William F. Martin • Miklós Müller (Editors)

# Origin of Mitochondria and Hydrogenosomes

William F. Martin
Miklós Müller (Editors)

# Origin of Mitochondria and Hydrogenosomes

With 31 Figures, 7 in Color and 6 Tables

Springer

Professor Dr. William F. Martin
Institut für Botanik III
Heinrich-Heine Universität Düsseldorf
Universitätsstr. 1
40225 Düsseldorf
Germany

Dr. Miklós Müller
The Rockefeller University
1230 York Avenue
New York, NY 10021-6399
USA
and
Collegium Budapest
1014 Budapest
Hungary

*Cover photo:* 'Hydrogenosomes 1974' by M. Müller and H. Shio

ISBN 978-3-642-07246-8 e-ISBN 978-3-540-38502-8



**Springer-Verlag is a part of Springer Science + Business Media**

springer.com



Editor: Dr. Sabine Schreck, Heidelberg
Desk Editor: Anette Lindqvist, Heidelberg

Cover Design: Design & Production, Heidelberg

Printed on acid-free paper 39/3152-HM 5 4 3 2 1 0

# Preface

At the time of their discovery over 30 years ago, it was thought that hydrogenosomes might be a novel form of either peroxisome, endosymbiotically derived organelle, or mitochondrion. For about two decades, this third possibility had seemed the least likely of all, probably because the concept of the mitochondrion was inextricably intertwined with the concept of oxidative phosphorylation, an attribute altogether lacking in all manifestations of the hydrogenosome ever found (so far). In recent years, however, it became clear that they are, indeed, a novel form of mitochondrion.

The realization that hydrogenosomes are mitochondria does not stem solely from experimental work specifically on hydrogenosomes. Instead, it stems from work on eukaryotic anaerobes in general: those that have hydrogenosomes, those that have mitosomes, and those that have anaerobically functioning mitochondria. The evolutionary significance of hydrogenosomes only became apparent in the context of a broader view of the molecular commonalities shared among those organisms and their organelles. In that sense, hydrogenosomes have helped to significantly broaden our concept of the mitochondrion.

As far as we are aware, this is the first monograph dedicated to hydrogenosomes. The views presented by the authors of the present volume are their own. We did not enter into the standard editing practice of negotiating content; all authors have had full opportunity to express their own views as they see fit. We hope that the reader finds this volume to be worthwhile. We feel that it is a uniquely rich source of information on eukaryotic anaerobes and their organelles and thank all contributors heartily for their hard work in preparing the chapters.

September, 2006

Bill Martin, Düsseldorf
Miklós Müller, Budapest

# Foreword

As Miklós Müller points out in his historical introduction, the evolutionary origins of hydrogenosomes have been the subject of considerable debate. From early days it was apparent that hydrogenosomes had evolved on multiple occasions in different eukaryotes, but from which progenitor organelle or endosymbiont was unresolved because of sparse and sometimes ambiguous data. Work from many different laboratories has contributed towards formulating the current hypothesis that hydrogenosomes and mitosomes, their even more reduced cousins, share common ancestry with mitochondria. These modern discoveries can be interpreted in terms of classical evolutionary theory. Hydrogenosomes, mitosomes and mitochondria are evolutionary homologues in the sense meant by Charles Darwin. Their shared similarities, for example their common mechanisms of protein import and their double membrane, can be explained by common ancestry, and their differences by descent with modification under contrasting lifestyles. The hypothesis that mitochondria, mitosomes and hydrogenosomes are homologues predicts that, as the organelles are studied more deeply, additional shared features will be revealed.

Understanding the origins of eukaryotic organelles and proteins is inextricably linked to efforts to understand eukaryotic relationships and to the development of more reliable phylogenetic methods. This is an area of endeavour that demands its own expertise, since overly simple analyses can produce strongly supported, but nevertheless incorrect, trees. Early efforts to make phylogenetic trees using gene sequences suggested that *Trichomonas*, a key model organism for understanding hydrogenosome evolution, separated from other eukaryotes before the mitochondrial endosymbiosis. This created a situation where it was deemed credible to suggest that the *Trichomonas* hydrogenosome had an origin distinct from that of other hydrogenosomes, through a unique endosymbiosis involving the ancestor of *Trichomonas* and an anaerobic hydrogen-producing bacterium. Today, there are widely accepted data for only two endosymbioses producing eukaryotic organelles: the $\alpha$-proteobacterial and cyanobacterial endosymbioses that gave rise to mitochondria and primary plastids, respectively. Any evidence purporting to support a third invasion of the eukaryotic cell producing hydrogenosomes was thus exciting and important. But extraordinary claims need extraordinary data, and recent work suggests that the phylogenetic position of *Trichomonas* is uncertain. By contrast, there is general agreement that *Trichomonas* contains genes derived from the mitochondrial endosymbiont. Lastly, the key enzymes pyruvate:ferredoxin oxidoreductase and [Fe] hydrogenase that have fuelled the separate origin hypothesis are not unique to the *Trichomonas* hydrogenosome. As such, they cannot logically be used to infer a singularly independent origin for the organelle. Surprisingly, these two proteins, in various forms and cellular compartments, have been found in diverse other eukaryotes with, and without, hydrogenosomes. The functional role/s of these proteins in organisms that do not produce hydrogen is/are still mostly unknown, but their conservation across diverse eukaryote genomes suggests they might be important.

The origins of eukaryotic [Fe] hydrogenase and pyruvate:ferredoxin oxidoreductase are part of a bigger question concerning the origins of the eubacterial-like genes that encode much of eukaryote metabolism. The apparent ubiquity of mitochondrial homologues among eukaryotes suggests that the mitochondrial endosymbiont is a prime candidate for the source of at least some of these genes, but how many and which ones is uncertain. There are also a number of imaginative alternative hypotheses to explain their presence: as the product of multiple lateral transfers from different prokaryotes, or the legacy of different eubacteria that participated in eukaryogenesis. Genomics coupled with more sophisticated phylogenetic analyses should in principle be able to identify eukaryote gene origins, but the search is likely to push both our data and our methods to their limits.

Which group of α-proteobacteria provided the endosymbiont is still contentious, because of the aforementioned difficulties of using phylogenetics to pinpoint ancient events. However, the apparent ubiquity of mitochondrial homologues among eukaryotes bears testament to the importance of the mitochondrial endosymbiosis in eukaryotic evolution. It also means that we can no longer be sure what came first – nucleus or mitochondrion. Put another way, prokaryote host models for the mitochondrial endosymbiosis can no longer be so comfortably dismissed, but must be judged on their merits as predictive hypothesis that can be tested.

Most of what is known about mitochondrial function and the importance of mitochondria for the eukaryotic cell is drawn from the study of yeast, mammal and plant mitochondria. Apart from oxidative phosphorylation, other important reactions include the Krebs cycle, haem biosynthesis, β-oxidation of fatty acids, amino acid biosynthesis and the formation and export of iron–sulphur clusters. New roles for mitochondria in health and disease are continually being discovered. It is already evident from biochemical and genomic data that hydrogenosomes and mitosomes can have retained only a limited subset of these reactions. It is also clear that parasites such as *Plasmodium* and *Cryptosporidium* have also greatly reduced the coding capacity and functions of organelles that are still called mitochondria. Mitochondrial homologues exhibit a spectrum of form and function, the breadth of which is still being uncovered because most parasites and anaerobic eukaryotes remain poorly studied. This raises important questions concerning the fundamental importance of this compartment of endosymbiotic ancestry for the eukaryotic cell, its biochemical flexibility, and the limits of organelle reduction. Comparative study of mitochondrial homologues in *all* their various guises is still needed to fully resolve these questions. However, it is already apparent from the contributions to this volume that identifying the genetic contribution to eukaryotes of the mitochondrial endosymbiosis and revealing the functions of its descendent organelles are key to understanding eukaryotic biology and evolution.

*Martin Embley*
*Newcastle, May 2006*

# Contents

# List of Contributors

CAROL A. ALLEN
School of Biological and Chemical Sciences, Queen Mary, University of London, Mile End Road, London E1 4NS, UK

JOHN F. ALLEN
School of Biological and Chemical Sciences, Queen Mary, University of London, Mile End Road, London E1 4NS, UK

MARIA JOSÉ BARBERÀ
Program in Evolutionary Biology, Canadian Institute for Advanced Research, Department of Biochemistry and Molecular Biology, Dalhousie University, Halifax, NS B3H 1X5, Canada

THOMAS CAVALIER-SMITH
Department of Zoology, University of Oxford, South Parks Road, Oxford OX1 3PS, UK

PAVEL DOLEŽAL
Department of Parasitology, Faculty of Science, Charles University in Prague, Vinicná 7, 128 44 Prague 2, Czech Republic

VICTOR V. EMELYANOV
Department of General Microbiology, Gamaleya Institute of Epidemiology and Microbiology, Gamaleya Street 18, 123098 Moscow, Russia

JOHANNES H.P. HACKSTEIN
Department of Evolutionary Microbiology, Faculty of Science, Radboud University Nijmegen, Toernooiveld 1, 6525 ED Nijmegen, The Netherlands

LAURA A. HUG
Program in Evolutionary Biology, Canadian Institute for Advanced Research, Department of Biochemistry and Molecular Biology, Dalhousie University, Halifax, NS B3H 1X5, Canada

MARTIJN HUYNEN
Nijmegen Centre for Molecular Life Sciences (NCMLS) and Centre for Molecular and Biomolecular Informatics, Toernooiveld 1, 6525 ED Nijmegen, The Netherlands

WERNER KOOPMAN
Microscopical Imaging Centre and Department of Biochemistry, Nijmegen Centre of Molecular Life Sciences (NCMLS), University Medical Centre, 6500 HB Nijmegen, The Netherlands

NICK LANE
Royal Free and University College Medical School, Pond Street, London NW3 2QG, UK

Jessica Leigh
Program in Evolutionary Biology, Canadian Institute for Advanced Research, Department of Biochemistry and Molecular Biology, Dalhousie University, Halifax, NS B3H 1X5, Canada

Miklós Müller
The Rockefeller University, 1230 York Avenue, New York, NY 10021, USA and Collegium Budapest, Szentháromság utca 2, 1014 Budapest, Hungary

Andrew J. Roger
Program in Evolutionary Biology, Canadian Institute for Advanced Research, Department of Biochemistry and Molecular Biology, Dalhousie University, Halifax, NS B3H 1X5, Canada

Iñaki Ruiz-Trillo
Program in Evolutionary Biology, Canadian Institute for Advanced Research, Department of Biochemistry and Molecular Biology, Dalhousie University, Halifax, NS B3H 1X5, Canada

Jan Sapp
Biology Department, Faculty of Science and Engineering, York University, 4700 Keele Street, Toronto, ON M3J 1P3, Canada

Jan Tachezy
Department of Parasitology, Faculty of Science, Charles University in Prague, Vinicná 7, 128 44 Prague 2, Czech Republic

Aloysius G.M. Tielens
Department of Biochemistry and Cell Biology, Faculty of Veterinary Medicine, Utrecht University, PO Box 80176, 3508 TD Utrecht, The Netherlands

Joachim Tjaden
Department of Plant Physiology, University of Kaiserslautern, Erwin-Schrödinger-Str., 67653 Kaiserslautern, Germany

Jorge Tovar
School of Biological Sciences, Royal Holloway University of London, Egham TW20 0EX, UK

Mark van der Giezen
School of Biological and Chemical Sciences, Queen Mary, University of London, Mile End Road, London E1 4NS, UK

Jaap J. van Hellemond
Department of Biochemistry and Cell Biology, Faculty of Veterinary Medicine, Utrecht University, PO Box 80176, 3508 TD Utrecht, The Netherlands

# 1 The Road to Hydrogenosomes

Miklós Müller

## 1.1 Introduction

Recent studies increasingly support the hypothesis that organelles originating from the mitochondrial endosymbiotc event comprise a family of diverse structures and not only the typical mitochondria, with their characteristic cristate structure functioning as aerobic powerhouses of cells. This large family is regarded today to encompass a broad, almost continuous spectrum of organelles, from the much-studied typical aerobic mitochondria, through anaerobic mitochondria and hydrogenosomes to the still-enigmatic mitosomes. It is becoming generally accepted that these organelles evolved from a common ancestral organelle that arose in a single endosymbiotic event sometime during the emergence of the eukaryotic cell, as we know it today. Contributions to this volume explore the diverse members of this organelle family in detail and most of them also give a glimpse of the intense debates on the evolutionary origins of these organelles and their current impact on general hypotheses on the origin and evolutionary diversification of eukaryotes.

One of the key contributions leading to the prevalent current views was the discovery of hydrogenosomes in certain "anaerobic" flagellates in the early 1970s. Initially they were regarded as an oddity or aberration and it took time before they were seriously discussed outside the community of protistologists. They remained an enigma until their mitochondrial relationships received strong support in the 1990s. Recent developments overshadow the early days of hydrogenosomes, as they indeed should. This volume, presenting an exciting collection of papers on mitochondrial diversity and evolution, is perhaps the appropriate place to retell briefly those early days by someone who was there. I eagerly accepted the suggestion of my friend, the coeditor of this volume, Bill Martin, that I do this. Needless to say, this will be an intensely personal history and certainly not an autobiography. Since the contributions to this volume provide a clear picture of later developments, I restrict my story to the very early days, approximately the first 12 years after the discovery of hydrogenosomes (Lindmark and Müller 1973).

Origin of Mitochondria and Hydrogenosomes (ed. by William F. Martin and Miklós Müller)

## 1.2 The Story

I start in *medias res*. After having spent some years in Budapest exploring protist food vacuoles and lysosomes with histochemical methods, I was given the opportunity of my life by Christian de Duve, who received the Nobel Prize in 1974 for the discovery of lysosomes and peroxisomes. In 1964 he invited me to join his laboratory at the Rockefeller University (still called Institute then) to explore these organelles with biochemical and cell fractionation methods in protists. My work in New York started well, although I was a novice to biochemistry at that time. Soon I was able characterize biochemically the lysosomes of the ciliate *Tetrahymena pyriformis* and also demonstrated classic peroxisomes in this species (Baudhuin et al. 1965). This finding showed the peroxisome to be broadly distributed in eukaryotes, since until then they had only been detected in mammalian tissues. After a year in New York, I spent a year at the Carlsberg Laboratories in Copenhagen under the guidance of Heinz Holter and Cicily Chapman-Andresen working on lysosomes and peroxisomes of amoebae. I returned to de Duve's laboratory in 1966 and continued working on *T. pyriformis*. With James Hogg of Queens College, we demonstrated that *T. pyriformis* peroxisomes unexpectedly contain two enzymes of the glyoxylate bypass of the glyoxylate cycle responsible for gluconeogenesis from acetyl-coenzyme A. The rest of the cycle in this ciliate was localized in the mitochondria, an interesting intracellular division of labor (Müller et al. 1968). About the same time, certain plant peroxisomes were found by others to harbor the glyoxylate cycle and were named glyoxysomes. In the laboratory we became interested in the distribution of peroxisomes in the living world, looking for new organisms to explore. Little did I suspect that this quest would lead to a "novel" organelle, the hydrogenosome. I have recounted these events in some detail in a Past President's address given to the Society of Protozoology (Müller 1985).

The theory of endosymbiotic origin of chloroplasts and mitochondria was in the air in the late 1960s. Lynn Margulis had just published her first paper on the serial endosymbiotic theory. As part of our studies on lysosomes and peroxisomes, the question of the evolutionary origin of these other organelles was also often discussed. The lysosomes could be relatively easily derived from the cell membrane, a conclusion that is still valid today. The origin of the peroxisomes, single membrane-bounded organelles that contain direct oxidases and catalase, was a more complicated question. de Duve (1969) suggested a modified serial endosymbiotic hypothesis (Fig. 1.1). The events envisaged by him invoked as a host cell an ancestral anaerobic phagocytic eukaryote without any respiratory organelles. In a first endosymbiotic event it was supposed to take up and integrate an aerobic bacterium containing direct oxidases and no cytochrome-mediated electron transport chain. This became the peroxisome. The uptake of a bacterium with a cytochrome-mediated respiratory system in a subsequent, second

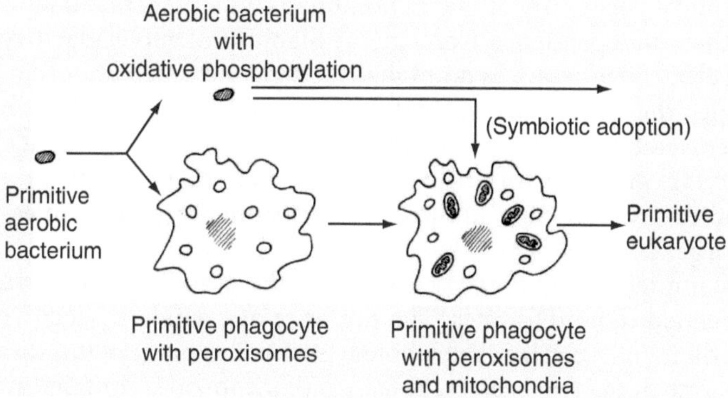


**Fig. 1.1.** Schematic representation illustrating de Duve's hypothesis concerning the evolutionary origin of peroxisome. (Figure 3 in de Duve 1969)

endosymbiotic event lead to the formation of a protomitochondrium, resulting in a complex eukaryotic cell with two respiratory organelles, the peroxisome and the mitochondrium.

An important corollary of this hypothesis was the possible existence of organisms that derived from an intermediary stage already containing peroxisomes but not yet mitochondria. The search for extant representatives of such organisms was an interesting challenge. One major motivation in trying to find these intermediates in eukaryogenesis was the hope that they could provide insight into the ancestral function of peroxisomes, which in the late 1960s were already known to vary significantly in their enzymatic composition from organism to organism. As proposed by de Duve:

"In a more speculative vein, our considerations of the role peroxisomes may have played in a primitive cell devoid of mitochondria could be relevant to the physiology of microroganisms that either do not have mitochondria ... or have deficient mitochondria.... Perhaps a survey of microorganisms, inspired by these considerations, may disclose the existence of eukaryotic organisms truly devoid of mitochondria, but possessing peroxisomes, in which the postulated ancestral function of these particles could be explored and assessed." (p. 380 in de Duve 1969)

While I continued my biochemical studies on *T. pyriformis*, I eagerly began a search for such protists, too.

Trichomonad flagellates seemed to be a good first choice. The human pathogen *Trichomonas vaginalis* and the cattle pathogen *Tritrichomonas foetus* were much-studied species available in bacterium-free cultures and were thus amenable to biochemical and cell fractionation studies, approaches extensively practiced by our group. The available physiological data showed that the respiration of these species was not of mitochondrial type, because it could not be inhibited with cyanide and other mitochondrial inhibitors. Furthermore no cytochromes were detected in these trichomonads (Ryley 1955).

A biochemical peculiarity of the trichomonads was their essentially anaerobic nature, their independence from the availability of molecular oxygen as a terminal electron acceptor, a point that soon gained much importance. At the same time, the ultrastructural data accumulated revealed the absence of typical christate mitochondria. Instead the well-known paracostal and paraxostylar granules of mitochondrial size turned out to resemble microbodies, by that time known to correspond to peroxisomes in many aerobic organisms.

We attempted to test the hypothesis that the granules seen in trichomonads are related to peroxisomes. The presence of cyanide-insensitive respiration and their characteristic morphology seemed to point in this direction. In addition, one of the two species, *T. foetus*, was known to contain catalase, the marker enzyme of peroxisomes. In 1971 I obtained cultures of this organism, verified the presence of catalase activity and performed a seemingly obvious experiment. I prepared a homogenate in isosmotic sucrose, and separated it by centrifugation into sediment and supernatant, with the expectation that the sediment containing the granules would contain most of the catalase activity. The results were disappointing: most of the activity was recovered in the supernatant. The total recovery of the activity was close to 100%, indicating no problems with the fractionation experiment itself. Two repeat experiments gave the same results. Since I well "knew" that catalase had to be in the organelles, I assumed that they had been broken during homogenization owing to their excessive fragility and decided that this approach would not lead me to the desired goal. I turned to other organisms with cyanide-insensitive respiration, work that did not lead to the desired breakthrough.

This disappointment did not leave me in peace for long, however. Reflecting on the data, I finally was willing to accept them at face value. OK, the organelles were not the subcellular location of catalase! But then what were they? A dominant structure occupying a significant portion of the cell volume, as readily seen in the electron microscope, surely has some important role to play. Since this was the only large, membrane-bounded entity seen in trichomonads, its relationship to either mitochondria or peroxisomes was still a possibility. The alternative assumption of having an independent origin remained open of course. But the immediate task was to define the *biochemical* nature of the organelle, the possible relationships to either mitochondria or peroxisomes only being important in suggesting enzymes and functions to explore.

Returning to the problem, I looked for typical mitochondrial or peroxisomal enzymes in *T. foetus* cell homogenates and fractions obtained by different centrifugation methods, but all results were negative (Müller 1973). It occurred to me then that this enigmatic organelle might be somehow connected to the anaerobic nature of the organisms already mentioned. A tempting possibility was that they could be somehow involved in the production of hydrogen by these organisms, a metabolic end product rarely found among eukaryotic organisms. Donald G. Lindmark, whose thesis work was on hydrogen production in anaerobic prokaryotes, joined our group and we set out to test our assumption.

We assayed homogenates of *T. foetus* for enzymes known to be involved in hydrogen formation in anaerobic bacteria and we detected the presence of pyruvate:ferredoxin oxidoreductase and hydrogenase, enzymes fundamentally different from the pyruvate-metabolizing enzymes of mitochondria. This confirmed the results of Bauchop (1971), who demonstrated that hydrogen production by *T. vaginalis* is biochemically analogous to processes in *Clostridium* species. Assays of subcellular fractions obtained by diverse centrifugation methods localized these activities to a large granule fraction that was shown by electron microscopy to be significantly enriched in granules, identifiable with the paracostal and paraxostylar granules. These results strongly suggested that the granules are indeed the subcellular site of hydrogen production. We published our first paper on these results in the *Journal of Biological Chemistry* in 1973, where we concluded:

"These findings underscore the unique nature of the microbody-like particles of *T. foetus*. In contrast to mitochondria or peroxisomes, in which electron transfer is directed toward molecular oxygen, they utilize protons as terminal electron acceptors and thus produce molecular hydrogen. We propose the term "hydrogenosome" to designate this new biochemically defined subcellular entity." (p. 7728 in Lindmark and Müller 1973)

I wish to stress the point that our definition was based on the biochemical properties of the organelle and did not extend to its morphological characteristics, leaving open the possibility that hydrogenosomes, if and when detected in other organisms, could display a morphology different from that observed in trichomonads.

Our immediate goals were twofold: first, to see whether hydrogenosomes are a peculiarity of *T. foetus* or are present also in other trichomonads as well as in other protist groups; second, to characterize the biochemistry of the newly discovered organelle in more detail. Along the first line of inquiry, we studied two additional trichomonad species with the same methods. The results were identical for the important human parasite *T. vaginalis* (Lindmark et al. 1975) and for *Monocercomonas* sp. (Lindmark and Müller 1974a), a species regarded as a primitive form at that time. The selection of these two species reflected our interest in the possible medical and evolutionary significance of the new organelle. The publication of our *T. vaginalis* results met unexpected and amusing obstacles, however. We submitted our detailed manuscript to the *Journal of Parasitology*, regarding it as the most appropriate venue to reach applied and general parasitologists alike. The paper was *not* accepted for publication because the reviewer stated that its content is self-evident and does not contribute any new information. This was at a time when practically no parasitologist or biologist was even aware of the existence of hydrogenosomes. Finally we summarized our data in a brief communication (Lindmark et al. 1975). We felt that the case for hydrogenosomes as a unique type of organelle in trichomonad flagellates had been made and decided not to expand our taxonomic sample further. The results strongly indicated that the hydrogenosomes were a general characteristic of trichomonad flagellates, a fact already listed as a property of these organisms in the 1980 Classification of the Protozoa, endorsed by the Society of Protozoologists (Levine et al. 1980). In 1980, I pointed to other groups that might harbor hydrogenosome-like organelles (Müller 1980), but the successful exploration of these was done later by others.

Along the second line of investigation, we looked for additional enzymes in the organelle to define its biochemical composition. We demonstrated the presence of an unusual acyltransferase, acetate:succinate coenzyme A transferase and succinylcoenzyme A synthetase (Lindmark 1976). These enzymes accounted for the production of acetate by the organism. Somewhat later we purified the electron transfer protein linking pyruvate:ferredoxin oxidoreductase and hydrogenase and showed that is a [2Fe–2S] ferredoxin (Marczak et al. 1983; Gorrell et al. 1984), belonging to the same class of ferredoxins as those found in mitochondria. We also devoted some effort to strengthen the evidence for the metabolic differences between hydrogenosomes and mitochondria by lowering the limit of detection of typical mitochondrial components, notably cytochromes and F1F0-ATP synthetase (Lloyd et al. 1979a, b). Neither of these was detectable with the methods then available to us.

Already by 1976, there were sufficient enzymatic data to propose a scheme for the main metabolic activities of trichomonad hydrogenosomes (Fig. 1.2a). Although much work was devoted subsequently to the exploration of this metabolism and new results were added to the picture, the originally proposed core pathway did not require substantial correction later (Fig. 1.2b). It needs to be stressed that the hydrogenosomal metabolic map was drawn up on the basis of enzymatic information and was verified only in 1988 by Steinbüchel and me by determination of the metabolic balance of isolated hydrogenosomes (Steinbüchel and Müller 1986; Fig. 1.3).

The nature of these organelles raised the curiosity of other investigators at about the same time. In contrast to us, they started from the assumption that these structures could be unusual forms of mitochondria. In Prague, the couple Jiři Čerkasov and Apolena Čerkasovová were studying the carbohydrate metabolism of trichomonads and decided to determine whether the cyanide-insensitive respiration was localized in the granules or the cytosol (Čerkasovová and Čerkasov 1974). They found part of the process connected with the granular fraction. They also showed that it is dependent on the presence of ADP in the incubation medium, suggesting a mitochondrial-type respiratory control (Čerkasovová and Čerkasov 1976; Čerkasov et al. 1978). The ADP effect was inhibited by atractyloside, an inhibitor of the mitochondrial ADP/ATP exchange transporter. In Clermont-Ferrand, Guy Brugerolle and Guy Metenier also explored possible similarities between the trichomonad granules and typical mitochondria. They selected as a marker for mitochondria the enzyme decarboxylating malate dehydrogenase ("malic enzyme") and found it to be localized in the granules (Brugerolle and Metenier 1973).

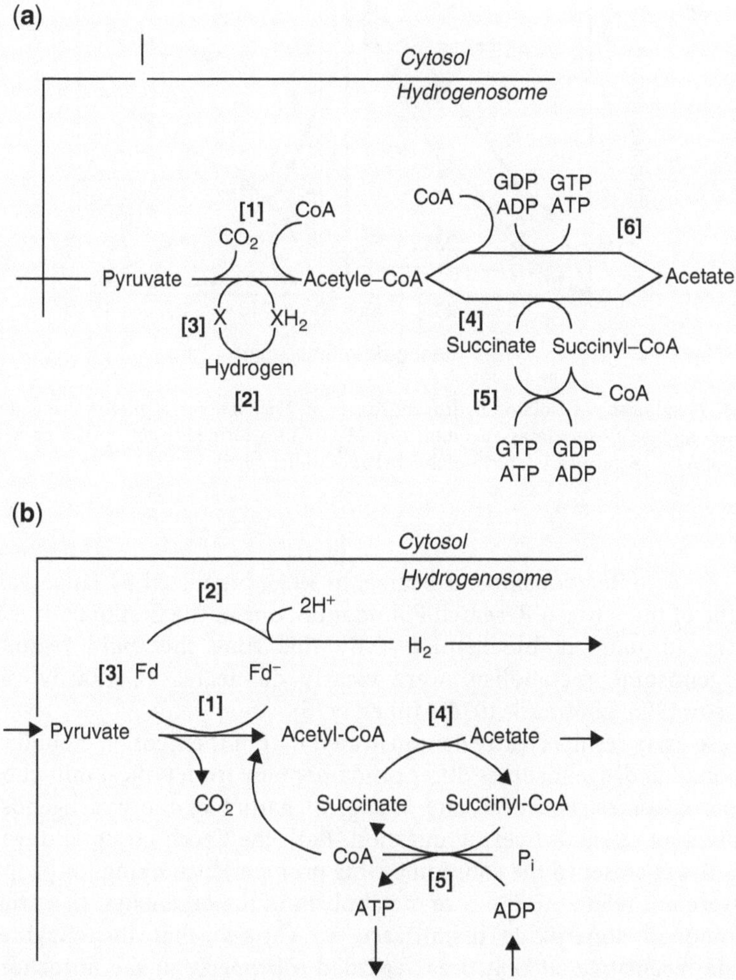


**Fig. 1.2.** Maps of pyruvate metabolism in trichomonad hydrogenosomes published in 1976 (**a**) and 2003 (**b**). Metabolism of malate omitted. *1* pyruvate:ferredoxin oxidoreductase, *2* hydrogenase, *3* electron transport protein (unidentified in 1976, known to be ferredoxin in 2003), *4* acetate:succinate coenzyme A (*CoA*) transferase, *5* succinyl-CoA synthase, *6* acetyl-CoA synthase (its presence was not confirmed later). (**a** From Fig. 2 in Müller 1976; **b** from Fig. 7.3 in Müller 2003)

They published their results but did not pursue their study further. These results hinted at similarities, and possible homologies, of the mitochondrial and hydrogenosomal metabolic machinery. The three groups had the opportunity to discuss their results at the Fifth International Congress of Protozoology in Clermond-Ferrand in 1973 establishing friendly contacts and subsequent sharing of information. Further intense discussions took place at the Second International Symposium on the Biochemistry of Parasites and

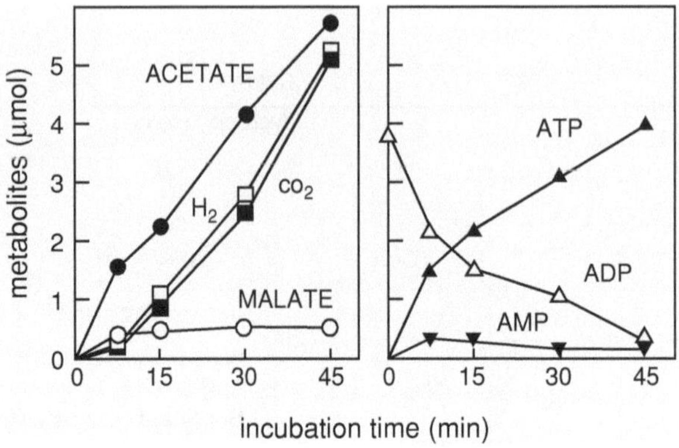


**Fig. 1.3.** Metabolism of isolated hydrogenosomes of *Trichomonas vaginalis* suspended in a pyruvate- and ADP-containing medium. *Left* Metabolites formed; *right* uptake of ADP and release of ATP. (Figure 3a, b from Steinbüchel and Müller 1986)

Host–Parasite Relationships in Beerse, in 1976, organized by Hugo van den Bossche of the Janssen Research Foundation, one of the first major meetings devoted to parasite biochemistry. By this time the main features of hydrogenosome metabolism were clearly delineated (Čerkasovová and Čerkasov 1976; Lindmark 1976; Müller 1976).

These early results clearly demonstrated that in their core metabolism trichomonad hydrogenosomes differ fundamentally from typical mitochondria and peroxisomes. However, the *biological* nature of the hydrogenosomes remained an open and vexing question. Both the Czech group and we were trying to get closer to the solution of this problem, by looking for properties that were not related to the core metabolism of the organelles. In a study on trichomonad superoxide dismutases we showed that the enzyme was cyanide-insensitive, at that time regarded a property of the mitochondrial and bacterial enzymes but not of the cytosolic enzyme. We concluded, "The occurrence of the bacterial and mitochondrial type of superoxide dismutase in hydrogenosomes is an interesting finding which may be a significant lead in attempts to elucidate the phylogenetic origin of this organelle." (p. 4637 in Lindmark and Müller 1974b). But only about one sixth of the total enzyme activity was in the organelles; thus, the evidence was not overwhelming. The trichomonad superoxide dismutases were later shown to be iron enzymes, related more closely to the α-proteobacterial ones and more distantly to the mitochondrial manganese enzymes. The Czech group obtained early results demonstrating the presence of a genome and cardiolipin in trichomonad hydrogenosomes (Čerkasovová et al. 1976), but these results could not be confirmed later (Turner and Müller 1983; Paltauf and Meingassner 1982). Essentially no convincing evidence linking the hydrogenosomes to mitochondria could be obtained in this early period. The interpretation that seemed to be the most consistent with all the data was that the organelles represent a unique structure, unrelated to either mitochondria or peroxisomes.

Whatley et al. (1979) in an overview of the origins of mitochondria and chloroplasts raised the possibility that hydrogenosomes arose through an independent endosymbiotic event. This idea agreed well with our thinking at the time and I embraced it wholeheartedly. In my review of 1980 I discussed this problem as follows:

"It is suggested that the hydrogenosome might be the anaerobic equivalent of the mitochondrion.... Although this suggestion is only a hypothesis at present, we can compare the known properties of the two organelles. Unfortunately, the available data are rather scanty and some or them need further confirmation. Clearly the most important point would be the unequivocal confirmation in hydrogenosomes of the presence or absence of gentic and protein-synthetic machinery.... Although the evolutionary origin of mitochondria and chloroplasts is still an open question it is possible that the hydrogenosome:*Clostridium* pair represents a further example in the list of organelle and 'its free living prokaryotic ancestor' pairs which includes the mitochondrion:aerobic bacterium and the chloroplast:cyanobacterium." (pp. 137–138 in Müller 1980)

In 1992 I formalized the separate origin in a symposium presentation entitled "Energy metabolism of ancestral eukaryotes: a hypothesis based on the biochemistry of amitochondriate parasitic protists" (Müller 1992). At that time this idea seemed to fit all available information best. However, in a few years convincing data started appearing that showed my hypothesis to be way off the mark. I did not hesitate long before accepting the contrary view, but I could not undo what had already been published. My 1992 paper is in the databases and references to it are still showing up from time to time in the literature.

Although my coworkers and I continued to make contributions to this field until I closed my laboratory recently, these represent only additional bricks for the now emerging edifice constructed by a number of investigators. The more recent advances are discussed in the following chapters and are no longer the results of the lonely adventures of a few in those first years. They have become part of the overall history of the exploration of mitochondrial diversity, a history still being written.

## 1.3 Conclusion

In retrospect, the discovery of hydrogenosomes was prompted by a search for an organism and an organelle that were not to be found. It led to the biochemical characterization of a "novel" organelle in certain protists, the trichomonads, started a hunt for similar organelles in other protists and finally opened up a Pandora's box full of diverse organelles in eukaryotes. Slow and steady increase in taxonomic, morphological, biochemical and sequence information since the original 1973 paper revealed some of the early attempts to understand the biological nature and evolutionary origins of the organelle to be misleading, albeit here and there pointing in the right direction. All work since then has essentially confirmed the functional interpretations presented in that first paper. Looking back at this history, it is apparent that in the absence of critical information, the repeated attempts made to find the place of this organelle in the "overall" picture of the eukaryotic cell had to lead to hypotheses now known to be in error. But without exaggeration it can be stated that they played a significant role in provoking further experimental work that "exploded" in the past two decades. This work, well presented in this volume, led to a consistent picture of the until recently unsuspected diversity of mitochondria and firmly placed the trichomonad hydrogenosomes in the big family of organelles derived from the ancestral "protomitochondrion." We have reached the stage of this story where these organelles cannot be swept under the rug anymore, as was done by most cell biologists until recently. Finally they can be seriously considered even in polite company.

*Acknowledgements:* First and foremost, I wish to thank Christian de Duve for willing to gamble on a young man from Hungary, offering him the possibility to start out in the New World, and providing friendship, inspiration, support, direction and criticism ever since. I wish to remember here Bronislav M. Honigberg, who generously shared with me his extensive knowledge and insight of trichomonad biology. I owe special acknowledgment to Donald G. Lindmark, who joined my efforts on trichomonads at the very beginning and was an equal partner in my early adventures in the world of "amitochondriate" eukaryotes. Many scientists, postdoctoral fellows and research associates participated in our work over the years and made critical contributions to this evolving story. The following were participants in the early stages of this work: Tom E. Gorrell, David Lloyd, Steve R. Mack, Regis Marczak, John McLaughlin, Vincent N'Seka, Alexander Steinbüchel, Geoff Turner and Nigel Yarlett. I express my sincere thanks to them here. Our research was generously supported by NIH and NSF, first by grants to Christian de Duve and since 1974 to me.

## References

Bauchop T (1971) Mechanism of hydrogen formation in *Trichomonas foetus*. J Gen Microbiol 68:27–33

Baudhuin P, Müller M, Poole B, de Duve C (1965) Non-mitochondrial oxidizing particles (microbodies) in rat liver and kidney and in *Tetrahymena pyriformis*. Biochem Biophys Res Commun 20:53–59

Brugerolle G, Metenier G (1973) Localisation intracellulaire et caracterisation de deux types de malate déshydrogénase chez *Trichomonas vaginalis* Donné. J Protozool 20:320–327

Čerkasov J, Čerkasovová A, Kulda J, Vilhelmová D (1978) Respiration of hydrogenosomes of *Tritrichomonas foetus*. I. ADP-dependent oxidation of malate and pyruvate. J Biol Chem 253:1207–1214

Čerkasovová A, Čerkasov J (1974) Location of the NADH oxidase activity in fractions of *Tritrichomonas foetus* homogenate. Fol Parasitol (Prague) 21:193–203

Čerkasovová A, Čerkasov J (1976) Some properties of the membrane of the hydrogenosome *Tritrichomonas foetus*. In: H. Van den Bossche (ed.), Biochemistry of parasites and host-parasite relationships. Elsevier/North Holland, Amsterdam, pp 23–30

Čerkasovová A, Čerkasov J, Kulda J, Reischig J (1976) Circular DNA and cardiolipin in hydrogenosomes, microbody like organelles in trichomonads. Fol Parasitol (Praha) 23:33–37

de Duve C (1969) Evolution of the peroxisome. Ann N Y Acad Sci 168:369–381

Gorrell TE, Yarlett N, Müller M (1984) Isolation and characterization of *Trichomonas vaginalis* ferredoxin. Carlsberg Res Commun 49:259–268

Levine ND et al. (1980) A newly revised classification of the Protozoa. J Protozool 27:37–58

Lindmark DG (1976) Acetate production by *Tritrichomonas foetus*. In: H Van den Bossche (ed) Biochemistry of parasites and host-parasite relationships. Elsevier/North Holland, Amsterdam, pp 15–21

Lindmark DG, Müller M (1973) Hydrogenosome, a cytoplasmic organelle of the anaerobic flagellate, *Tritrichomonas foetus*, and its role in pyruvate metabolism. J Biol Chem 248:7724–7728

Lindmark DG, Müller M (1974a) Biochemical cytology of trichomonad flagellates. II. Subcellular distribution of oxidoreductases and hydrolases in *Monocercomonas* sp. J Protozool 21:374–378

Lindmark DG, Müller M (1974b) Superoxide dismutase in the anaerobic flagellates, *Tritrichomonas foetus* and *Monocercomonas* sp. J Biol Chem 249:4634–4637

Lindmark DG, Müller M, Shio H (1975) Hydrogenosomes in *Trichomonas vaginalis*. J Parasitol 61:552–554

Lloyd D, Lindmark DG, Müller M (1979a) Respiration of *Tritrichomonas foetus*: absence of detectable cytochromes. J Parasitol 65:466–469

Lloyd D, Lindmark DG, Müller M (1979b) Adenosine triphosphatase of *Tritrichomonas foetus*. J Gen Microbiol 115:301–307

Marczak R, Gorrell TE, Müller M (1983) Hydrogenosomal ferredoxin of the anaerobic protozoon, *Tritrichomonas foetus*. J Biol Chem 258:12427–12433

Müller M (1973) Biochemical cytology of trichomonad flagellates. I. Subcellular localization of hydrolases, dehydrogenases, and catalase in *Tritrichomonas foetus*. J Cell Biol 57:453–474

Müller M (1976) Carbohydrate and energy metabolism of *Tritrichomonas foetus*. In: H Van den Bossche (ed.), Biochemistry of parasites and host-parasite relationships. Elsevier/North Holland, Amsterdam, pp 3–14

Müller M (1980) The hydrogenosome. Symp Soc Gen Microbiol 30:127–142

Müller M (1985) Search for cell organelles in protozoa. J Protozool 32:559–563

Müller M (1992) Energy metabolism of ancestral eukaryotes: a hypothesis based on the biochemistry of amitochondriate parasitic protists. BioSystems 28:33–40

Müller M (2003) Energy metabolism. Part I: Anaerobic protozoa. In: Marr J, Nilsen T, Komuniecki R (eds) Molecular medical parasitology. Academic, London, pp 125–139

Müller M, Hogg JF, de Duve C (1968) Distribution of tricarboxylic acid cyle enzymes and of glyoxylate cycle enzymes between mitochondria and peroxisomes of *Tetrahymena pyriformis*. J Biol Chem 243:5385–5395

Paltauf F, Meingassner JG (1982) The absence of cardiolipin in hydrogenosomes of *Trichomonas vaginalis* and *Tritrichomonas foetus*. J Parasitol 68:949–950

Ryley JF (1955) Studies on the metabolism of the protozoa. 5. Metabolism of the parasitic flagellate *Trichomonas foetus*. Biochem J 59:361–369

Steinbüchel A, Müller M (1986) Anaerobic pyruvate metabolism of *Tritrichomonas foetus* and *Trichomonas vaginalis* hydrogenosomes. Mol Biochem Parasitol 20:57–65

Turner G, Müller M (1983) Failure to detect extranuclear DNA in *Trichomonas vaginalis* and *Tritrichomonas foetus*. J Parasitol 69:234–236

Whatley JM, John P, Whatley FR (1979) From extracellular to intracellular: the establishment of mitochondria and chloroplasts. Proc R Soc Lond B Biol Sci 204:165–187

# 2 Mitochondria: Key to Complexity

Nick Lane

## 2.1 Introduction

All known eukaryotic cells either have, or once had, and later lost, mitochondria (which is to say, the common ancestor of mitochondria and hydrogenosomes; Gray et al. 1999, 2001; Embley et al. 2003; Tielens et al. 2002; Boxma et al. 2005; Gray 2005). If this statement is upheld (discussed elsewhere in this volume), then possession of mitochondria could have been a sine qua non of the eukaryotic condition. That cannot be said of any other organelle. The eukaryotic cell apparently evolved only once – all modern eukaryotes are descended from a single common ancestor – and that ancestor had mitochondria (Martin 2005; Lane 2005; Martin and Müller 1998).

By definition, there are no eukaryotic cells without a nucleus, but it is a surprise that there are no eukaryotes that did not have mitochondria in their past. The implications have not yet been properly digested. Why was this? What advantage did the mitochondria offer? Whatever the advantage, it was not trivial. Bacteria and archaea ruled the Earth for three billion years (Knoll 2003). During this time, they evolved a dazzling wealth of biochemical variety, making the eukaryotes look impoverished (Martin and Russell 2003). Yet the prokaryotes failed to evolve greater morphological complexity: although some bacteria might best be thought of as multicellular organisms, their degree of organisation falls far short of eukaryotic attainments (Kroos 2005; Velicer and Yu 2003). In general, bacteria today seem to be no more complex than in the earliest known fossils (Knoll 2003; Maynard Smith and Szathmáry 1995). Such lack of 'progress' seems to be true of their biochemistry too: all the most important geochemical cycles were apparently in place by 2.7 billion years ago, implying that prokaryotes had already by then evolved oxygenic photosynthesis, sulphate reduction, fermentation, oxidative phosphorylation, methanogenesis, denitrification and nitrification (Martin et al. 2003; Lane 2002; Anbar and Knoll 2002; Nisbet and Sleep 2001; Canfield et al. 2000; Castresana and Moreira 1999; Canfield 1998).

The traditional long list of differences between prokaryotic and eukaryotic cells has been gradually eroded as exceptions are found to each. There are prokaryotes with structures resembling a nucleus, straight chromosomes, cytoskeleton, giant size, internal membranes, multiple replicons, introns, mitotic-like apparatus, gene regulation, inter-cellular signalling, genetic

Origin of Mitochondria and Hydrogenosomes (ed. by William F. Martin and Miklós Müller)

recombination, and indeed with endosymbionts (Margolin 2005; Jones et al. 2001; van den Ent et al. 2001; Vellai and Vida 1998; Fonstein and Haselkorn 1995). There has been a revolution in our perception of the prokaryotic state (Gitai 2005). But if all these supposedly 'eukaryotic' traits are found in prokaryotes, why do we not see a continuum in complexity between prokaryotes and eukaryotes? Some might argue that we do, indeed that such a spectrum of eukaryotic traits in prokaryotes is already good evidence of a continuum. Even so, there is still a void – although there is some degree of overlap in cell biology, no prokaryotes ever gave rise directly (without endosymbiosis) to organisms even of the complexity of protozoa.

The distinction between prokaryotes and eukaryotes is essentially one of degree. Prokaryotes made a start with most aspects of molecular organisation, then stopped short. The eukaryotes took up the baton and ran. Picoeukaryotes, discovered at the turn of the millennium to thrive in surprising abundance in extreme environments, such as iron-rich rivers (Amaral Zettler et al. 2002) and deep oceans (López-García et al. 2001), are similar to prokaryotes in their size and complexity. Even so, they have a nucleus, straight chromosomes, and organelles including tiny mitochondria (Baldauf 2003). In general, however, eukaryotes are 10,000–100,000 times the size of prokaryotes, and have genomes to match. Including non-coding DNA, no known prokaryote has a genome larger than about 10 Mb (megabases), whereas eukaryotes have expanded their genome size up to an extraordinary 670,000 Mb in *Amoeba dubia* (200 times larger than the human genome), a range of more than 4 orders of magnitude (Cavalier-Smith 1985, 2005; Gregory 2001; Cavalier-Smith and Beaton 1999).

This distinction is usually ascribed to nuclear factors, such as straight chromosomes, or to the invention of meiotic sex, which has the potential to postpone mutational meltdown, and so enable an expansion in genome size (Kondrashov 1988; Ridley 2001). On the other hand, if recombination can indeed prevent mutational meltdown (and there is little evidence to show it does; Keightley and Eyre-Walker 2000; Elena and Lenski 1997), then so too in principle could lateral gene transfer in bacteria. Despite the ubiquity of lateral gene transfer, indeed perhaps because of it, as I will argue, bacteria still do not expand their genome size (Konstantinidis and Tiedje 2004; Kunin and Ouzounis 2003; Gregory 2002). Similarly, if straight chromosomes were all that was needed to enable multiple origins of replication, then why could bacteria like *Borrelia burgdorferi* and several species of *Streptomyces*, which have straight chromosomes (Fonstein and Haselkorn 1995; Baril et al. 1989), not expand their own genome sizes up to eukaryotic proportions? Indeed, other bacteria, such as *Pseudomonas cepacia* and *Rhodobacter sphaeroides*, do have multiple replicons, but still do not expand their genome size (Vellai et al. 1998; Fonstein and Haselkorn 1995). Likewise, if the accumulation of DNA in eukaryotes was linked to the proliferation of selfishly replicating elements like transposable elements and retroviruses, as some have argued (Doolittle and Sapienza 1980; Orgel and Crick 1980), then why are bacteria virtually immune to this? If the duplication of selfish DNA has the power to expand genome size, and this is an advantage (Zuckerkandl 2002), why did prokaryotes not take advantage?

## 2.2 Size

In short, the size, complexity and genomic expansion of eukaryotic cells cannot obviously be explained in terms of one of the two characteristic eukaryotic traits, the nucleus. Might it then be explained in terms of the other key trait, the mitochondria? Traditionally, mitochondria have been thought of as an optional extra for eukaryotic cells, an efficient power supply, but little more (Keeling 1998). Such a view is intuitive, and perhaps befits their status as an organelle, but not their newfound status as co-progenitors of the eukaryotic cell (Martin and Müller 1998; Martin 1999; Martin et al. 2001). The emasculated conception of mitochondrial significance was reinforced by Cavalier-Smith's depiction of the archezoa as primitive amitochondriate eukaryotes (Cavalier-Smith 1987, 1989), implying that mitochondria were an asset rather than a necessity. But this view is not compatible with more recent evidence that mitochondria were a necessity, a sine qua non of the eukaryotic condition; and Cavalier-Smith (2002) no longer holds that the archezoa are primitively amitochondriate. Might it be instead that mitochondria, by compartmentalising ATP generation, enabled greater cell volume; and that genomic expansion and reorganisation were a consequence of larger cell volume?

The idea that the genomic expansion of the eukaryotes was made possible by the acquisition of mitochondria has been argued by Vellai and Vida (Velai et al. 1998; Vellai and Vida 1999), and they make several important points. In particular, they argue that energetic compartmentalisation overcame the ATP subsaturation characteristic of bacteria. They imply that it did so by facilitating phagocytosis, which requires (1) a large ATP investment to change shape by altering cytoskeletal structure dynamically, (2) loss of the cell wall, permitting vigorous changes in shape and (3) specialisation of the plasma membrane for invagination into food vacuoles, which perhaps could only be achieved after it had been freed from its commitment to chemiosmotic proton pumping. According to Vellai and Vida, this ensemble of changes is only made possible by compartmentalisation of ATP production, effectively giving eukaryotic cells multibacterial power, analogous to multiple horsepower. This is almost certainly true, and might well help explain why bacteria do not practise phagocytosis.

But there are two problems with this argument as an explanation for the origin of eukaryotes, rather than the process of phagocytosis itself. First, the key assumption is that prokaryotes are typically subsaturated with ATP, whereas eukaryotic cells are somehow not – that compartmentalisation was enough to improve the efficiency of respiration and cure the ATP subsaturation. Perhaps phagocytosis did, because it enabled predation, which in turn meant that selection for the speed of replication was no longer overwhelming: eukaryotes could now compete by eating the competition, rather than racing against it. However, as Vellai and Vida (1999) themselves point out, phagocytosis is complex, requiring a dynamic cytoskeleton, loss of the cell wall, an endomembrane system, and the targeting of digestive enzymes to food vacuoles. These did not evolve overnight, and might well never have evolved at all in fungi (Martin et al. 2003; Martin 2005), which are nonetheless morphologically complex. In the meantime, we are faced with the question: how did energy compartmentalisation, in itself, solve ATP subsaturation, to permit genomic expansion? The problem is all the more acute, as eukaryotic cells are larger, and presumably have a higher demand for ATP to complete cell division. They have mitochondria, certainly, but with no better access to food they have no advantage, just a lot of hungry mitochondrial mouths to feed. There is no evidence that bacteria are any less saturated with ATP than are eukaryotes. It seems that the argument from horsepower puts the cart before the horses.

## 2.3 Compartments

The second problem relates to the nature of energy compartmentalisation: if it was such a big advantage, why did bacteria not solve it for themselves? There are good examples of energy compartmentalisation in bacteria, such as *Nitrosococcus* and *Nitrosomonas* (Madigan et al. 2002), which do have a eukaryotic 'look' about them. Furthermore, the speed and efficiency of bacteria at generating ATP often massively outstrips that of eukaryotic cells. For example, the speed of ATP production by means of oxidative phosphorylation in *Azotobacter* is 5,000 times faster than in ourselves per gram weight (Schatz 2003). Fermentation is also typically faster, giving fermenting bacteria an edge in terms of replicative speed from the same resources (Pfeiffer et al. 2001; Cox and Bonner 2001). If bacteria outstrip eukaryotes in ATP production, they should be less subsaturated in ATP. And if they can compartmentalise their own ATP production, as some certainly can, then surely they have every advantage. Why did they not use them, and expand in genome size and complexity as the eukaryotes did?

The answer I shall put forward in this chapter lies in the need for tight *genetic* control over electron transport linked to chemiosmotic proton pumping, in particular to the requirement for co-localisation of genes with the respiratory chains to control redox poise in mitochondria, as argued by John Allen in Chap. 3 of this volume. I shall take this hypothesis a step further, to propose that only endosymbiosis could compartmentalise the correct core contingent of genes in the energy-producing organelles of eukaryotes (both mitochondria and chloroplasts) and that the ability to regulate electron transport and chemiosmotic proton pumping over a wide area of internal membranes led directly to the greater complexity of the eukaryotes. I shall make four key points:

1. Bacteria are limited in their size and genetic complexity by heavy selection for fast replication and small genome size. The selection pressure to lose genes is set against the ubiquity of gene duplication and lateral gene transfer, which mitigate the potential calamity of losing necessary genes during short periods of disuse.
2. Because chemiosmotic proton pumping takes place across the plasma membrane, a 2D sheet, ATP generation scales with the 2/3 power of cell volume. Unless bacteria alter the density of the respiratory complexes, or the speed of electron transport, their respiratory efficiency (ATP production per unit mass) tails away with increasing size. Eukaryotes escape this fate by compartmentalisation of energy generation in mitochondria, enabling larger size without a corresponding loss of energetic efficiency (assuming that cell surface area is sufficient for nutrient absorption).
3. Bacteria were unable to compartmentalise themselves in this way because a core contingent of genes is necessary to maintain redox poise. Without endosymbiosis, bacteria have no means of allocating the correct contingent of genes to regulate redox poise across large areas of bioenergetic membranes. All mitochondria have retained the same core contingent of genes (Gray 1999; Allen 1993, 2003; Allen and Raven 1996; Race et al. 1999). In this chapter, I postulate that the only way that eukaryotic cells could do so was by way of endosymbiosis.
4. Unlike bacteria, larger eukaryotic cells are energetically more efficient, according to the biological scaling laws of metabolic rate. Further, the optimal karyoplasmic ratio means that larger cells have larger nuclei. Large nuclei accumulate more DNA, as argued by Cavalier-Smith (1985, 2005), which provides raw material for greater complexity (Zuckerkandl 2002). Thus, bacteria are under a heavy selection pressure to remain small and lose DNA, while eukaryotes are under a selection pressure to become larger and to expand their genomes, giving them the raw materials necessary for greater complexity.

## 2.4 Dynamics of Gene Gain and Gene Loss in Bacteria

Reductive evolution is now thought to play a pervasive role in bacteria. The classic examples are obligate intracellular parasites such as *Rickettsia prowazekii* (Frank et al. 2002; Andersson and Kurland 1998), but the rule is general. Many facets of prokaryotic genome organisation attest to selection for small genomes, including the prevalence of haploid genomes, the low proportion of non-coding DNA, the rarity of introns, the existence of a single copy most genes, and the organisation of functionally related genes into operons, lowering the costs of transcriptional regulation (Vellai and Vida 1999; Vellai et al. 1998; Fonstein and Haselkorn 1995).

The strength of the selection pressure to lose genes in free-living bacteria is hard to measure, as it can obviously be difficult to know exactly which genes are lost without trace. However, the tendency to gene loss can be estimated on the basis of the dynamic balance between gene loss and gene gain. Most prokaryotic genomes are between 0.6 and 9.5 Mb in size, a range of barely more than tenfold, despite the fact that bacteria regularly take up genes by lateral gene transfer (Cavalier-Smith 2005; Doolittle et al. 2003). In contrast, eukaryotic genome sizes vary by more than 4 orders of magnitude, from around 12 Mb in *Saccharomyces cerevisiae* up to 670,000 Mb in *A. dubia* (Cavalier-Smith 1985; Gregory 2001, 2002, 2005). Given that bacterial genomes do not expand upwards in a eukaryotic fashion, there must be a dynamic balance between gene gain and gene loss. Genes can be gained by gene genesis, in particular via duplication and divergence of genes and even whole genomes, or by lateral gene transfer from other bacteria (Kunin and Ouzounis 2003).

From an evolutionary point of view, lateral gene transfer is mainly considered in terms of its potential to confound gene trees (Doolittle et al. 2003; Feil et al. 2001). Within particular species, lateral transfer and recombination is indeed the predominant form of genetic variety, accounting for as much as 97% of variation in *Neisseria gonorrhoeae*, for example (Maynard Smith et al. 1993, 2000; Spratt et al. 2001). In this case, even the gene for 16S ribosomal RNA, often thought of as a true guide to phylogeny, part of the small, unshifting, core bacterial gene set, is in fact passed around by lateral transfer (Smith et al. 1999; Spratt et al. 2001). The scale of lateral gene transfer in other bacterial species is illustrated by two different strains of *Escherichia coli*, which differ more radically in their gene content (a third of their genomes, or nearly 2,000 different genes) than all mammals, perhaps even all vertebrates (Doolittle et al. 2003).

The extent to which lateral transfer confounds evolutionary trees is uncertain. In one study, Kunin and Ouzounis (2003) analysed the distribution of 12,762 protein families on a phylogenetic tree derived from the entire genomes of 41 species of bacteria and ten species of archaea. They examined the presence or absence of genes on this tree. If a gene was consistently present in a clade, they concluded the corresponding gene had been present in the ancestor of that clade; occasional absence was ascribed to gene loss, while fragmented distribution across distantly related species was ascribed to lateral gene transfer. The authors concluded that gene loss was the dominant driving force in prokaryotic evolution, while gene duplication was twice as common as lateral transfer. The number of protein families involved in lateral gene transfer was estimated to be 25–39% of the protein families examined, while 60% of families appeared to have been inherited vertically.

This is good news from the point of view of the veracity of gene trees – they are not completely garbled by lateral gene transfer across species and clades, and vertical inheritance apparently still shines through – but the study still corroborated the importance of lateral gene transfer as one of the major driving forces of prokaryotic evolution, partially balancing the strong tendency to gene loss. In fact, the importance of lateral transfer is likely to be greater than suggested by this study, as the estimates of its prevalence were insensitive to transfers within single species or closely related species, which are presumably more common than lateral transfer across clades (because the genes are closely related, so homologous recombination is likely to be successful). Despite these shortcomings, this large study nonetheless illustrates well the dynamic balance between gene loss and gene gain in bacteria.

The benefits of a dynamic balance can best be understood in the context of a heavy pressure to lose genes, combined with fluctuating environmental conditions. The pressure to lose genes derives from the time and energy cost of replicating large genomes. Under optimal conditions, cell division in *E. coli* takes just 20 min, even though it takes 40 min to replicate the full genome (O'Farrell 1992). The only way *E. coli* can complete its cycle of cell division in 20 min is by initiating the second round of genome replication before the first round is complete (O'Farrell 1992; Donachie and Blakely 2003). Clearly, the smaller the genome, the less time it takes to replicate. This applies to cells growing in a single population, rather than to different species, which replicate at disparate speeds that are not necessarily related to their genome size (Vellai, personal communication). In a striking study, Vellai et al. demonstrated the importance of genome size in *E. coli* by engineering three plasmid vectors, each of different size (6, 11.8 and 15.5 kb), but all of them encoding a gene conferring resistance to the antibiotic ampicillin. Cells containing vector DNA were cultured in the presence or absence of ampicillin, and growth curves were constructed. In the absence of ampicillin, all three cultures grew at a similar rate, but under the restrictive phase (when the culture medium begins to be exhausted), the proportion of viable cells retaining their vector DNA declined in proportion to the size of the vector. In other words, in the restrictive phase, equivalent to natural conditions, cells were under a heavy selection pressure to jettison unnecessary DNA. When cells were grown in the presence of ampicillin, the outcome was very different. Cells losing their vector were rendered vulnerable to ampicillin, so all viable cells had retained the vector. Now, in the restrictive phase, the cells with the smallest plasmids (6 kb) outgrew the cells with the largest plasmids (15.5 kb) by an order of magnitude over a 12-h period ($10^{10}$ versus $10^{9}$ cells). So, for *E. coli*, larger genomes are penalised by natural selection, especially under restricted conditions, when cells are subsaturated with ATP.

These studies are revealing, for they imply that there is a strong pressure to lose superfluous genes from a bacterial population, a pressure which is presumably balanced by the ability to regain genes by lateral gene transfer, either from within the species or elsewhere, whenever conditions change again. Plasmid transfer could easily balance gene loss from vectors – this is well known to account for the swift rise of antibiotic-resistant strains among pathogens (Maynard Smith et al. 2000; Spratt et al. 2001) – but presumably the same process also occurs in bacterial chromosomes, if more slowly, given a dynamic balance between gene loss and gene gain, and the roughly constant size of bacterial genomes. Gene loss is random, and may cause cell death (if an essential gene is deleted), or decrease growth rates (if functioning but non-essential genes are deleted); any clones that are affected in this way will be eliminated by natural selection. On the other hand, DNA loss can be advantageous if it eliminates unused sequences, for this speeds replication, if only slightly, as shown by Vellai et al. (1998). Importantly, it takes a prolonged period, relative to environmental fluctuations (such as seasonal fluctuations in substrate concentration) to lose all copies of any particular gene from a population or species, in any one environment. Furthermore, bacteria can recombine gene fragments to regenerate a functional gene by homologous recombination (Maynard Smith et al. 1993; Vellai, personal communication). Indeed, DNA sequencing has revealed that individual bacterial genes have a mosaic structure that could only have arisen by homologous recombination, even in bacteria not previously thought to show natural transformation (Maynard Smith et al. 2000; Spratt et al. 2001). It is therefore likely that, if conditions fluctuate, then functional copies of deleted genes can be regenerated within a population as a whole. Despite the heavy rates of gene loss over evolutionary time, it must be relatively rare, in terms of the dynamics of population turnover, to lose a gene from a population altogether. As Vellai (personal communication) put it, "Natural bacterial populations contain dispersed genomes, which reflects a community-like organization of the genetic material."

Dispersed genomes are the optimal stable strategy for fluctuating environmental conditions, as only cells that can *draw on* large genomes can survive most exigencies, while at the same time bacteria that invariably *retain* large genomes are outcompeted by bacteria with smaller genomes, which are viable *at that particular time*. The combination of gene loss with lateral transfer enables speedy replication and genetic resourcefulness: bacteria that economise and load genes dynamically thrive in place of bacteria with large, metabolically flexible genomes, or bacteria with small but inflexible genomes. If conditions are relatively stable, however, the rules shift. If lean resources are the rule, bacteria can tolerate larger genomes, as the metabolic flexibility can enable ATP to be produced more effectively than competitors. This tendency is borne out by an examination of 115 fully sequenced bacterial genomes by Konstantinidis and Tiedje (2004). They found that the bacteria that have the largest genomes dominate in environments where resources are scarce but diverse, and where there is little penalty for slow growth, such as soil. Soil bacteria produce on average three generations in a year, so there is less selection for speed than for any replication whatsoever (Konstantinidis and Tiedje 2004). Under such conditions an ability to take advantage of scant resources is important, which requires more genes to encode the metabolic flexibility needed (Stêpkowski and Legocki 2001). It is no accident that *Streptomyces avermitilis*, ubiquitous in the soil, is metabolically versatile with a big genome to match. Even so, there is still selection for small size in relation to other versatile bacteria, and presumably this sets the bacterial genome ceiling of about 9.5 Mb. In other words, even the most versatile bacteria have small genomes compared with those of the eukaryotic cells living in the same environment. So how were the eukaryotes released from a selection pressure that stifles even the most versatile bacteria?

## 2.5 ATP Regulation of Bacterial Replication

For both bacteria and eukaryotes, ATP subsaturation is the norm. On a global scale it is clear that this is the case. In optimal conditions, *E. coli* can divide every 20 min, or 72 times in a day. A single *E. coli* bacterium weighs about $10^{-12}$ g, so 72 cell divisions in a day corresponds to an amplification of $2^{72}$ ($= 10^{72} \times \log 2 = 10^{21.6}$), or an increase in weight from $10^{-12}$ g to 4,000 t (O'Farrell 1992). In 2 days the mass of exponentially doubling *E. coli* would be 2,664 times larger than the mass of the Earth (which weighs $5.977 \times 10^{21}$ t). The fact that this does not happen shows that nutrient supply is an overriding limiting factor.

The speed of bacterial replication is closely tied to nutrient availability. DNA replication is controlled directly by ATP availability, through the formation of orisomes (Cunningham and Berger 2005; Leonard and Grimwade 2005). These create a small bubble of unwound DNA within the replication origin, which recruits the helicase. A critical first step is the stable unwinding of the chromosomal replication origin, *oriC*, by the multiprotein orisome complexes, comprising the initiator DnaA and modulator proteins that bend DNA (Leonard and Grimwade 2005). About 20 DnaA molecules bind to at least five 'DnaA boxes' in the *oriC* region of the chromosome (McGarry et al. 2004; Messer 2002). To bind to the DnaA boxes, the DnaA proteins must also bind to ATP or ADP (Messer 2002). The DnaA protein binds either nucleotide with equal avidity, but it seems that only the ATP-DnaA is active in initiating replication, at least in critical boxes (McGarry et al. 2004). ATP binding promotes an allosteric modification, and does not provide energy for the unwinding reaction, as non-hydrolysable ATP analogues are equally effective (Leonard and Grimwade 2005).

Presumably, such a mechanism calibrates the balance of ATP to ADP within the cell. During active replication, some 1,000–2,000 DnaA proteins are present in the cell, and occasionally as many as 10,000 (Donachie and Blakely 2003). These bind to either ATP or ADP, which in turn enables thermodynamic control over bacterial replication. This is because the Gibbs free energy of ATP hydrolysis depends on the ratio of ATP to ADP, not their total concentration (Jensen et al. 1995). The energy of ATP resides not in a 'high-energy' bond, as is often stated (the bond is actually quite low in energy, which is why it is so easy to break, and thus useful), but in the displacement of the chemical equilibrium towards ATP from ADP (Harold 1986, 2001). Increasing the ATP concentration is only helpful if it is at the expense of ADP concentration. The cell must have a mechanism to determine whether ATP supplies are sufficient at the outset to complete cell division, and this depends on the thermodynamic driving force – the ratio of ATP to ADP. The cell can monitor the driving force simply by means of a molecular switch that has an equal binding affinity for ATP or ADP (reflecting their concentration ratio), but which is unequally activated by nucleotide binding. To be genuinely sensitive to the ratio also requires a large statistical 'sample size' (just as political polls are only reliable if they question a representative section of the population) and this may account for the surprisingly high number of DnaA proteins, totalling several thousand.

Such thermodynamic control of cell replication is also evident in DNA supercoiling (Jensen et al. 1995; Koedoed et al. 2002; Snoep et al. 2002). Negative supercoiling facilitates the opening of the double helix, which is necessary for growth-related processes such as transcription and replication. This is achieved with the aid of enzymes like DNA gyrase. In vitro studies with isolated DNA gyrase from *E. coli* show that the activity of the enzyme is sensitive to the ratio of ATP and ADP, whereas the absolute concentrations of the nucleotides have no effect on the enzyme activity (Westerhoff et al. 1988; Westerhoff and van Workum 1990). When cells run out of growth substrates (or are in the presence of respiratory uncouplers), the changes in DNA supercoiling stimulate expression of some genes, while repressing expression of others, thereby adjusting the pattern of gene expression to the new conditions (Jensen et al. 1995). In vivo, in *E. coli* at least, when the ratio of ATP to ADP falls to less than 1, there is a general inhibition of transcription and replication, and so of cell division (Jensen et al. 2001). Thus, it seems there are a number of ways in which cells can use the thermodynamic indicator of the ATP to ADP ratio to control cell division.

Another control over bacterial division is cell size, which is also regulated by the ATP to ADP ratio (Donachie and Blakely 2003). Initiation of replication takes place at a fixed cell volume, the 'initiation volume', which is constant under a wide variety of growth conditions, despite the fact that cell size at division varies greatly at different growth rates (Donachie 1968). Once DNA replication has been initiated, the cell then enters an 'eclipse period', during which time it cannot start a second cycle of DNA replication. The end of the eclipse period is related to cell volume, by way of a mechanism that involves the binding of DnaA proteins to various DnaA-binding sequences in DNA, including the *oriC* boxes mentioned before, as well as a number of other sites, such the as *datA* locus (Donachie and Blakely 2003; Christensen et al. 1999). The latter may bind as many as 370 DnaA proteins (Kitagawa et al. 1996; Ogawa et al. 2002). During the eclipse period, most of the DnaA proteins are bound to ADP rather than to ATP, though exactly how the shift is brought about is unknown (Donachie and Blakely 2003). One plausible factor is that the DnaA protein has a weak ATPase activity, which is needed for it to dissociate from DNA: the DnaA protein hydrolyses the bound ATP, leaving it free in the inactive ADP-bound form (Messer 2002). However, such a mechanism could only account for a rather small proportion of perhaps 2,000 DnaA proteins, so there is more to it than that (Donachie and Blakely 2003). Regardless of the precise mechanism, the eclipse period is linked to the ATP to ADP ratio. It begins shortly after initiation of DNA replication and persists for about a third of the cell cycle, during which time the DnaA-ATP to DnaA-ADP ratio reaches a minimum (Donachie and Blakely 2003). Thereafter, the ratio climbs again with de novo DnaA-ATP production, at a rate that parallels protein synthesis. This implies that the ratio of DnaA-ATP to DnaA-ADP increases in parallel with cell size, until it reaches a critical value, at which the active form can compete successfully with the inactive form for binding at *oriC*, and thus initiate a new round of DNA replication (Donachie and Blakely 2003). Such an interpretation is corroborated by mutants lacking the *datA* locus, which show asynchronous initiation of replication (Ogawa et al. 2002), and by mutants with additional *datA* copies, which have a higher initiation volume (a larger cell size at the initiation of DNA replication), consistent with delayed initiation of replication (Kitagawa et al. 1996).

All these various mechanisms tie bacterial replication to ATP production. Clearly, the more efficiently and quickly a cell produces ATP, the more quickly it can replicate. The balance between speed and efficiency is an interesting one, for fermentation is faster than oxidative phosphorylation, but has a substantially lower yield of ATP, for it only partially oxidises the substrate. This leaves substrate for cells capable of oxidative phosphorylation. Thus, cells growing by fermentation typically outcompete respirers, at least in the short term (Pfeiffer et al. 2001; Cox and Bonner 2001) Afterwards, their growth is rapidly curtailed by the exhaustion of substrate. If anaerobic bacteria excrete waste products such as ethanol into their surroundings, then slowly replicating aerobic bacteria may ultimately outgrow them by oxidising the ethanol. The most able replicators might therefore be bacteria capable of facultative anaerobic respiration, exactly those proposed by Martin and Müller (1998) in the hydrogen hypothesis to be the free-living ancestors of the mitochondria, such as *Rhodobacter* (see also Müller and Martin 1999; Tielens et al. 2002). Interestingly, an analysis by Pfeiffer et al. (2001) suggests that obligate aerobic cells, which have a high yield but low rate of ATP production, might do best when grouped in clusters, hiving off substrate for later use. Such collaborative behaviour could have been an incentive towards the origin of multicellular colonies in eukaryotes (Pfeiffer et al. 2001; Cox et al. 2001). In prokaryotes, however, obligate aerobic respiration is most commonly found in obligate intracellular parasites such as *Rickettsia* (which may gain by maximising ATP yield in their constricted intracellular environment) and some soil bacteria, where growth is slow, owing to scarcity of nutrients, so ATP yield is critical (Stêpkowski and Legocki 2001).

In considering the efficiency and speed of ATP production in bacteria, size matters. Bacteria respire over their plasma membrane, pumping protons into the periplasm (in Gram-negative bacteria, or the equivalent space between the plasma membrane and the cell wall in the other prokaryotes; Vellai et al. 1999; Harold 2001; Lane 2005). This means that the efficiency of ATP production depends on the surface-area to volume ratio. As bacteria become larger, their chemiosmotic ATP production depends on the surface area of the plasma membrane, which increases with the square of cell dimensions, whereas the volume of the cell increases with the cube of its dimensions. Overall, as cell size increases, the metabolic rate scales with the 2/3 power of cell volume, all other factors being equal. Excluding vacuoles (which, for example, explain the giant size of *Thiomargarita namibiensis*, with its diameter of up to 300 μm; Schulz et al. 1999), larger bacteria are composed of more structural macromolecules. These in turn consume more ATP for their synthesis, so larger bacterial cells must have a higher requirement for ATP, coupled with a relatively lower capacity for producing that ATP. Presumably, the ATP to ADP ratio takes longer to arrive at the critical threshold, such that the eclipse period is longer and cell replication is accordingly slower. As bacteria become larger, their energetic efficiency tails away, a serious cost liable to be penalised by selection.

There are of course many adaptations that bacteria could make, which influence the speed or efficiency of ATP production. Fermentation, for example, does not depend on the surface area and so is not affected by size. Relatively simple changes in cell shape affect the surface-area to volume ratio: rods have a larger surface area relative to their volume than spheres, perhaps conferring a slight selective advantage on bacilli compared with cocci. The kinetic efficiency of respiratory enzymes could be optimised, as can their packing density within the membrane. All these changes have costs attached, however. Fermentation has a low ATP yield, which probably restricts cell size, as smaller cells can replicate more times given limited substrate; complex morphology requires more genes to encode it, and macromolecules to build it; optimising the efficiency of respiratory enzymes requires greater selectivity for their substrates or electron acceptors, to the exclusion of others that may at times be more abundant. Increasing packing density of respiratory complexes must affect the density of other membrane proteins needed for nutrient absorption, homeostasis, motility, or sensing the concentration gradient of various substrates or toxins. And if the packing density of all membrane proteins rises, then the fluid properties of the lipid phase must change, again with potentially detrimental consequences. While species adapted to different environments might benefit from any of these adaptations, within a single population of cells competing for the same resources, the most important factor is cell size. If all else is roughly equal, large cells are penalised by natural selection, as they produce ATP less efficiently than small cells, and so are slower to replicate. As a result, prokaryotes are under a heavy selection pressure to remain small in size, just as they are under a heavy pressure for small genomes (Lane 2005).

## 2.6 Redox Poise Across Bioenergetic Membranes

There is one change that bacteria could make to enable larger size without a loss of energetic efficiency, and that is internalising the bioenergetic membranes. If the speed of respiration in the cell depends on the surface area of bioenergetic membranes, then obviously the more membranes the faster the ATP production. What is more, internalising respiration frees up the plasma membrane to specialise in other tasks, such as the absorption of nutrients, tasks that no longer need to compete for space with the respiratory or ATPase complexes. This is of course exactly what the eukaryotes did, for the surface area of bioenergetic membranes in eukaryotic cells depends on the density of cristae and the total number of mitochondria, but is no longer tied to the external surface area of the cell. The question is: why did bacteria not internalise their own bioenergetic membranes? In some cases, such as *Nitrosomonas* and *Nitrosococcus*, they did (Madigan et al. 2002), so the question is again really one of degree: why did the bacteria once again make a start, and then stop short?

The answer is to be found not in the bacteria themselves, but in the mitochondria. The mitochondria were, after all, once bacteria, and their respiratory prowess is bacterial (Berry 2003). Within the context of the cell as a whole, what are the differences between the internal membranes of bacteria like *Nitrosocossus* and the cristae of mitochondria? The answer, surely, is *genes*. All known mitochondria in all known eukaryotes have retained a small genome, closely associated with the inner mitochondrial membrane (Gray et al. 1999, Burger et al. 2003). In contrast, the internal membranes of *Nitrosococcus* are not associated with any core contingent of genes (Madigan et al. 2002). Why mitochondria retained their own genes is not generally agreed, but Allen (1993, 2003) has put forward a convincing case that both chloroplasts and mitochondria retain their genomes to control redox poise during electron transport coupled to chemiosmotic proton pumping. If so, then perhaps bacteria cannot expand their internal membranes because they cannot localise the correct contingent of genes with their bioenergetic membranes, except by way of endosymbiosis.

Allen's theory of redox poise, and the evidence supporting it, are discussed in Chap. 3 of this volume. Here, I want to make a few general observations on necessity and workability. Each mitochondrion needs a genome because the speed of electron flow down the respiratory chains depends not just on supply and demand (concentration of NADH, $O_2$, ADP and inorganic phosphate) but also on the number and redox state of respiratory complexes (Allen 1993,

2003; Allen and Raven 1996; Race et al. 1999; Otten et al. 1999; Blackstone 2000). If there are too few complexes, electrons pass slowly down the respiratory chains from NADH to oxygen, even if demand is high and substrates and electron acceptors are abundant. Conversely, if there are too many complexes, electrons accumulate in the chain, perhaps resulting in a higher free-radical leak (Race et al. 1999; Allen 1993, 2003; Lane 2005). Thus, it is necessary to regulate the turnover of respiratory complexes within individual mitochondria to correspond with demand, maintaining redox poise and minimising free-radical leak. Respiration is fastest when a balanced number of electron carriers are balanced in their redox state, which is to say 50% are reduced and 50% oxidised (redox poise; Allen 2003).

The difficulty is a matter of targeting. If one mitochondrion needs more cytochrome oxidase, it could signal its deficiency to the nucleus, stimulating transcription of nuclear genes; however, if there are several hundred mitochondria in a cell, then there is a problem targeting de novo cytochrome oxidase to the correct mitochondrion. If the new protein is delivered to all the mitochondria, 99% will receive too much, and 1% will receive too little. Redox poise is lost in all the mitochondria. In contrast, if a core of genes is retained in individual mitochondria, then the signal is sent only as far as the local genome. It responds by synthesising more cytochrome oxidase on site, and restores the correct balance. Such a rapid local response could take place in any of the cell's mitochondria and might in principle operate quite differently in different mitochondria in the same cell at the same time. Redox poise is retained in all mitochondria. In other words, by having a small genome within each mitochondrion, despite the costs (such as maintenance costs of numerous copies of the requisite genetic machinery, and a fast mutation rate of mitochondrial DNA), the cell as a whole can maintain redox poise across a wide area of bioenergetic membranes (Allen and Raven 1996).

Exactly how such signalling works is not known. In fact, however, a simple mechanism exists that depends on no more than the biophysics of electron flow down the respiratory chains, as argued by Blackstone (Blackstone and Green 1999; Blackstone 2000). In essence, the rate of free-radical leak does not depend on the speed of respiration, but rather on the redox state of the respiratory complexes: a high reduction state of complex I (above about 90% reduced) leads to a high free-radical leak (Turrens 2003; Barja 1998; Brand et al. 2004; Skulachev 1998, 2004). This means that the signal might simply be free-radical leak: a rise in free-radical leak alters gene activity through the action of redox-sensitive transcription factors, or differential stability of transcripts (of which there are many examples; Allen 2003). So, for example, if cytochrome oxidase (complex IV) is deficient, the redox state of complex I will rise, and with it free-radical leak. One question is how the cell interprets this signal to 'know' that more cytochrome oxidase is needed. Free-radical leak also rises if there is a low demand for ATP: the redox state of complex I then rises as electrons accumulate in the chain, but this leak could not be relieved by building new complexes. In principle, though, the mitochondria could detect ATP levels and so combine two signals: 'high ATP' and 'high free radicals'. An appropriate response would now be to dissipate the proton gradient, to maintain electron flow, and there is indeed evidence that this happens (Brand et al. 2004). In contrast, if there were not enough respiratory complexes, ATP levels would decline and electrons would again accumulate in the respiratory chains. Now the signal would combine 'low ATP' with 'high free-radical leak'. This system could in theory discriminate the need for more respiratory complexes from low ATP demand. Similar signalling systems operate at a cellular level to signal apoptosis in eukaryotes ('high free radicals', 'low ATP'; Zamzami et al. 1995, Richter et al. 1996; Ott et al. 2002) and homologous recombination in simple eukaryotes like *Volvox carteri* and *S. cerevisiae* ('high free radicals', 'high ATP'; Brennan and Schiestl 1998; Nedelcu et al. 2004). In the case of *V. carteri*, a twofold rise in free-radical production activates the sex genes, leading to the formation of new gametes (Nedelcu et al. 2004). Importantly, this effect can be induced by inhibitors of the respiratory chain, which increase the rate of free-radical leak (Nedelcu et al. 2004).

A problem with this model in mitochondria is that the respiratory complexes are constructed from a large number of subunits, as many as 48 separate proteins in complex I. Mitochondrial genes encode a handful of these subunits (seven in the case of mammalian complex I), but nuclear genes encode the great majority. How, then, could the mitochondrial genes dominate? They could do so if the respiratory complexes assemble themselves around a few core subunits; and there is evidence that this happens (Ugalde et al. 2004; Vogel et al. 2004; Bai et al. 2004; Cardol et al. 2002). If the mitochondrial genes encoded these critical or rate-limiting subunits, then they would control the assembly of new complexes. Effectively, mitochondria would make a construction decision, and plant a flag in the membrane; and the nuclear components then assemble around the flag. Again, there is evidence that this is indeed what happens (Chomyn 2001). Given that the nucleus serves hundreds of mitochondria at once, the total number of flags in the cell at any one time might remain fairly constant. There would be no need to change the overall rate of nuclear transcription to compensate for fluctuations in individual mitochondria, but the effect would be to keep a tight grip on the rate of respiration in all the mitochondria in a cell at once. This is indeed a plausible mechanism, for in both the mitochondria and the chloroplasts, the organellar genome encodes the core subunits of complexes (Chomyn 2001; Race et al. 1999; Allen 2003).

If these arguments are correct, then they can explain why bacteria are limited in their internal membrane systems: they are unable to retain redox control over a larger area of membranes because they cannot localise the correct core contingent of genes. In the light of sections 2.4 and 2.5, it is salutary to consider the penalties for redox control. The simplest way would be to copy a subset of genes and delegate it to regulate the extra membranes – but how are the necessary genes chosen? There is no way that I can think of that does not involve foresight, and evolution has none. The only way that such delegation could work would be to replicate the entire genome, and whittle away at one of the two genomes until all superfluous genes were deleted or transferred to the host cell (as happened in mitochondria). But which genome should lose its genes? Both must be active for redox control to work. In the meantime, however, the bacterium has two active genomes, each under a heavy selection pressure to lose superfluous genes. Either genome might be expected to lose some genes, but two competing genomes in the same cell raises the spectre of genomic conflict, certainly not aiding it in competition with other cells (Cosmides and Tooby 1981; Hurst and Hamilton 1992; Partridge and Hurst 1998; Ridley 2001).

Genomic conflict might be prevented if it were possible to demark the sphere of influence of each genome. The eukaryotes solved the problem by sealing off the mitochondrial genome in a double membrane. This is not possible in bacteria, however. If the spare set of genes were sealed off, there would be no way of extracting ATP. The ADP/ATP carriers do not exist in bacteria, with the exception of obligate intracellular parasites such as *Rickettsia*, whose ADP/ATP carriers bear some similarities to plastid carriers and appear to branch deeply in the eukaryotic tree (Amiri et al. 2003; Andersson et al. 2003); but exporting ATP is of course a suicidal trait for free-living bacteria. The mitochondria ADP/ATP carriers, along with the family of 150 mitochondrial transport proteins to which they belong, are purely a eukaryotic invention: their gene sequences are closely related in all eukaryotes, but there are no similar bacterial genes (Löytynoja and Milinkovitch 2001; Santamaria et al. 2004). This implies that the ADP/ATP carriers had evolved in the last common ancestor of all the eukaryotes, before the divergence of the major groups.

The eukaryotes had time to evolve ADP/ATP carriers because the metabolic syntrophy between the two partners of the original chimeric eukaryote was stable, providing ample time and stability for evolutionary changes to take place (Martin and Müller 1998). In the case of bacteria evolving by natural selection (rather than symbiosis), however, there is no corresponding stability. Simply duplicating a gene set and sealing it off in a membrane could in itself provide no advantages in the interim. Far from it: maintaining extra genes and membranes without any benefit is costly, and would be eliminated by natural selection. Whichever way we look at it, selection pressure is likely to jettison the burdensome additional genes needed for respiratory control over a wide area of membranes in bacteria. The most stable state is always a small cell that respires across the plasma membrane, using a periplasmic space to generate a chemiosmotic proton gradient – in short, a bacterium. Because selection is probabilistic, small cells will almost invariably be favoured over larger, inefficient, free-radical-leaking competitors. And that is why bacteria are still bacteria.

## 2.7 Allometric Scaling of Metabolic Rate and Complexity

Metabolic syntrophy, such as posited by the hydrogen hypothesis, explains why only endosymbiosis is stable enough over evolutionary time to enable multiple core contingents of genes controlling redox poise to be co-localised with a wide area of bioenergetic membranes. Although such endosymbiosis is common in eukaryotes, it is far less common in bacterial communities, which tend to live together but not inside one another. A major obstacle to endosymbiosis in bacteria is the absence of phagocytosis among prokaryotes, and indeed Cavalier-Smith (2002) has argued that the first eukaryote was necessarily phagocytic. There are nonetheless examples of bacteria living inside other bacteria, without a hint of phagocytosis, so clearly it is possible, if uncommon, as postulated by the hydrogen hypothesis (von Dohlen et al. 2001; Hoffmeister and Martin 2003; see also Chap. 4 by Sapp in this volume). If the ideas in this chapter are correct, then the advantages of eukaryotic energy generation in mitochondria can only be gained by endosymbiosis in bacteria. Because only a handful of scenarios explain how an endosymbiosis could have led to the rise of a eukaryotic cell, this helps to explain why the eukaryotes only arose once in the history of life on Earth.

There is still a question about the rise of complexity to answer, however, and it is this. Once a small, if incipiently complex, eukaryotic cell was established, what 'drove' the eukaryotes on to greater size and complexity? Gould (1997) and others have argued that there is no inherent direction in evolution, that the rise of complexity was little more than a random walk stopping off at vacant niches on route. Because the bacteria and archaea already occupied all the simplest niches, the eukaryotes had no option but to evolve in the direction of greater complexity. Nonetheless, bacteria and archaea dominated the Earth for three billion years, yet never showed any tendency towards greater morphological complexity in that time, nor indeed towards greater biochemical complexity, if the major geochemical cycles were all in place 2.7 billion years ago (Martin et al. 2003; Lane 2002; Anbar and Knoll 2002; Nisbet and Sleep 2001; Canfield et al. 2000; Castresana and Moreira 1999; Canfield 1998). Yet once the eukaryotes had evolved, there was a gearshift. In less than half the time open to bacteria, the eukaryotes developed elaborate endomembrane systems, complex cell cycles, specialised organelles, mitosis, meiosis, huge genomes, phagocytosis, predatory activity, multicellularity, differentiation, large size, and spectacular feats of mechanical engineering: flight, sight, hearing, echolocation, brains, sentience. Insofar as this progression happened over time, it can reasonably be plotted as a ramp of ascending complexity. So we are faced with the prokaryotes, with nearly unlimited biochemical diversity but no drive towards complexity, and eukaryotes, which have little biochemical diversity, but a marvellous flowering in the realm of bodily design. Is there anything about the evolution of mitochondria that made the evolution of larger size and greater complexity not just possible, but probable?

The answer is yes: the scaling of metabolic rate with size (Kleiber 1961). Allometric scaling laws in biology have undergone a renaissance in interest since West et al. (1997) proposed a fractal model to explain the 'three quarters law' of metabolism, the apparent scaling of metabolic rate with mass to the power of 3/4 across an extraordinary 27 orders of magnitude, from respiratory complexes within mitochondria to the blue whale. In essence, the fractal model argues that metabolic rate is determined by the fractal properties of the supply network, i.e. fractal geometry constrains energy metabolism (West et al. 1997, 1999, 2002; Weibel 2002). This theory has been criticised from various points of view (Lane 2005), notably (1) three-quarter scaling is an illusion; in fact metabolic rate more commonly scales with $m^{2/3}$, where $m$ is mass (Heusner 1991; Dodds et al. 2001; White and Seymour 2003); (2) the theory predicts that the fractal supply network constrains the resting metabolic rate, whereas in fact it can only constrain maximal metabolic rate, which scales with 0.88, not 0.75 (Bishop 1999); (3) in mammals, the scaling of the supply network generally reflects tissue demand, so capillary density scales with $m$ in muscle tissue; (4) different tissues scale differently with body mass (Bennett 1988; Rolfe and Brown 1997; Porter 2001). Bone, for example, is metabolically inert. Because bone strength depends on cross-sectional area, and the weight it bears depends on body mass, a greater proportion of body mass must be composed of metabolically inert bone with increasing size; and this must affect the scaling of metabolic rate with mass. This last point was the basis of the allometric cascade model of Hochachka et al. (Hochachka et al. 2003; Darveau et al. 2002) using metabolic control analysis . Their model demonstrated that different tissues scale differently with body mass, according to demand and the economies of scale. They showed that, in animals, larger size is metabolically more efficient, i.e. the specific metabolic rate falls with size as tissue demand falls, and the aerobic capacity correspondingly increases. As a result, both organ size and mitochondrial density in organs fall with rising body size. Importantly, this is not a constraint of greater size, but an opportunity, which may, for example, have given rise to endothermy in mammals and birds (Lane 2005).

There has been far less work at the level of individual cells, but the scaling of metabolic rate with cell size has been studied in various unicellular organisms. Despite claims that metabolic rate scales with $m^{0.75}$ in single cells, a re-evaluation by Prothero (1986) and other groups declared the relation to be "generally not at all persuasive" (Dodds et al. 2001). Metabolic rate actually scales with an exponent of between 0.3 and 1, depending on which groups are considered (Prothero 1986). The fact that the exponent is generally less than 1 means that metabolic rate falls relative to mass as cell size increases (a cell of twice the volume has a metabolic rate that is less than double); but it is unclear whether that is an opportunity or constraint of size. In bacteria, metabolic rate often scales with $m^{2/3}$ because bacteria respire over the plasma membrane. This is a constraint of size, and helps to explain why bacteria remain small (Vellai et al. 1998; Vellai and Vida 1999; Lane 2005). However, in animals, Hochachka et al. (2003) showed that the metabolic rate falls with size as a result of the economies of scale: an opportunity of size, not a constraint. So which of these two possibilities applies to unicellular eukaryotes? The answer is not known, but I suspect that even unicellular eukaryotes do gain from the economy of scale. As in society, the benefits depend on set-up costs, operation costs and distribution costs, and these impose upper limits on the economies of scale. But within these limits, the benefits ought to apply widely. This is because living organisms are highly conservative in their operational principles. In particular, their organisation is invariably modular. Both unicellular and multicellular organisms are composed of a mosaic of modular units. Modular functions in unicellular eukaryotes include transcription and translation, protein synthesis, glycosylation, ionic homeostasis, ribosomal assembly, membrane synthesis, vacuolar or lysosomal digestion, signalling, chemotaxis, ATP generation, motility, molecular trafficking, and so on. I imagine the economies of scale apply as much to these modular aspects of single cells as to multicellular organisms. Within the limits set by the supply network, single-celled eukaryotes become more efficient as they grow larger: the set-up costs for manufacturing 100 ribosomes are higher than for manufacturing 1,000, so fewer mitochondria are needed to provide the ATP necessary.

This idea brings us back to the question of genome size that I touched on at the start of the chapter. Large cells are energetically more efficient, but large cells usually have a larger nucleus (Cavalier-Smith 1985, 2005; Gregory 2001, 2002, 2005). Balanced growth during the cell cycle apparently requires that the ratio of nuclear volume to cell volume (the karyoplasmic ratio) is roughly constant, for reasons that are disputed (discussed in detail by Cavalier-Smith 2005). Essentially, over evolution, the nuclear size and DNA content adjusts to changes in cell volume and metabolic rate (Kozłowski et al. 2003; Lane 2005). As cells grow larger, they adjust by developing a larger nucleus, with more DNA, even if this extra DNA does not necessarily code for more genes. This explains the C-value paradox, and is why cells like *A. dubia* have 200 times more DNA than a human being, albeit coding for fewer genes. The extra DNA may be partly structural, as argued by Cavalier-Smith (1985, 2005), but it can also be called upon to serve useful purposes, from forming the structural scaffolding of chromosomes, to providing binding sites for transcriptional regulation (Zuckerkandl 2002; Mattick and Makunin 2005; Mattick 2001). It also forms the raw material for new genes and regulatory RNAs, building the foundations of complexity (Zuckerkandl 2002; Mattick and Makunin 2005; Mattick 2001). The sequences of many genes betray their ancestry as selfishly replicating elements (Zuckerkandl 2002). Thus, as soon as eukaryotic cells became powered with mitochondria, there was a selective advantage to them being bigger. Bigger cells need more DNA, and with that they had the raw material needed for greater complexity.

This reverses the selection pressures on bacteria: whereas bacteria are oppressed by heavy selection pressure to lose genes, eukaryotes are under pressure to gain them.

## 2.8 Conclusions

Bacteria are under heavy selection pressure to lose any unnecessary genes, regardless of whether they may prove necessary again in the future (Vellai and Vida 1999; Kunin and Ouzounis 2003). This pressure is balanced by the capacity to regain functional copies of deleted genes by lateral transfer from within the same population, or other species, clades, or even domains (Kunin and Ouzounis 2003). Natural bacterial populations therefore effectively have dispersed genomes, combining fast replication with the genetic versatility necessary for adapting to fluctuating conditions (Vellai, personal communication).

The speed of replication is tied to cell size and ATP availability by the thermodynamic driving force of ATP to ADP ratio (Donachie and Blakely 2003). Both the rate and the yield of ATP generation are important, favouring facultative anaerobic respiration in bacteria (Pfeiffer et al. 2001; Cox et al. 2001). Size matters because the rate of ATP generation by way of a proton gradient across the plasma membrane means that metabolic rate scales with cell mass to the power of 2/3. All else being equal, respiratory efficiency (ATP production per unit mass) declines with rising cell volume. Respiratory efficiency can be maintained in larger cells by internalising the bioenergetic membranes, and some bacteria do have specialised bioenergetic membranes enclosing an extensively folded periplasmic space (Madigan et al. 2002); however, the surface area of such membranes falls orders of magnitude short of the cristae area of energetic eukaryotic cells (Lane 2005).

The surface area of internal membrane systems in bacteria is restricted by the need to maintain tight redox poise during electron transport. This is achieved in mitochondria by retaining a small subset of genes that encode core subunits of the respiratory complexes (Allen 1993, 2003; Allen and Raven 1996; Race et al. 1999). However, neither bacteria nor mitochondria can possibly 'know' in advance which is the correct subset of genes to control redox poise: they were retained in the mitochondria by selection. The ancient mitochondrial genome was gradually whittled away by gene loss and transfers to the nucleus, but if any of the critical genes necessary for maintaining redox poise were lost from the mitochondria, the host cell and its mitochondria would die. Such a gradual whittling process requires stability over evolutionary time that can only be achieved by endosymbiosis. Any other method, such as duplicating the bacterial genome, is liable to be penalised by the heavy selection for small genome size in bacteria. Thus, eukaryotic cells were released from the constraints of a limited bioenergetic surface area by a rare endosymbiotic event between prokaryotic cells, probably some form of metabolic syntrophy, such as that postulated by the hydrogen hypothesis, which ultimately gave rise to the mitochondria (Martin and Muller 1998). Such endosymbiosis is rare but not unknown in prokaryotes. In contrast, without endosymbiosis, bacteria remained under heavy selection pressure for small size and small genome.

Once mitochondria were established, larger cell size became not just possible but probable. This was because metabolic rates scale allometrically with body mass, with an exponent of less than 1, so that doubling the mass does not double metabolic rate (Kleiber 1961; Hochachka et al. 2003). The metabolic rates of unicellular organisms scale with cell size in a similar fashion, probably owing to the economies of scale in modular operative units (Prothero 1986). Larger cell size is therefore more cost-effective in eukaryotic cells. Larger cells usually have a larger nucleus, with more DNA, giving an optimal karyoplasmic ratio during the cell cycle (Cavalier-Smith 2005). So the selection pressure favouring larger cell size also drove the eukaryotic tendency to accumulate more DNA, and indirectly, greater complexity. Thus, mitochondria made larger size and greater complexity probable, rather than staggeringly unlikely, inverting the constrained world of prokaryotes (Lane 2005).

Biological complexity was only made possible by the energetic and genomic reorganisation emerging from a confederacy of endosymbionts, which ceded most of their powers to central government (the host cell nucleus), but retained flexible regional governance of ATP generation. The hypothesis might be known, if with tongue in cheek, as the federal power hypothesis.

## References

Allen JF (1993) Control of gene expression by redox potential ad the requirement for chloroplast and mitochondrial genes. J Theor Biol 165:609–631

Allen JF (2003) The function of genomes in bioenergetic organelles. Philos Trans R Soc Lond B Biol Sci 358:19–38

Allen JF, Raven JA (1996) Free-radical-induced mutation vs redox regulation: costs and benefits of genes in organelles. J Mol Evol 42:482–492

Amaral Zettler LA, Gómez F, Zettler E, Keenan BG, Amils R, Sogin ML (2002) Eukaryotic diversity in Spain's River of Fire. Nature 417:137

Amiri H, Karlberg O, Andersson SG (2003) Deep origin of plastid/parasite ATP/ADP translocases. J Mol Evol 56:137–150

Anbar AD, Knoll AH (2002) Proterozoic ocean chemistry and evolution: a bioinorganic bridge? Science 297:1137–1142

Andersson SG, Karlberg O, Canback B, Kurland CG (2003) On the origin of mitochondria: a genomics perspective. Philos Trans R Soc Lond B Biol Sci 358:165–177

Andersson SGE, Kurland CG (1998) Reductive evolution of resident genomes. Trends Microbiol 6:263–268

Bai Y, Hu P, Park JS, Deng JH, Song X, Chomyn A, Yagi T, Attardi G (2004) Genetic and functional analysis of mitochondrial DNA-encoded complex I genes. Ann N Y Acad Sci 1011:272–283

Baldauf SL (2003) The deep roots of eukaryotes. Science 300:1703–1706
Baril C, Richaud C, Baranton G, Saint-Girons IS (1989) Linear chromosome of Borrelia burgdorferi. Res Microbiol 140:570–576
Barja G (1998) Mitochondrial free radical production and aging in mammals and birds. Ann N Y Acad Sci 854:224–238
Bennett AF (1988) Structural and functional determinates of metabolic rate. Amer Zool 28:699–708
Berry S (2003) Endosymbiosis and the design of eukaryotic electron transport. Biochim Biophys Acta 1606:57–72
Bishop CM (1999) The maximum oxygen consumption and aerobic scope of birds and mammals: getting to the heart of the matter. Proc R Soc Lond B Biol Sci 266:2275–2281
Blackstone NW (2000) Redox control and the evolution of multicellularity. BioEssays 22:947–953
Blackstone NW, Green DR (1999) The evolution of a mechanism of cell suicide. BioEssays 21:84–88
Boxma B, de Graaf RM, van der Staay GW, van Alen TA, Ricard G, Gabaldon T, van Hoek AH, Moon-van der Staay SY, Koopman WJ, van Hellemond JJ, Tielens AG, Friedrich T, Veenhuis M, Huynen MA, Hackstein JH (2005) An anaerobic mitochondrion that produces hydrogen. Nature 434:74–79
Brand MD, Buckingham JA, Esteves TC, Green K, Lambert AJ, Miwa S, Murphy MP, Pakay JL, Talbot DA, Echtay KS (2004) Mitochondrial superoxide and aging: uncoupling-protein activity and superoxide production. Biochem Soc Symp 71:203–213
Brennan RJ, Schiestl RH (1998) Chloroform and carbon tetrachloride induce intrachromosomal recombination and oxidative free radicals in Saccharomyces cerevisiae. Mutat Res 397:271–278
Burger G, Gray MW, Lang BF (2003) Mitochondrial genomes: anything goes. Trends Genet 19:709–716
Canfield DE (1998) A new model of Proterozoic ocean chemistry. Nature 396:450–452
Canfield DE, Habicht KS, Thamdrup B (2000) The Archean sulfur cycle and the early history of atmospheric oxygen. Science 288:658–661
Cardol P, Matagne RF, Remacle C (2002) Impact of mutations affecting ND mitochondria-encoded subunits on the activity and assembly of complex I in Chlamydomonas. Implication for the structural organization of the enzyme. J Mol Biol 319:1211–1221
Castresana J, Moreira D (1999) Respiratory chains in the last common ancestor of living organisms. J Mol Evol 49:453–460
Cavalier-Smith T (1985) The evolution of genome size. Wiley, Chichester
Cavalier-Smith T (1989) Archaebacteria and Archezoa. Nature 339:100–101
Cavalier-Smith T (1987) Eukaryotes with no mitochondria. Nature 326:332–333
Cavalier-Smith T (2002) The phagotrophic origin of eukaryotes and phylogenetic classification of Protozoa. Int J Syst Evol Microbiol 52:297–354
Cavalier-Smith T (2005) Economy, speed and size matter: evolutionary forces driving nuclear genome miniaturization and expansion. Ann Bot 95:147–175
Cavalier-Smith T, Beaton MJ (1999) The skeletal function of non-genic nuclear DNA: new evidence from ancient cell chimaeras. Genetica 106:3–13
Chomyn A (2001) Mitochondrial genetic control of assembly and function of complex I in mammalian cells. J Bioenerg Biomembr 33:251–257
Christensen BB, Atlung T, Hansen FG (1999) DnaA boxes are important elements in setting the initiation mass of Escherichia coli. J Bacteriol 181:2683–2688
Cosmides LM, Tooby J (1981) Cytoplasmic inheritance and intragenomic conflict. J Theor Biol 89:83–129
Cox E, Bonner J (2001) Ecology. The advantages of togetherness. Science 292:448-449
Cunningham EL, Berger JM (2005) Unraveling the early steps of prokaryotic replication. Curr Opin Struct Biol 15:68–76
Darveau CA, Suarez RK, Andrews RD, Hochachka PW (2002). Allometric cascade as a unifying principle of body mass effects on metabolism. Nature 417:166–170

Dodds PS, Rothman DH, Weitz JS (2001) Re-examination of the "3/4-law" of metabolism. J Theor Biol 209:9–27
Donachie WD (1968) Relationship between cell size and time of initiation of DNA replication. Nature 219:1077–1079
Donachie WD, Blakely GW (2003) Coupling the initiation of chromosome replication to cell size in Escherichia coli. Curr Opin Microbiol 6:146–150
Doolittle WF, Sapienza C (1980) Selfish genes, the phenotype paradigm and genome evolution. Nature 284:601–603
Doolittle WF, Boucher Y, Nesbo CL, Douady CJ, Andersson JO, Roger AJ (2003) How big is the iceberg of which organellar genes in nuclear genomes are but the tip? Philos Trans R Soc Lond B Biol Sci 358:39–58
Elena SF, Lenski RE (1997) Test of synergistic interactions among deleterious mutations in bacteria. Nature 390:395–398
Embley TM, van der Giezen M, Horner DS, Dyal PL, Foster P (2003) Mitochondria and hydrogenosomes are two forms of the same fundamental organelle. Philos Trans R Soc Lond B Biol Sci 358:191–201
Feil EJ, Holmes EC, Bessen DE, Chan MS, Day NP, Enright MC, Goldstein R, Hood DW, Kalia A, Moore CE, Zhou J, Spratt BG (2001) Recombination within natural populations of pathogenic bacteria: short-term empirical estimates and long-term phylogenetic consequences. Proc Natl Acad Sci USA 98:182–187
Fonstein M, Haselkorn R (1995) Physical mapping of bacterial genomes. J Bacteriol 177:3361–3369
Frank AC, Amiri H, Andersson SGE (2002) Genome deterioration: loss of repeated sequences and accumulation of junk DNA. Genetica 115:1–12
Gitai Z (2005) The new bacterial cell biology: moving parts and subcellular architecture. Cell 120:577–86
Gray MW (2005) Evolutionary biology: the hydrogenosome's murky past. Nature 434:29–31
Gray MW, Burger G, Lang BF (1999) Mitochondrial evolution. Science 283:1476–1481
Gray MW, Burger G, Lang BF (2001) The origin and early evolution of mitochondria. Genome Biol 2: Rev1018.1–1018.5
Gregory TR (2001) Coincidence, coevolution, or causation? DNA content, cell size and the C-value enigma. Biol Rev 76:65–101
Gregory TR (2002) Genome size and developmental complexity. Genetica 115:131–146
Gregory TR (2005) The C-value enigma in plants and animals: a review of parallels and an appeal for partnership. Ann Bot 95:133–146
Gould SJ (1997) Full house. Random House, New York
Harold FM (2001) The way of the cell. Molecules, organisms and the order of life. Oxford University Press, New York
Harold FM (1986) The vital force: a study of bioenergetics. Freeman, New York
Heusner A (1991) Size and power in mammals. J Exp Biol 160:25–54
Hochachka PW, Darveau CA, Andrews RD, Suarez RK (2003) Allometric cascade: a model for resolving body mass effects on metabolism. Comp Biochem Physiol A 134:675–691
Hoffmeister M, Martin W (2003) Interspecific evolution: microbial symbiosis, endosymbiosis and gene transfer. Environ Microbiol 5:641–649
Hurst L, Hamilton WD (1992) Cytoplasmic fusion and the nature of sexes. Proc R Soc Lond B Biol Sci 247:189–194
Jensen PR, Loman L, Petra B, van der Weijden C, Westerhoff HV (1995) Energy buffering of DNA structure fails when Escherichia coli runs out of substrate. J Bacteriol 177:3420–3426
Jones LJ, Carballido-Lopez R, Errington J (2001) Control of cell shape in bacteria: helical, actin-like filaments in Bacillus subtilis. Cell 104:913–922
Keeling PJ (1998) A kingdom's progress: archezoa and the origin of eukaryotes. Bioessays 20:87–95
Keightley PD, Eyre-Walker A (2000) Deleterious mutations and the evolution of sex. Science 290:331–333

Kitagawa R, Mitsuki H, Okazaki T, Ogawa T (1996) A novel DnaA protein-binding site at 94.7 min on the Escherichia coli chromosome. Mol Microbiol 19:1137–1147
Kleiber M (1961) The fire of life. Wiley, New York
Knoll AH (2003) Life on a young planet: the first three billion years of evolution on earth. Princeton University Press, Princeton
Koedoed S, Otten MF, Koebmann BJ, Bruggeman FJ, Bakker BM, Snoep JL, Krab K, van Spanning RJM, van Verseveld HW, Jensen PR, Koster JG, Westerhoff HV (2002) A turbo engine with automatic transmission? How to marry chemicomotion to the subtleties and robustness of life. Biochim Biophys Acta 1555:75–82
Kondrashov AS (1988) Deleterious mutations and the evolution of sexual reproduction. Nature 336:435–440
Konstantinidis KT, Tiedje JM (2004) Trends between gene content and genome size in prokaryotic species with larger genomes. Proc Natl Acad Sci USA 101:3160–3165
Kozłowski, J, Konarzewski M, Gawelczyk AT (2003) Cell size as a link between noncoding DNA and metabolic rate scaling. Proc Natl Acad Sci USA 100:14080–14085
Kroos L (2005) Eukaryotic-like signaling and gene regulation in a prokaryote that undergoes multicellular development. Proc Natl Acad Sci USA 102:2681–2682
Kunin V, Ouzounis CA (2003) The balance of driving forces during genome evolution in prokaryotes. Genome Res 13:1589–1594
Lane N (2002) Oxygen: The molecule that made the world. Oxford University Press, Oxford
Lane N (2005) Power, sex, suicide: mitochondria and the meaning of life. Oxford University Press, Oxford
Leonard AC, Grimwade JE (2005) Building a bacterial orisome: emergence of new regulatory features for replication origin unwinding. Mol Microbiol 55:978
López-García P, Rodríguez-Valera F, Pedrós-Alió C, Moreira D (2001) Unexpected diversity of small eukaryotes in deep-sea Antarctic plankton. Nature 409:603–610
Löytynoja A, Milinkovitch MC (2001) Molecular phylogenetic analyses of the mitochondrial ADP-ATP carriers: the plantae/fungi/metazoa trichotomy revisited. Proc Natl Acad Sci USA 98:10202–10207
Madigan MT, Martinko JM, Parker J (2002) Brock biology of microorganisms. Prentice Hall, Englewood Cliffs
Margolin W (2005) Bacterial mitosis: actin in a new role at the origin. Curr Biol 15:R259–261
Martin W (1999) A briefly argued case that mitochondria and plastids are descendents of endosymbionts but that the nuclear compartment is not. Proc R Soc Lond B Biol Sci 266:1387–1395
Martin W (2005) Woe is the tree of life. In: Sapp J (ed) Microbial phylogeny and evolution: concepts and controversies. Oxford University Press, New York
Martin W, Müller M (1998) The hydrogen hypothesis for the first eukaryote. Nature 392: 37–41
Martin W, Russell MJ (2003) On the origins of cells: a hypothesis for the evolutionary transitions from abiotic geochemistry to chemoautotrophic prokaryotes, and from prokaryotes to nucleated cells. Philos Trans R Soc Lond B Biol Sci 358:59–85
Martin W, Hoffmeister M, Rotte C, Henze K (2001) An overview of endosymbiotic models for the origins of eukaryotes, their ATP-producing organelles (mitochondria and hydrogenosomes) and their heterotrophic lifestyle. Biol Chem 382:1521–1539
Martin W, Rotte C, Hoffmeister M, Theissen U, Gelius-Dietrich G, Ahr S, Henze K (2003) Early cell evolution, eukaryotes, anoxia, sulfide, oxygen, fungi first (?), and a tree of genomes revisited. IUBMB Life 55:193–204
Mattick JS (2001) Non-coding RNAs: the architects of eukaryotic complexity. EMBO Rep 2:986–991
Mattick JS, Makunin IV (2005) Small regulatory RNAs in mammals. Hum Mol Genet 14 Spec No 1:R121–132
Maynard Smith J, Szathmáry E (1995) The major transitions of evolution. Freeman, San Francisco

Maynard Smith J, Smith NH, O'Rourke M, Spratt BG (1993) How clonal are bacteria? Proc Natl Acad Sci USA 90:4384–4388

Maynard Smith J, Feil EJ, Smith NH (2000) Population structure and evolutionary dynamics of pathogenic bacteria. BioEssays 22:1115–1122

McGarry KC, Ryan VT, Grimwade JE, Leonard AC (2004) Two discriminatory binding sites in the Escherichia coli replication origin are required for DNA strand opening by initiator DnaA-ATP. Proc Natl Acad Sci USA 101:2811–2816

Messer W (2002) The bacterial replication initiator DnaA. DnaA and oriC, the bacterial mode to initiate DNA replication. FEMS Microbiol Rev 26:355–374

Müller M, Martin W (1999) The genome of Rickettsia prowazekii and some thoughts on the origin of mitochondria and hydrogenosomes. BioEssays 21:377–381

Nedelcu AM, Marcu O, Michod RE (2004) Sex as a response to oxidative stress: a two-fold increase in cellular reactive oxygen species activates sex genes. Proc R Soc Lond B Biol Sci 271:1591–1592

Nisbet EG, Sleep NH (2001) The habitat and nature of early life. Nature 409:1083–1091

O'Farrell PH (1992) Cell cycle control: many ways to skin a cat. Trends Cell Biol 2:159–163

Ogawa T, Yamada Y, Kuroda T, Kishi T, Moriya S (2002) The datA locus predominantly contributes to the initiator titration mechanism in the control of replication initiation in Escherichia coli. Mol Microbiol 44:1367–1375

Orgel LE, Crick FH (1980) Selfish DNA: the ultimate parasite. Nature 284:604–607

Ott M, Robertson JD, Gogvadze V, Zhitotovsky B, Orrenius S (2002) Cytochrome c release from mitochondria proceeds by a two-step process. Proc Natl Acad Sci USA 99:1259–1263

Otten MF, Reijnders WN, Bedaux JJ, Westerhoff HV, Krab K, Van Spanning RJ (1999) The reduction state of the Q-pool regulates the electron flux through the branched respiratory network of Paracoccus denitrificans. Eur J Biochem 261:767–774

Partridge L, Hurst LD (1998) Sex and conflict. Science 281:2003–2008

Pfeiffer T, Schuster S, Bonhoeffer S (2001) Cooperation and competition in the evolution of ATP-producing pathways. Science 292:504–507

Porter RK (2001) Allometry of mammalian cellular oxygen consumption. Cell Mol Life Sci 58:815–822

Prothero J (1986) Scaling of energy metabolism in unicellular organisms: a re-analysis. Comp Biochem Physiol A 83:243–248

Race HL, Herrmann RG, Martin W (1999) Why have organelles retained genomes? Trends Genet 15:364–370

Richter C, Schweizer M, Cossarizza A, Franceschi C (1996) Control of apoptosis by the cellular ATP level. FEBS Lett 378:107–110

Ridley M (2001) Mendel's demon: gene justice and the complexity of life. Phoenix, London

Rolfe DFS, Brown GC (1997) Cellular energy utilization and molecular origin of standard metabolic rate in mammals. Physiol Rev 77:731–758

Santamaria M, Lanave C, Saccone C (2004) The evolution of the adenine nucleotide translocase family. Gene 333:51–59

Schatz G (2003) The tragic matter. FEBS Lett 536:1–2

Schulz HN, Brinkhoff T, Ferdelman TG, Hernández Mariné M, Teske A, Jørgensen BB (1999) Dense populations of a giant sulfur bacterium in Namibian shelf sediments. Science 284:493–495

Skulachev VP (2004) Mitochondria, reactive oxygen species and longevity: some lessons from the Barja group. Aging Cell 3:17–19

Skulachev VP (1998) Uncoupling: new approaches to an old problem of bioenergetics. Biochim Biophys Acta 1363:100–124

Smith NH, Holmes EC, Donovan GM, Carpenter GA, Spratt BG (1999) Networks and groups within the genus Neisseria: analysis of argF, recA, rho, and 16S rRNA sequences from human Neisseria species. Mol Biol Evol 16:773–783

Snoep JL, van der Weijden CC, Andersen HW, Westerhoff HV, Jensen PR (2002) DNA supercoiling in Escherichia coli is under tight and subtle homeostatic control, involving

gene-expression and metabolic regulation of both topoisomerase I and DNA gyrase. Eur J Biochem 269:1662–1669
Spratt BG, Hanage WP, Feil EJ (2001) The relative contributions of recombination and point mutation to the diversification of bacterial clones. Curr Opin Microbiol 4:602–606
Stêpkowski T, Legocki AB (2001) Reduction of bacterial genome size and expansion resulting from obligate intracellular lifestyle and adaptation to soil habitat. Acta Biochim Polonica 48:367–381
Tielens AG, Rotte C, van Hellemond JJ, Martin W (2002) Mitochondria as we don't know them. Trends Biochem Sci 27:564–572
Turrens JF (2003) Mitochondrial formation of reactive oxygen species. J Physiol 552:335–344
Ugalde C, Vogel R, Huijbens R, Van Den Heuvel B, Smeitink J, Nijtmans L (2004) Human mitochondrial complex I assembles through the combination of evolutionary conserved modules: a framework to interpret complex I deficiencies. Hum Mol Genet 13:2461–2472
van den Ent F, Amos LA, Lowe J (2001) Prokaryotic origin of the actin cytoskeleton. Nature 413:39–44
Velicer GJ, Yu YT (2003) Evolution of novel cooperative swarming in the bacterium Myxococcus xanthus. Nature 425:75–78
Vellai T, Vida G (1999) The origin of eukaryotes: the difference between prokaryotic and eukaryotic cells. Proc R Soc Lond B Biol Sci 266:1571–1577
Vellai T, Takács K, Vida G (1998) A new aspect to the origin and evolution of eukaryotes. J Mol Evol 46:499–507
Vogel R, Nijtmans L, Ugalde C, van den Heuvel L, Smeitink J (2004) Complex I assembly: a puzzling problem. Curr Opin Neurol 17:179–186
von Dohlen CD, Kohler S, Alsop ST, McManus WR (2001) Mealybug beta-proteobacterial endosymbionts contain gamma-proteobacterial symbionts. Nature 412:433–436
Weibel ER (2002) The pitfalls of power laws. Nature 417:131–132
West GB, Brown JH, Enquist BJ (1997) A general model for the origin of allometric scaling in biology. Science 276:122–126
West GB, Brown JH, Enquist BJ (1999) The fourth dimension of life: fractal geometry and allometric scaling of organisms. Science 284:1677–1679
West GB, Woodruff WH, Brown JH (2002) Allometric scaling of metabolic rate from molecules and mitochondria to cells and mammals. Proc Natl Acad Sci USA 99:2473–2478
Westerhoff HV, van Workum M (1990) Control of DNA structure and gene expression. Biomed Biochim Acta 49:839–853
Westerhoff HV, O'Dea MH, Maxwell A, Gellert M (1988) DNA supercoiling by DNA gyrase. A static head analysis. Cell Biophys 12:157–181
White CR, Seymour RS (2003) Mammalian basal metabolic rate is proportional to body mass$^{2/3}$. Proc Natl Acad Sci USA 100:4046–4049
Zamzami N, Marchetti P, Castedo M, Decaudin D, Macho A, Hirsch T, Susin SA, Petit PX, Mignotte B, Kroemer G (1995) Sequential reduction of mitochondrial transmembrane potential and generation of reactive oxygen species in early programmed cell death. J Exp Med 182:367–377
Zuckerkandl E (2002) Why so many noncoding nucleotides? The eukaryote genome as an epigenetic machine. Genetica 115:105–129

# 3 Origin, Function, and Transmission of Mitochondria

Carol A. Allen, Mark van der Giezen, John F. Allen

## 3.1 Introduction

Mitochondria have existed for more than a billion years, but it was not until the middle of the nineteenth century that they were actually recognised in cells, at first as a grainy appearance in the cell cytoplasm when observed by light microscopy. The anatomist Kölliker (1856) observed mitochondria in muscle cells in the 1850s, while Altman (1890) suggested that his "bioblasts" (granules microscopically observable throughout the cell) were symbionts, something Schimper (1883) had suggested for chloroplasts 7 years earlier, and this idea was taken further by Mereschkowsky (1905). The name mitochondrion was first coined by Benda (1898), and it comes from two Greek words, *mitos* (thread) and *chondros* (granule), which describes the appearance of mitochondria during spermatogenesis. In the following years, many people speculated on the role of mitochondria in the cell, with Warburg (1913) recognising the particulate nature of cell respiration and Keilin (1925) associating the cytochrome system with cellular structures. The first direct evidence for this functional association depended on isolation of the mitochondria from the rest of the cell, which became possible in the 1930s. The first isolations of mitochondria by cell fractionation were made by Bensley and Hoerr (1934), and, following this breakthrough, the path opened for study of the biochemical reactions occurring in mitochondria.

As the sites of energy conversion and cellular respiration, mitochondria became regarded as the "powerhouses" of the cell. However the possible origin of mitochondria was not looked at for some time, not really until the 1950s. It was in the early 1950s that Ephrussi (1950) and Mitchell and Mitchell (1952) observed that mitochondrial replication in yeast cells was controlled by non-Mendelian genetic factors, and slightly later that McLean et al. (1958) observed that mitochondria synthesise proteins. The discovery of mitochondrial DNA followed in the early 1960s, when a number of different groups (Luck and Reich 1964; Nass and Nass 1963a,b; Schatz et al. 1964) published their findings of mitochondrial and chloroplast DNA.

While the endosymbiotic origin of mitochondria had been considered since the time of Mereschkowsky (Martin and Kowallik 1999), the advances in biochemical techniques in the 1960s led to a revival of the idea, and a new and enthusiastic following for it. The driving force behind this renewed

Origin of Mitochondria and Hydrogenosomes (ed. by William F. Martin and Miklós Müller)

interest was, of course, the discovery of organelle DNA, and the non-Mendelian inheritance of organelles. Margulis (1970) published a reformulation of the endosymbiotic theory in 1970, and endosymbiosis has subsequently become the accepted view of the origin of mitochondria. There is still disagreement about how this endosymbiosis arose, and which organisms it involved, but there is a general agreement now that the mitochondrion has descended ultimately from a free-living bacterium.

In the course of a little over 100 years, scientists have gone from the first observations of mitochondria to an understanding of their structure, function, inheritance, and origin. The mitochondrial genomes of over 250 different species are now known (Tsang and Lemire 2003), as are the effects of mutations in many mitochondrial genes. There are, however, basic questions still left to be answered. Which organisms contributed to the first eukaryotic cell? Why do mitochondria retain a genome? Can mitochondria still function without their own genomes? Why are only certain mitochondria passed on to the next generation?

## 3.2 Origins of Mitochondria

All developments seem to progress from simple to more complex forms. Whether this is true or just an imaginary chain of events that fits more comfortably with our way of thinking remains to be seen. The anthropocentric view comes naturally to us. Cars, for example, "evolved" from simple horseless carriages to high-performance automotive vehicles. This "evolution", according to some proponents, is similar to the evolution of living organisms, including parameters such as natural selection. Nonetheless, the evolution of life is often thought to have occurred in a smooth and gradual manner. The first group of organisms on our planet, the prokaryotes, are generally the simplest. In principle, prokaryotes are "nothing more" than membrane-enclosed bags of enzymes capable of some biochemical trickery. In contrast, take eukaryotes, with ourselves as the glorifying example of how complex life can be. Clearly, "higher life" evolved from lowly creatures such as bacteria by gradually improving their simple architecture into more elaborate cells which include organelles such as nuclei and mitochondria.

The gradual transformation of a prokaryote into a primitive anucleate eukaryote is still considered the logical chain of events in many textbooks. Accordingly, this primitive eukaryote at one stage took up a free-living bacterium which converted into our modern-day mitochondrion. Such endosymbiosis theories for the evolution of eukaryotes at one stage involved amitochondriate (i.e. without mitochondria) eukaryotes. This hypothetical group received recognition in the now defunct kingdom of the Archezoa (Cavalier-Smith 1987). All studied members of this group have been shown to contain mitochondria of some sort (van der Giezen et al. 2005). This raises the question as to why true amitochondriate eukaryotes, which would have been direct descendants of this anucleate eukaryote, do not seem to exist nowadays. Normally, one would expect intermediary stages of evolutionary development to be capable of producing a lineage of descendants even if the ancestors themselves become extinct. So, why do we not see truly amitochondriate eukaryotes nowadays? This could be for two reasons: either they never existed or they lost the battle of the "survival of the fittest". The latter scenario suggests that, although these organisms did evolve in a particular environmental niche, they no longer occupy this niche, either because it does not exist anymore or, again, because its former occupants lost out to more competitive eukaryotes. An easier way to explain the absence of intermediate forms is to suggest they never existed in the first place. Although this might run in the face of our convenient way of ordering things in a gradual progression from simple to more complex, it actually explains our observations without invoking subsequent events (selective culling of the amitochondriates). So, perhaps the origin of the eukaryotes evolved in a "big bang"-like fashion; with a momentous event.

Let us consider a counterintuitive but satisfying proposal for this event, and one that explains several key aspects in the evolution of eukaryotes. Firstly, the origin of eukaryotes and mitochondria was the same event. In addition, in contrast to general belief, the eubacterial organism that gave rise to the mitochondrion was not an obligate aerobe, far from it. Finally, again in contrast to general belief, the reason for the establishment of the mitochondrion was not energy production. The name of this heretical hypothesis? The hydrogen hypothesis (Martin and Müller 1998), which suggests that hydrogen, and not oxygen or energy, was the currency for the establishment of the mitochondrial endosymbiont. This suggests that the host was able to metabolise hydrogen.

Eukaryotic genome analyses have indicated that almost all informational genes (i.e. involved in genetics) are archaebacterial in origin (Rivera et al. 1998). In contrast, all operational genes, i.e. those involved in metabolism, are eubacterial in origin (Rivera and Lake 2004). Various analyses, for example cytochrome phylogenies, had already indicated that the origin of the mitochondrion might be sought amongst the α-proteobacteria (Schwartz and Dayhoff 1978). So, the players involved are an archaebacterial methanogenic host and an α-proteobacterial endosymbiont (Fig. 3.1). *Rickettsia* has been put forward as the α-proteobacterium which would be most closely related to the original endosymbiont. One reason is the similarity of its aerobic respiration to mitochondrial respiration. But here is a problem: methanogens are one of the most oxygen-intolerant prokaryotes, and cannot produce any energy in the presence of oxygen. So, in a last-ditch attempt, the mitochondrial endosymbiont is put forward as a saviour of the oxygen-sensitive host (Kurland and Andersson 2000). But why put an oxygen scavenger inside the host it is supposedly protecting from harm? One would not put the knights on the courtyard but put them up on the walls to fend off any enemy.

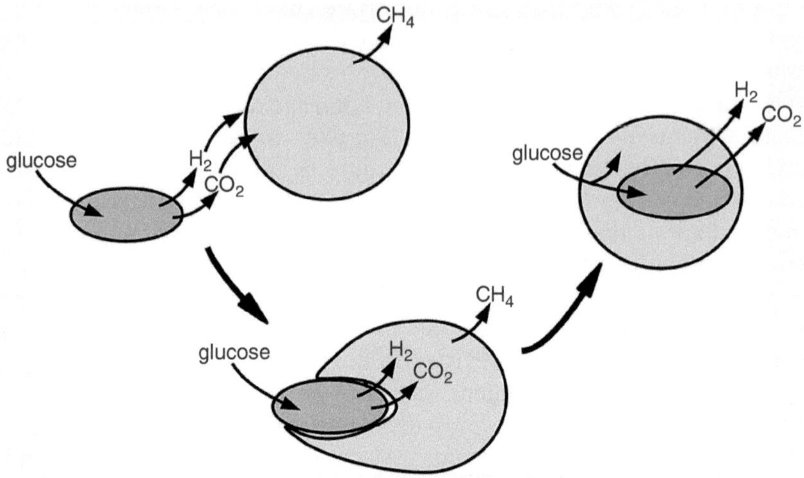


**Fig. 3.1.** The hydrogen hypothesis of Martin and Müller. The events leading to the establishment of the mitochondrial endosymbiont. *Top left*: A facultative anaerobic α-proteobacterium (*dark grey*) produces hydrogen which is taken up by an autotrophic methanogen (*light grey*). *Middle*: A more intimate relationship results into a larger surface area that can be used for interspecies hydrogen transfer. *Top right*: After eventually becoming fully incorporated, the proteobacterium initially kept producing hydrogen and in return received reduced organic compounds. (Martin and Müller 1998)

So, an α-proteobacterial host with a methanogenic endosymbiont would make more sense if oxygen protection was the reason that forged the symbiosis. The hydrogen hypothesis does present the symbiosis as an interspecies hydrogen transfer gone too far. The endosymbiont remains an α-proteobacterium, but this time something more similar to *Rhodobacter*, capable of aerobic and anaerobic metabolism. This endosymbiont offered the methanogen molecular hydrogen and carbon dioxide, and the autotrophic host returned reduced organic compounds, which geared the endosymbiont's metabolism to new heights. Subsequent gene transfers forged the symbiosis for eternity.

It has been argued that anaerobic metabolism could not have been the driving force in times when atmospheric oxygen concentrations were rising (Kurland and Andersson 2000). The concentration of atmospheric oxygen around the time of the endosymbiosis (about 2,000 million years ago; Martin et al. 2003) was about 3%, or about 7 times less then the present-day concentration (Nisbet and Sleep 2001). Perhaps more importantly, large parts of ocean waters around these times were anoxic (Canfield 1998), and it is thought that these important evolutionary events would have taken place in the sea and not on the land as perhaps commonly thought. So, oxygen seems to have been an extremely unlikely factor to have influenced the establishment of the mitochondrial endosymbiont and hydrogen seems more important then ever imagined.

One problem discussing mitochondrial function is that there does not seem to be a typical mitochondrion. Mitochondria evolved over a period of 2,000 million years in an huge variety of organisms living under an enormous range of environmental conditions (van der Giezen and Tovar 2005). Mitochondria range from archetypal aerobic mitochondria, via various anaerobic versions and hydrogenosomes, to the most derived forms, mitosomes (Tielens et al. 2002). Nonetheless, currently we know of at least one function found in all mitochondrial varieties; iron–sulphur cluster assembly (Lill and Muhlenhoff 2005). This essential pathway produces iron–sulphur co-factors for both mitochondrial and cytosolic enzymes involved in electron transport, enzyme catalysis, and regulation of gene expression. The most aerobic of mitochondria are involved in oxidative phosphorylation using oxygen as terminal electron acceptor, while more anaerobic versions use alternative electron acceptors such as nitrate (Tielens et al. 2002). Hydrogenosomes, similarly to aerobic mitochondria, convert pyruvate to acetyl coenzyme-A, however not using pyruvate dehydrogenase but by means of the oxygen-sensitive pyruvate:ferredoxin oxidoreductase (Embley et al. 2003). All these mitochondrial variants produce energy, be it by means of harvesting the electrochemical gradient generated via the respiratory chain or by substrate-level phosphorylation. Mitosomes on the other hand are not known to be directly involved in energy generation; currently, their function seems exclusively tied to iron–sulfur cluster assembly (van der Giezen et al. 2005)

## 3.3 Mitochondrial Genomes

As discussed by van der Giezen and Tovar (2005), mitochondria are an enormously diverse set of various organelles. Even if one is not willing to include the anaerobic varieties as being mitochondrial, the vast biochemical repertoire present in aerobic mitochondria alone is staggering. In addition to this biochemical heterogeneity, there exists a genetic heterogeneity as well. There is no such thing as a standard-issue mitochondrial genome. This genome can be as small as 5,967 base pairs in the case of *Plasmodium falciparum* (Feagin et al. 1991) and only code for three proteins (cytochrome b, cytochrome oxidase I and III) or as large as 490,000 base pairs for rice (Notsu et al. 2002). Strangely enough, although the rice mitochondrial genome is 80 times larger than the *Plasmodium* one, it does not contain 80 times as many genes. Although plant mitochondrial genomes tend to be the largest, the mitochondrial genome which actually contains the most genes is the one from the freshwater protozoon *Reclinomonas americana*, which contains 97 genes (Lang et al. 1997). The median is therefore something around 45 genes. If one takes a present-day $\alpha$-proteobacterium (*Rhodobacter sphaeroides*, for example) which contains almost 4,000 genes, it becomes obvious that many genes of the original endosymbiont have been lost as a consequence of the symbiotic interaction. These genes have either been lost owing to redundancy (the host already contained homologous genes) or been transferred to the host genome (Timmis et al. 2004). It has been estimated that up to 75% of a eukaryotic genome could actually originate from the endosymbiont (Esser et al. 2004). The remaining mitochondrial genes are involved in a limited set of functions; always respiration and translation (as evident in the case of *P. falciparum*), and occasionally also in transcription, RNA maturation, and protein import (Burger et al. 2003). The partially sequenced hydrogenosomal genome from the ciliate *Nyctotherus ovalis* does indeed code for parts of a mitochondrial electron transport chain (Boxma et al. 2005). Other hydrogenosomes and

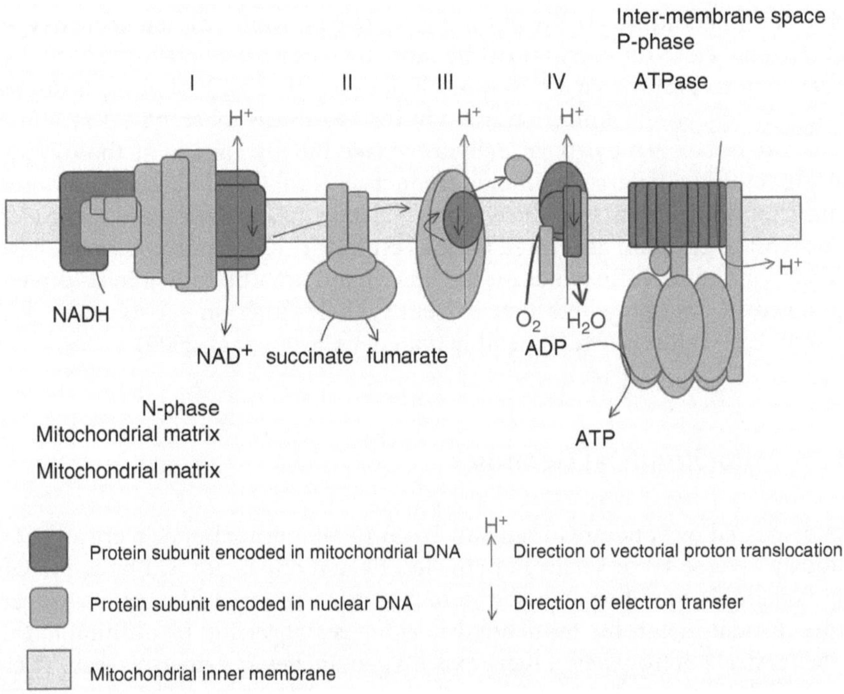


**Fig. 3.2.** Elements of energy transduction in respiration and oxidative phosphorylation in mitochondria. The mitochondrial inner membrane is shown in *yellow*. The principal complexes involved in energy transduction are complex I (NADH dehydrogenase), complex II (succinate dehydrogenase), complex III (the cytochrome *b*–cytochrome $c_1$ complex), complex IV (cytochrome *c* oxidase), and the coupling ATPase. Vectorial electron transfer is depicted as *thin, dark-blue arrows*. Proton (hydrogen ion; $H^+$) translocation is depicted as *thin, red arrows*. Other chemical conversions are given *black arrows*. The major, variable environmental input is oxygen ($O_2$), shown in *blue*. Subunits of protein complexes are coloured according to the location of the genes encoding them. Mitochondria are usually pink or reddish-brown, the colour of cytochromes and iron–sulphur proteins, so *reddish-brown subunits* have genes in the mitochondrion and are synthesised in the mitochondrial matrix; *light-brown subunits* have genes in the nucleus, and are imported from the cytosol as precursors. The depiction of sites of synthesis is schematic only and corresponds roughly to the arrangement in vertebrates. (Adapted from Allen 1993a)

mitosomes as well do not seem to have kept their mitochondrial genome (van der Giezen et al. 2005), indicating that the mitochondrial genome's core function is respiration and oxidative phosphorylation (Fig. 3.2).

## 3.4 The Mitochondrial Theory of Ageing

Reactive oxygen species, generated largely by the mitochondrial electron transport chain, damage the mitochondrial proteins and DNA, and the mitochondrial theory of ageing, simply put, states that this damage leads to ageing and its associated degenerative diseases (Fig. 3.3). This theory was first put forward by Harman (1956), although earlier observations had linked life span to metabolic rate: the higher the metabolic rate, the shorter the life span (Pearl 1928). Although Harman's theory has been around for 50 years, and there is a lot of circumstantial evidence to support it, there remain many

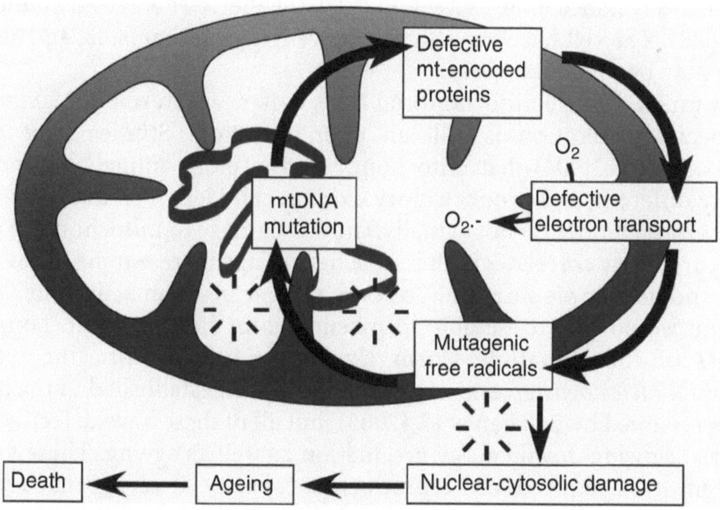


**Fig. 3.3.** Why we grow old and die: the mitochondrial theory of ageing. Free radicals (whose reactions are symbolised by a *star*), including the superoxide anion radical, $O_2^{\cdot-}$, are produced at a low frequency as by-products of respiratory electron flow in oxidative phosphorylation. Free-radical mutagenesis of mitochondrial DNA (*mtDNA*) then impairs the structure and function of respiratory chain proteins, in turn increasing the frequency of free-radical production. Univalent reduction of oxygen by semiquinone anion radicals may be an important initial step, since ubisemiquinone is an intermediate in protonmotive Q-cycles in oxidative phosphorylation, and readily reduces oxygen to the superoxide anion radical, $O_2^{\cdot-}$. Other oxygen free radicals and sites in the respiratory chain may also be involved. Direct damage to proteins and membranes may accelerate the cycle and initiate somatic degeneration. Mitochondria may minimise, but never eliminate, mutagenic electron transfers. For animal cells, this positive feedback loop, or "vicious circle", has been proposed as the primary cause of ageing. (Adapted from Allen 1996)

uncertainties. When Harman first put forward his hypothesis, it had not actually been shown that cells generated free radicals, and it was only in 1969, with the discovery of superoxide dismutase (McCord and Fridovich 1969), that this question was satisfactorily answered. It is now known that cells generate reactive oxygen species at many sites, the majority of these being within the mitochondria. The two major sites are believed to be complexes I and III of the respiratory chain. Experiments increasing the redox potential of either site I or site III increase the rate of generation of free radicals (Chen et al. 2003; Kushnareva et al. 2002). Both of these complexes reduce ubiquinone (ubiquinol is also oxidised by complex III), and univalent reduction of oxygen probably occurs by electron transfer from the ubisemiquinone free radical, an intermediate in ubiquinone–ubiquinol oxidation and reduction, as follows.

$$UQ^{\cdot -} + O_2 \rightarrow UQ + O_2^{\cdot -}.$$

It is not known how much of the oxygen consumption of the cell is turned over to generating reactive oxygen species, but the figure is thought to be between 2% (Chance et al. 1979) and 0.2% (St-Pierre et al. 2002; Staniek and Nohl 2000). The cell has very efficient scavenging mechanisms, and so these figures may be underestimates.

How much damage mitochondrial DNA suffers as a result of reactive oxygen species generation is still an open question. Studies have shown (Shigenaga et al. 1994) that mitochondria from older animals are morphologically different, and produce more oxidants and less ATP than those from younger ones, but we do not actually know if damage to mitochondria causes ageing, or merely correlates with it. The field of ageing research – what causes ageing and how do we stop, slow, or even reverse it – is an active one. Almost everyone would like to be able to extend his or her life span. Long-lived mutants of the nematode worm *Caenorhabditis elegans*, the fruit fly *Drosophila melanogaster*, and even mice have been established in the laboratory, as reviewed by Balaban et al. (2005), but all of these have defective mitochondria, slowing down energy production as well as ageing. These animals also seem to have a reduced reproductive capacity. It seems that reducing generation of reactive oxygen species does indeed slow ageing, but at what cost? These animals can survive under laboratory conditions, but it is unlikely that they could survive in nature. Perhaps our mortality is the price we have to pay for survival in the short term, and our immortality has been secured by reproduction. Mitochondrial DNA is kept in the most hostile environment in the cell. While the vast majority of genes from the original endosymbiont have been transferred to the nucleus or lost, a small core of genes persist in the mitochondrial matrix. There is strong evidence that damage to mitochondrial DNA by reactive oxygen species generated during oxidative phosphorylation contributes to ageing and death of an organism, and so it is reasonable to assume that there must be a very compelling reason for the organism to continue to keep DNA there.

## 3.5 Why Are There Genes in Mitochondria?

Mitochondria have descended, in evolution, from free-living bacteria (Gray and Doolittle 1982; Gray 1992; Martin et al. 2001). Before the bacterial origin of mitochondria was generally appreciated, there were attempts to account for mitochondrial biogenesis in terms of sequestration of nuclear DNA in the cytoplasm. These need not detain us. However, there is a more recent dogma: that mitochondria retain genes and genetic systems *because* they are descended from bacteria. This statement, while correct, is not a complete explanation. For one thing, there are clearly subcellular organelles, hydrogenosomes and mitosomes, which are also derived from bacteria, and which no longer possess their own, internal genetic systems (van der Giezen et al. 2005). Another objection to this otherwise reasonable first guess – mitochondria happen to be stuck with bacterial genes – is as follows: many mitochondrial proteins with homology to bacterial proteins are now encoded in the cell nucleus, and are successfully imported, post-translationally, as precursors, prior to processing and assembly into functional complexes (Schatz 1998). Indeed, the major respiratory chain complexes are hybrids as regards the location of the genes for their subunits (Fig. 3.2), and there is no indication that their nuclearly encoded subunits are any less bacterial in origin than the mitochondrially encoded ones.

Thus, even granted the endosymbiotic origin of mitochondria, the persistence of mitochondrial genes and genomes requires explanation: if most ancestral, bacterial genes have been successfully relocated to the cell nucleus, then why not all? What is it about mitochondrial genes, or their gene products, that has prevented their successful removal to the nucleus?

The textbook *The Molecular Biology of the Cell* (Alberts et al. 1994) states the problem very clearly, and the following quotation has been retained, unchanged, from the first edition (1983).

"Why do mitochondria and chloroplasts require their own separate genetic systems when other organelles that share the same cytoplasm, such as peroxisomes and lysosomes, do not? ... The reason for such a costly arrangement is not clear, and the hope that the nucleotide sequences of mitochondrial and chloroplast genomes would provide the answer has proved unfounded. We cannot think of compelling reasons why the proteins made in mitochondria and chloroplasts should be made there rather than in the cytosol".

There seems to be no explicit proposal for the most widely held hypothesis for the persistence of mitochondria genomes, but the hypothesis is implicit in many discussions of mitochondrial structure and function. For example, and in contrast to the open question posed by Alberts et al., *Cell and Molecular Biology Concepts and Experiments* (Karp 2002) provides what is probably still the current consensus view.

"Mitochondrial DNA is a relic of ancient history. It is a legacy from a single aerobic bacterium that took up residence in the cytoplasm of a primitive cell that ultimately became an ancestor of all eukaryotic cells. Most of the genes of this ancient symbiont were either lost or transferred over the course of evolution to the nucleus of the host cell, leaving only a handful of genes to encode some of the most hydrophobic proteins of the inner mitochondrial membrane".

Thus, according to this "hydrophobicity hypothesis", proteins that are encoded and synthesised within organelles are characterised by shared, and extreme, hydrophobicity - all are intrinsic membrane proteins (Claros et al. 1995; Popot and de Vitry 1990; Von Heijne 1986).

This view amounts to mitochondrial genes being stuck where they are because of an insuperable difficulty if translocating hydrophobic proteins between subcelluar compartments. Yet there seems to be no evidence that hydrophobicity presents a particular barrier to protein import. For example, mitochondrial ADP–ATP carriers (AACs) are intrinsic to the mitochondrial inner membrane, have six transmembrane helices, and yet are encoded in the nucleus (Saraste and Walker 1982; van der Giezen et al. 2002).

## 3.6 Co-location of Gene and Gene Product Permits Redox Regulation of Gene Expression

This hypothesis states that mitochondria and chloroplasts contain genes whose expression must be under the direct, regulatory control of the redox state of their gene products, or of electron carriers with which their gene products interact (Fig. 3.4). These genes comprise a primary subset of organellar genes. The requirement for redox control of each gene then confers a selective advantage upon location of that gene within the organelle instead of in the cell nucleus. Mitochondrial and chloroplast genomes also contain genes for components of the their own, distinct, genetic systems. These genes comprise a secondary subset of organellar genes: genetic system genes. Retention of genetic system genes is necessary for the operation of redox control of expression of genes in the primary subset: bioenergetic genes. Without genes in the primary subset, the function of genetic system genes is eventually lost, and organelles lose their genomes.

This hypothesis of co-location for redox regulation of gene expression, CORR, was first outlined, in general terms, in a review on protein phosphorylation in regulation of photosynthesis (Allen 1992). The hypothesis was put forward in two articles (Allen 1993a, b), where the function of the location of organellar genes was proposed as redox regulation of gene expression. The term CORR was introduced more recently (Allen 2003a, b).

CORR applies equally to mitochondria and chloroplasts, and accounts for the fact that both of these organelles possess membrane-intrinsic electron transport systems along with discrete, extranuclear genetic systems. CORR rests on ten assumptions, or principles, as follows:

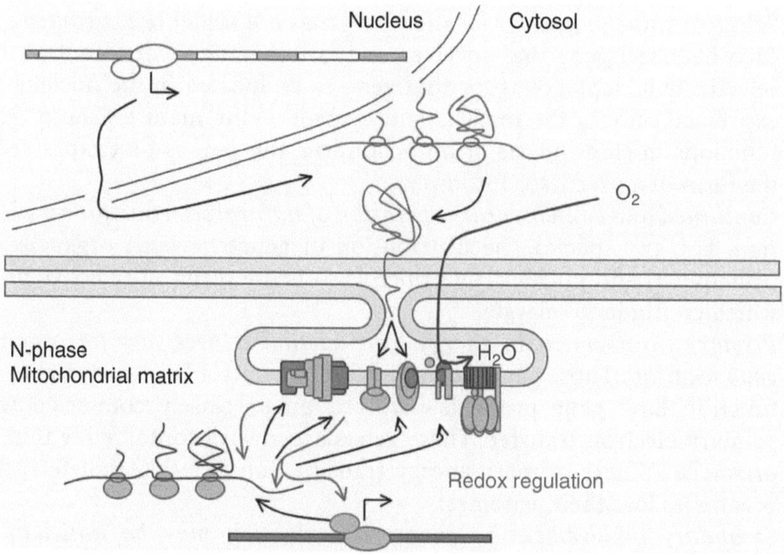


**Fig. 3.4.** Gene expression and principal pathways of biosynthesis of subunits of protein complexes involved in respiration and oxidative phosphorylation in animal mitochondria. *Reddish-brown DNA, RNA, and protein subunits* are located and synthesised in the mitochondrial matrix; *light-brown protein subunits* have genes (also *light brown*) in the nucleus, and are imported from the cytosol as precursors. *White genes* and ribosomal and protein subunits are nuclear-cytoplasmic and of archaebacterial origin. *Reddish-brown and light-brown genes* and ribosomal and protein subunits are of bacterial origin. The major, variable environmental input is oxygen (*blue*). It is proposed that it is beyond the ability of the nuclear-cytoplasmic system to respond rapidly and directly to changes in oxygen concentration or partial pressure, and so redox regulation of gene expression (*red arrows*) has been retained from the ancestral, bacterial endosymbiont. This redox regulation requires co-location of certain genes, with their gene products, within the mitochondrion. (Adapted from Allen 2003)

1. *Endosymbiotic origin.* As now generally agreed, mitochondria and chloroplasts evolved from free-living bacteria.
2. *Unselective gene transfer.* Gene transfer between the symbiont or organelle may occur in either direction and is not selective for particular genes.
3. *Unselective protein import.* There is no barrier to the successful import of any precursor protein, nor to its processing and assembly into a functional, mature form.
4. *Evolutionary continuity of redox control.* Direct redox control of expression of certain genes was present in the bacterial progenitors of mitochondria and chloroplasts, and was vital for selectively advantageous cell function before, during, and after the transition from bacterium to organelle. The mechanisms of this control have been conserved.
5. *Selective value of redox control.* For each gene under redox control (principle 4), it is selectively advantageous for that gene to be retained and expressed only within the organelle.

6. *Selective value of nuclear location for genes not under redox control.* For each bacterial gene that survives and is not under redox control, it is selectively advantageous for that gene to be located in the nucleus and expressed only in the nucleus and cytosol. If the mature gene product functions in chloroplasts or mitochondria, the gene is first expressed in the form of a precursor for import.
7. *Continued and contemporary operation of natural selection for gene location.* For any species, the distribution of genes between organelle (by principle 5) and nucleus (by principle 6) is the result of selective forces which continue to operate.
8. *Primary involvement in energy transduction is necessary for organelle gene location.* Those genes for which redox control is always vital to cell function have gene products involved in, or closely connected with, primary electron transfer. These genes are always contained within the organelle. Where primary energy transduction is lost completely, then organelles lose their genomes.
9. *Secondary involvement in energy transduction may be sufficient for organelle gene location.* Genes whose products contribute to the organelle genetic system itself, or whose products are associated with secondary events in energy transduction, may be contained in the organelle in one group of organisms, but not in another, depending on the physiology and biochemistry of respiration and photosynthesis in the species concerned.
10. *Nuclear encoding of redox-signalling components.* Components of the redox-signalling pathways upon which co-location for redox regulation depends are themselves not involved in primary electron transfer, and so their genes have been relocated to the nucleus, in accordance with principle 6.

At present, direct evidence for the redox control of organellar gene expression, as predicted CORR, is stronger for chloroplasts than for mitochondria (Pfannschmidt et al. 1999). Redox effects on mitochondrial gene expression in vitro are largely confined, at present, to protein synthesis (Allen et al. 1995; Galvis et al. 1998). The search for a direct signalling pathway from the respiratory chain to mitochondrial DNA is likely to be an active area of future research (Allen et al. 2005; Lane 2005).

## 3.7 Maternal Inheritance of Mitochondria

As discussed previously, the mitochondrial theory of ageing rests on the observation that mitochondrial DNA is exposed to high levels of reactive oxygen species when the mitochondrion is performing its redox chemistry. These reactive oxygen species cause mutation. These mutations accumulate, gradually damaging the mitochondrion's ability to function. This happens in all the cells of an organism, leading to the symptoms we know as ageing, and eventually to death. One of the problems facing any eukaryotic organism is how to ensure the next generation does not inherit the damage that it had suffered to its mitochondria. We can see that each generation starts off with new, healthy mitochondria, but how does this happen?

The theory that two sexes are derived from a division of labour between male and female germ line mitochondria was first proposed in 1996 (Allen 1996). Here, it was proposed that the mitochondria of the female germ line have a repressed bioenergetic function, and so they escape the damage to their DNA caused by mutagens generated by respiratory electron transport. The mitochondria are therefore able to replicate and pass to the next generation with minimal change. The male gametes need to have functional mitochondria, in order to generate the energy needed to reach the egg cell. The hypothesis proposes that these bioenergetically active, and therefore damaged, mitochondria will be prevented from entering the germ line.

As far as the mitochondria go, the most significant difference between the male and female is that the male mitochondria have to produce a lot of ATP to propel the sperm as quickly as possible towards the egg, and the female mitochondria do not really have to do much at all. If free radicals generated by an active electron transport chain are responsible for damage to mitochondrial DNA, then it is reasonable to assume that sperm mitochondria are likely to be more damaged than those in the egg. To take this idea one step further, perhaps the mitochondria in the egg cells are prevented from carrying out oxidative phosphorylation at all, in order to preserve as accurate a copy as possible of the DNA to pass to the offspring.

In the systems that have been studied, it can be seen that cells destined to become the female germ line are identified and set aside very early in development. In the female, these cells will go on to form the eggs, and will contain the mitochondria that will pass to the next generation. When one of these eggs is fertilised, the cells that will form the gametes for the next generation are set aside, and so on.

In the majority of organisms, and all mammals, the mitochondria are inherited from only one parent, and that is the mother (Law and Hutson 1992). It is thought that there are different mechanisms for excluding the male mitochondria: some organisms prevent entry to the egg cell, but in mammals it is an active destruction process. Sutovsky et al. (2000) have shown that the sperm cell mitochondria in cattle do get into the egg cell, and are subsequently targeted and destroyed, mostly between the four-cell and the eight-cell stage of the embryo. The male sperm are ubiquitinated within the oocyte cytoplasm, which provides a target for proteolysis. The germ line is determined at a very early stage of embryology. In the nematode worm *C. elegans* (Seydoux et al. 2001), the fate of every cell from the zygote onwards has been studied, and even from the four-cell stage, the germ line can be distinguished. In mammals, the germ line is segregated very early, before implantation in fact. By the blastocyst stage, the germ line has already been segregated, and this is before the embryo becomes aerobic.

From these observations, it is possible to see how mitochondria could be passed from one generation to the next without ever having to fulfil their role as generators of cellular energy, but only having to act as templates. The energy for replication could come from the egg's helper cells, and once any mitochondria have an active respiratory chain they become unsuitable to act as a template for future generations (Fig. 3.5). If such mitochondria did get

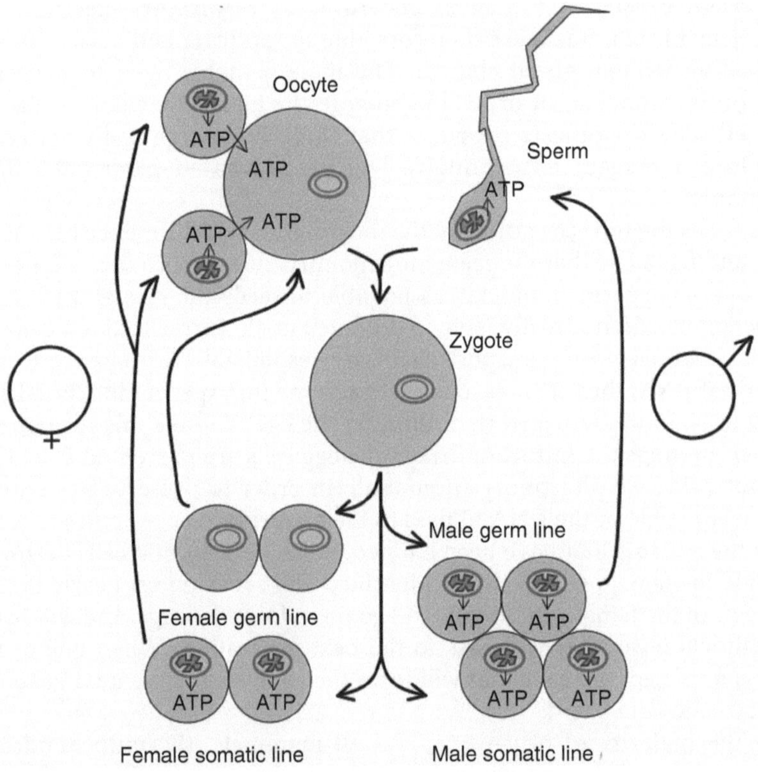


**Fig. 3.5.** Why our genes do not die with us: differentiation of male and female gametes for motility and for fidelity of mitochondrial genome replication. The probability of encounters of two gametes is ideally the same whether one or both are motile; therefore, one sex (male) may produce gametes that sacrifice the mitochondrial genome in favour of oxidative phosphorylation. The other sex (female) is then free to produce immobile gametes in which oxidative phosphorylation is repressed in promitochondria, and through which the mitochondrial genome is thus transmitted with increased fidelity. Promitochondria are sequestered early in development in the female germ line. Female oocytes obtain ATP from oxidative phosphorylation in the differentiated mitochondria of ancillary somatic cells (follicle cells in animals). Promitochondria persist in plants in meristematic cells, prior to differentiation of somatic and germ cells. In contrast, any ancillary germ cells (nurse cells in invertebrates) will also require imported ATP, since they share the oocyte's cytoplasm. Gamete differentiation may likewise rescue the chloroplast genomes of plants. Mitochondria and chloroplasts are thus maternally inherited. (Adapted from Allen 1996)

into the egg cell, and so form the basis of the inherited mitochondria, one would expect that the offspring would be born at a more advanced cellular age than normal. Although the data are sparse, it is thought that early reproductive cloning of mammals was achieved by fusing a somatic cell with an egg cell whose DNA had been destroyed. "Dolly" the sheep was the first cloned mammal, and her nuclear DNA was derived from a 6-year-old ewe. She showed signs of ageing very early on, and died at the age of 5 (usual life span would be 11 years), of a disease more usually associated with old age (Allen and Allen 1999).

## 3.8 Conclusions

From the time that mitochondria were first seen in cells, they have been an interesting enigma. The more we learn about them, the more questions are raised. There would seem to be enough unanswered questions in the field of mitochondrial function, genetics, and origins to engage researchers for a long time to come.

## References

Alberts B, Bray D, Lewis J, Raff M, Roberts K, Watson JD (1994) The molecular biology of the cell. Garland, New York

Allen CA, Håkansson G, Allen JF (1995) Redox conditions specify the proteins synthesized by isolated chloroplasts and mitochondria. Redox Report 1:119–123

Allen JF (1992) Protein phosphorylation in regulation of photosynthesis. Biochim Biophys Acta 1098:275–335

Allen JF (1993a) Control of gene-expression by redox potential and the requirement for chloroplast and mitochondrial genomes. J Theor Biol 165:609–631

Allen JF (1993b) Redox control of gene-expression and the function of chloroplast genomes – an hypothesis. Photosynth Res 36:95–102

Allen JF (1996) Separate sexes and the mitochondrial theory of ageing. J Theor Biol 180:135–140

Allen JF (2003) The function of genomes in bioenergetic organelles. Philos Trans Roy Soc Lond B Biol Sci 358:19–38

Allen JF, Allen CA (1999) A mitochondrial model for premature ageing of somatically cloned mammals. IUBMB Life 48:369–372

Allen JF, Puthiyaveetil S, Ström J, Allen CA (2005) Energy transduction anchors genes in organelles. BioEssays 27:426–435

Altmann R (1890) Die Elementarorganismen und ihre Beziehungen zu den Zellen. Veit, Leipzig

Balaban RS, Nemoto S, Finkel T (2005) Mitochondria, oxidants, and aging. Cell 120:483–495

Benda C (1898) Weitere Mitteilungen über die Mitochondria. Verh Dtsch Physiol Ges 376–383

Bensley RR, Hoerr N (1934) Studies on cell structure by the freezing-drying method VI. The preparation and properties of mitochondria. Anat Rec 60:449–455

Boxma B, de Graaf RM, van der Staay GW, van Alen TA, Ricard G, Gabaldon T, van Hoek AH, Moon-van der Staay SY, Koopman WJ, van Hellemond JJ, Tielens AG, Friedrich T, Veenhuis M, Huynen MA, Hackstein JH (2005) An anaerobic mitochondrion that produces hydrogen. Nature 434:74–79

Burger G, Gray MW, Franz Lang B (2003) Mitochondrial genomes: anything goes. Trends Genet 19:709–716
Canfield DE (1998) A new model for proterozoic ocean chemistry. Nature 396:450–453
Cavalier-Smith T (1987) Eukaryotes with no mitochondria. Nature 326:332–333
Chance B, Sies H, Boveris A (1979) Hydroperoxide metabolism in mammalian organs. Physiol Rev 59:527–605
Chen Q, Vazquez EJ, Moghaddas S, Hoppel CL, Lesnefsky EJ (2003) Production of reactive oxygen species by mitochondria – central role of complex III. J Biol Chem 278:36027–36031
Claros MG, Perea J, Shu Y, Samatey FA, Popot JL, Jacq C (1995) Limitations to in vivo import of hydrophobic proteins into yeast mitochondria. The case of a cytoplasmically synthesized apocytochrome b. Eur J Biochem 228:762–771
Embley TM, van der Giezen M, Horner DS, Dyal PL, Foster P (2003) Mitochondria and hydrogenosomes are two forms of the same fundamental organelle. Philos Trans R Soc Lond B Biol Sci 358:191–204
Ephrussi B (1950) The interplay of heredity and environment in the synthesis of respiratory enzymes in yeast. Harvey Lect 46:45–67
Esser C, Ahmadinejad N, Wiegand C, Rotte C, Sebastiani F, Gelius-Dietrich G, Henze K, Kretschmann E, Richly E, Leister D, Bryant D, Steel MA, Lockhart PJ, Penny D, Martin W (2004) A genome phylogeny for mitochondria among alpha-proteobacteria and a predominantly eubacterial ancestry of yeast nuclear genes. Mol Biol Evol 21:1643–1660
Feagin JE, Gardner MJ, Williamson DH, Wilson RJ (1991) The putative mitochondrial genome of *Plasmodium falciparum*. J Protozool 38:243–345
Galvis MLE, Allen JF, Håkansson G (1998) Protein synthesis by isolated pea mitochondria is dependent on the activity of respiratory complex II. Curr Genet 33:320–329
Gray MW (1992) The endosymbiont hypothesis revisited. Int Rev Cytol 141:233–357
Gray MW, Doolittle WF (1982) Has the endosymbiont hypothesis been proven? Microbiol Rev 46:1–42
Harman D (1956) Aging: a theory based on free radical and radiation chemistry. J Gerontol 11:298–300
Karp G (2002) Cell and molecular biology concepts and experiments, 4th edn. Wiley, New York
Keilin D (1925) On cytochrome, a respiratory pigment, common to animals, yeast, and higher plants. Proc R Soc Lond B Biol Sci 98:312–339
Kölliker A (1857) Einige Bemerkungen über die Endigungen der Hautnerven und deu Bau der Muskelzellen. Z Wiss Zool 8:311–325
Kurland CG, Andersson SG (2000) Origin and evolution of the mitochondrial proteome. Microbiol Mol Biol Rev 64:786–820
Kushnareva Y, Murphy AN, Andreyev A (2002) Complex i-mediated reactive oxygen species generation: Modulation by cytochrome c and nad(p)(+) oxidation-reduction state. Biochem J 368:545–553
Lane N (2005) Power, sex, suicide. Mitochondria and the meaning of life. Oxford University Press, Oxford
Lang BF, Burger G, O'Kelly CJ, Cedergren R, Golding GB, Lemieux C, Sankoff D, Turmel M, Gray MW (1997) An ancestral mitochondrial DNA resembling a eubacterial genome in miniature. Nature 387:493–497
Law R, Hutson V (1992) Intracellular symbionts and the evolution of uniparental cytoplasmic inheritance. Proc Roy Soc Lond B Biol Sci 248:69–77
Lill R, Mühlenhoff U (2005) Iron-sulfur-protein biogenesis in eukaryotes. Trends Biochem Sci 30:133–141
Luck DJL, Reich E (1964) DNA in mitochondria of neurospora crassa. Proc Natl Acad Sci USA 52:931–938
Margulis L (1970) Origin of eukaryotic cells. Yale University Press, New Haven
Martin W, Kowallik KV (1999) Annotated English translation of Mereschkowsky's 1905 paper "Über Natur und Ursprung der Chromatophoren im Pflanzenreiche." Eur J Phycol 34:287–295

Martin W, Müller M (1998) The hydrogen hypothesis for the first eukaryote. Nature 392:37–41
Martin W, Hoffmeister M, Rotte C, Henze K (2001) An overview of endosymbiotic models for the origins of eukaryotes, their ATP-producing organelles (mitochondria and hydrogenosomes), and their heterotrophic lifestyle. Biol Chem 382:1521–1539
Martin W, Rotte C, Hoffmeister M, Theissen U, Gelius-Dietrich G, Ahr S, Henze K (2003) Early cell evolution, eukaryotes, anoxia, sulfide, oxygen, fungi first (?), and a tree of genomes revisited. IUBMB Life 55:193–204
McCord JM, Fridovich I (1969) Superoxide dismutase. An enzymic function for erythrocuprein (hemocuprein). J Biol Chem 244:6049–6055
McLean JR, Cohn GL, Brandt IK, Simpson MV (1958) Incorporation of labeled amino acids into the protein of muscle and liver mitochondria. J Biol Chem 233:657–663
Mereschkowsky CS (1905) Über Natur and Ursprung der Chromatophoren im Pflanzenreiche. Biol Zentr 25:593–604
Mitchell MB, Mitchell HK (1952) A case of maternal inheritance in neurospora crassa. Proc Natl Acad Sci USA 38:442
Nass MM, Nass S (1963a) Intramitochondrial fibers with DNA characteristics. I. Fixation and electron staining reactions. J Cell Biol 19:593–611
Nass S, Nass MM (1963b) Intramitochondrial fibers with DNA characteristics. Ii. Enzymatic and other hydrolytic treatments. J Cell Biol 19:613–629
Nisbet EG, Sleep NH (2001) The habitat and nature of early life. Nature 409:1083–1091
Notsu Y, Masood S, Nishikawa T, Kubo N, Akiduki G, Kadowaki K, Nakazono M, Hirai A (2002) The complete sequence of the rice (oryza sativa l.) mitochondrial genome: Frequent DNA sequence acquisition and loss during the evolution of flowering plants. Mol Genet Genomics 268:434–445
Pearl R (1928) The rate of living. University of London Press, London
Pfannschmidt T, Nilsson A, Allen JF (1999) Photosynthetic control of chloroplast gene expression. Nature 397:625–628
Popot JL, de Vitry C (1990) On the microassembly of integral membrane proteins. Annu Rev Biophys Biophys Chem 19:369–403
Rivera MC, Lake JA (2004) The ring of life provides evidence for a genome fusion origin of eukaryotes. Nature 431:152–155
Rivera MC, Jain R, Moore JE, Lake JA (1998) Genomic evidence for two functionally distinct gene classes. Proc Natl Acad Sci USA 95:6239–6244
Saraste M, Walker JE (1982) Internal sequence repeats and the path of polypeptide in mitochondrial adp/atp translocase. FEBS Lett 144:250–254
Schatz G (1998) Protein transport – the doors to organelles. Nature 395:439–440
Schatz G, Haslbrunner E, Tuppy H (1964) Deoxyribonucleic acid associated with yeast mitochondria. Biochem Biophys Res Commun 15:127–132
Schimper AFW (1883) Über die Entwicklung der Chlorophyll Körner und Farbkörner. Bot Zeit 41:105–114
Schwartz RM, Dayhoff MO (1978) Origins of prokaryotes, eukaryotes, mitochondria, and chloroplasts. Science 199:395–403
Seydoux G, Schedl T (2001) The germline in C. elegans: origins, proliferation, and silencing. Int Rev Cytol 203:139–185
Shigenaga MK, Hagen TM, Ames BN (1994) Oxidative damage and mitochondrial decay in aging. Proc Natl Acad Sci USA 91:10771–10778
Staniek K, Nohl H (2000) Are mitochondria a permanent source of reactive oxygen species? Biochim Biophys Acta 1460:268–275
St-Pierre J, Buckingham JA, Roebuck SJ, Brand MD (2002) Topology of superoxide production from different sites in the mitochondrial electron transport chain. J Biol Chem 277: 44784–44790
Sutovsky P, Moreno RD, Ramalho-Santos J, Dominko T, Simerly C, Schatten G (2000) Ubiquitinated sperm mitochondria, selective proteolysis, and the regulation of mitochondrial inheritance in mammalian embryos. Biol Reprod 63:582–590

Tielens AGM, Rotte C, van Hellemond JJ, Martin W (2002) Mitochondria as we don't know them. Trends Biochem Sci 27:564–572

Timmis JN, Ayliffe MA, Huang CY, Martin W (2004) Endosymbiotic gene transfer: Organelle genomes forge eukaryotic chromosomes. Nat Rev Genet 5:123–135

Tsang WY, Lemire BD (2003) The role of mitochondria in the life of the nematode, Caenorhabditis elegans. Biochim Biophys Acta 1638:91–105

van der Giezen M, Slotboom DJ, Horner DS, Dyal PL, Harding M, Xue GP, Embley TM, Kunji ERS (2002) Conserved properties of hydrogenosomal and mitochondrial adp/atp carriers: A common origin for both organelles. EMBO J 21:572–579

van der Giezen M, Tovar J (2005) Degenerate mitochondria. EMBO Rep 6:525–530

van der Giezen M, Tovar J, Clark CG (2005) Mitochondria-derived organelles in protists and fungi. Int Rev Cytol 244:175–225

Von Heijne G (1986) Why mitochondria need a genome. FEBS Lett 198:1–4

Warburg O (1913) Über Sauerstoffatmende Körnchen aus Leberzellen und über Sauerstoffatmung in Berkefeld-Filtralen Wässriger Leberextrakte. Arch Gesamte Physiol 154:599–617

# 4 Mitochondria and Their Host: Morphology to Molecular Phylogeny

JAN SAPP

## 4.1 Introduction

We often envision the history of ideas about mitochondrial symbiosis to be that of a radical concept persisting on the margins of science, sometimes ridiculed and ignored, until rediscovered many decades later when techniques become available to assess it fairly. Legends of such lost and found discoveries are common in science. We think of Mendel's laws and Chatton's prokaryote–eukaryote dichotomy, both said to have been overlooked for more than three decades until they were "rediscovered". Yet, on closer inspection, such accounts have been found to be somewhat illusory (Sapp 2003, 2005b). Long-neglect stories often result from a specific form of historical myopia, a nearsightedness caused by looking only for ("correct") present views in the past ideas, and ignoring the rest: effectively "seeing what one wants to see."

When we situate past ideas about mitochondrial symbiosis in their own contexts, a more textured history comes into relief – of theories markedly different from those of today. Not only have ideas about the nature of mitochondria changed, so too have analogies, evidence, and requirements of proof. The symbiotic origin of mitochondria was resolved when new molecular methods were developed for bacterial classification previously based on comparative morphology and physiology. While the exogenous origin of mitochondria is no longer disputed today, there is no consensus on the nature of the host and how that symbiosis occurred. The search for the protomitochondrial host over the past decade has again entailed replacing the morphological classification of eukaryotes by molecular evolutionary methods to distinguish convergent evolution from phylogeny.

## 4.2 Alternative Visions

Seeing only what one wants to see is a problem in microscopy too. The story of mitochondrial research begins in the late nineteenth century, blinded by the brilliant cytological research on the chromosomes and their

Origin of Mitochondria and Hydrogenosomes (ed. by William F. Martin and Miklós Müller)

elaborate movements during cell division. Methods for fixing and staining chromosomes often bleached the cytoplasm. With attention concentrated on the nucleus, the aim of making fixatives was to reveal nuclear detail. Mixtures containing alcohol, chloroform or acetic acid were used for this purpose because of their rapid penetration; however, these substances were later found to kill mitochondria. The more one examined the dance of the chromosomes, it seemed, the more one could see nothing else (Cowdry 1916, 1918).

To see differently, to create other visions with different techniques, was far from straightforward. One of the main problems was to sort out visual facts from artefacts; that is, determining whether images represented natural objects or abnormalities due to fixation and staining procedures. The reality of many cytoplasmic structures, including centrioles (Sapp 1998), spindle fibers (Schrader 1953), and the Golgi apparatus (Mazzerello and Bentivoglio 1998), was doubted for decades. So too the reality of mitochondria was questioned.

In his book of 1890 *Die Elementarorganismen*, Richard Altmann (1852–1900) at Leipzig proposed that bodies, he called "bioblasts", which he observed using a special staining technique with fuchsin, reproduced by division and built up the cytoplasm of the cell. Bioblasts were "elementary organisms" which, he believed, secreted various cell substances, including fat, glycogen, and pigments, and could be transformed into, or produce, various rods and fibers. Altmann compared them to free-living bacteria and suggested that the nucleated cell had originated in the remote past when separate bioblasts formed a colony with a membrane constructed around it. This would be much like zoögloea, which secrete an enclosing envelop or capsule of gelatinous or gummy material around them.

Altmann's theory was rejected by his contemporaries. It conflicted with the growing conviction that the nucleus was the sole source of hereditary determination directing the formation of organisms. The USA's premier cell biologist, E.B. Wilson (1896, p. 327) stated this concept clearly in the late nineteenth century, when he wrote that "the nucleus alone suffices for the *inheritance* of specific possibilities of development". The reality of Altmann's bioblasts was also rejected – as artefacts of the staining techniques he used. Only later, in the early twentieth century, were they recognized as mitochondria (Cowdry 1916, Sapp 1994), at which time leading mitochondrial researchers argued that Altmann's identification of these bodies with bacteria had only discredited their importance (Gulliermond 1914, Cowdry 1918). The reality of mitochondria was established beginning in 1898 when Carl Benda (1857–1933) introduced crystal violet as a stain for the granules with greater certainty and brilliancy. Benda (1898) named them "mitochondria" from the Greek *mitos* (thread) and *khondrion* (little granule), since they seemed to exist as both threads and granules.

## 4.3 Before the Word

A symbiotic theory for the origin of all nucleated cells from two phylogenetically distinct kinds of organisms had been proposed before the reality of mitochondria was established. Although he is typically overlooked today, Japanese zoologist Shôsaburô Watasé (1862–1929) was perhaps the first to advocate it (Sapp 1994). Watasé had visited the USA as a student in 1886 and he completed his Ph.D. in 1890 under William Keith Brooks at Johns Hopkins University. Between 1890 and 1899, he was a member of C.O. Whitman's staff, first at Clark University, then at the University of Chicago and at the Marine Biological Laboratory at Woods Hole, where he studied the cleavage of the egg of squid.

In a lecture given before the Biology Club of the University of Chicago, Watasé (1893) noted that the behaviour of the cytoplasmic thread or network suggested that it was "formed of a group of small, living particles, each with the power to assimilate, to grow and multiply by division." The chromosome, was itself "a colony of minute organisms of another kind", each endowed with similar attributes of vitality. That the cell as a whole assimilates, grows, and divides he wrote (p. 86), "is ultimately due to the fact that the minute particles which compose the cytoplasm and chromosome are endowed with these functions." His argument was based on three premises: (1) neither nucleus nor cytoplasm arose de novo by differentiation one from the other – both arose only through division of pre-existing entities of the same kind; (2) analogous symbioses based on physiological complementarities were known; and (3) the same kind of argument had been applied to the chloroplasts of plant cells.

Watasé insisted that any theory of cellular organization had to recognize both the profound physiological interdependence between nucleus and cytoplasm, their reciprocal interchange of metabolic products, and their morphological independence. "The doctrine of symbiosis, first propounded by De Bary, just fulfils these requirements, in as much as it means now, in a more restricted sense, *the normal fellowship or the consortial union of two or more organisms of dissimilar origin, each of which acts as the physiological complement to the other in the struggle for existence*" (p. 101 in Watasé 1893). He also considered the possible symbiotic origin of the "centrosome". It had not been definitively established that centrosomes also arose only by division. But if this were shown to be the case, Watasé insisted that his arguments would apply to them as well.

Demonstrations of the dual nature of lichens led by Simon Schwendener and Anton de Bary in the 1860s provided the exemplar for how an organism could be created by the establishment of an intimate physiological relationship between two dissimilar organisms (Sapp 1994). Watasé also pointed to papers suggesting that chloroplasts of plants similarly arose as symbionts, and to classic papers describing "animal chlorophyll" in translucent animals such as hydra and sea anemones as symbiotic algae. In these latter cases, bodies at first thought to be organs derived from the differentiation of the germ were shown to be symbiotic algae. What were once called organs from a physiological point of view were in reality organisms in themselves.

The symbiotic nature of chloroplasts was suggested by de Bary's former student Andreas Schimper (1883, 1885) at the University of Bonn when he demonstrated that these cell organs arose only by division of pre-existing bodies of the same kind. Schimper (1883, pp. 112–113) first suggested they might be symbionts in the very paper in which he coined the term "chloroplast":

"Should it be definitively proven that the plastids are not formed anew in the egg cells, then their relationship to the organism that contains them would more or less remind us of a symbiosis. It is possible that the green plants indeed owe their origin to the union of a colourless organism with one evenly stained with chlorophyll".

The analogy between chloroplasts and algae living in the cells of translucent animals was clear. Studies of the algal symbionts of the flatworm *Convoluta roscoffensis* (Keeble 1910) by Gottlieb Haberlandt (1891), a former student of Schwendener, led him to suggest that these algae might eventually lose their protoplasm and nucleus, and evolve into simple chlorophyll corpuscles like those of higher green plants. In the early twentieth century, the symbiotic nature of plastids was generally considered with mitochondria; indeed plastids themselves were often understood to be modified mitochondria.

## 4.4 *Les Symbiotes*

During the first decades of the twentieth century, mitochondria were intensely investigated by histologists, especially in France, Germany, and Belgium, under various aliases, including chondriosomes, plastosomes, ergastidions (little workers), eclectosomes, vacuolides, and plastidules. Mitochondria were reported in the cells of plants, animals, and protists. Zoologists identified them in both eggs and sperm, and many leading cytologists ascribed to them the power of independent growth, and division. Those who investigated mitochondria considered them to be fundamentally important for tissue development, cellular differentiation, and as "bearers of heredity" (Duesberg 1913, 1919; Meves 1918; Cowdry 1918). This was a time when leading embryologists argued that the cytoplasm of the egg contained "organ-forming substances" and was responsible for the "fundamental characteristics" of the organism (Sapp 1987, 2003). These were developmental features which distinguished the higher taxonomic groups: phyla, classes, orders, and families. Accordingly, Mendelian genes in the cell nucleus would be responsible only for trivial characteristics which topped off the organism: differences between individuals or species.

According to the eclectosome theory of Claude Regaud (1909) at the Pasteur Institute, mitochondria select substances from the surrounding cytoplasm and condense and transform them into diverse products and structures. That mitochondria were formative granules was also supported by studies in plants led by Alexandre Guilliermond (1914) in Lyon, who followed the trajectory of mitochondria in the epidermal cells of the iris and argued that mitochondria could elaborate diverse materials and directly transform themselves into chloroplasts, amyloplasts, and chromoplasts. These concepts of mitochondria as formative granules permeated every theory about mitochondrial symbiosis before the Second World War.

Paul Portier (1866–1962), working at the Institut Océanographique de Monaco, was the first to write explicitly about mitochondria as symbionts. In his book *Les Symbiotes* (p. viii in Portier 1918 ), he developed an elaborate theory of symbiosis as a fundamental aspect of life when he argued that mitochondria were symbionts, transformed over eons by their intracellular existence:

"All living beings, all animals from Amoeba to Man, all plants from Cryptogams to Dicotyledons are constituted by an association, the '*emboitement*' of two different beings. Each living cell contains in its cytoplasm formations which histologists call 'mitochondria'. These organelles are, for me, nothing other than symbiotic bacteria, which I call 'symbiotes'".

Portier's interest in symbiosis had begun with studies of bacteria in the gut of cellulose-eating insects such as termites, which he argued were necessary for digesting cellulose, supplied the host with essential vitamins, and played fundamental roles in tissue differentiation and development of their hosts. Portier's *symbiote* was an organ of synthesis. He adopted Regaud's eclectosome theory that mitochondria selected substances in the cytoplasm and transformed them into diverse products, including chloroplasts. What they elaborated depended on the physicochemical milieu in which they found themselves in the course of development.

Portier's arguments about the symbiotic nature of mitochondria were based on similar staining properties, comparative physiology and morphology, as well as natural history. There was one one fundamental *symbiote*, in his view, which was prevalent in nature, extremely resistant to physical and chemical agents, and had extensive morphological and physiological plasticity. It was this that populated the cells of all organisms. He further postulated that bacteria entering with food fused with and rejuvenated the mitochondria. To demonstrate that mitochondria were *symbiotes* he pointed to bacteria in the root nodules of legumes, which he referred to as promitochondria. One was able to culture them independently of their host. Portier further claimed to have cultured bacteria from the healthy tissue of animals and he identified those bacteria as mitochondria.

Portier's critics in France rejected all of his ideas. Effectively pasteurizing his *symbiotes*, they insisted that the identification of mitochondria with bacteria based on morphology and staining properties was purely illusory (Sapp 1994). Symbionts in the cells of plant tissue was one thing, but it was simply too much to accept that bacteria lived in healthy animal tissue, or that mitochondria were ever free-living bacteria. Portier's claim to have cultured bacteria from the healthy tissue of animals led to a showdown with bacteriologists at the Institut Pasteur who dismissed his results as due to contamination, a charge to which Portier astutely responded was unscientific because it was impossible to refute: there were no such experimental proofs that could not be attributed to contamination.

Russian botanist Constantin Mereschkowsky (1855–1921) who coined the word "symbiogenesis" for the synthesis of new organisms by symbiosis, and who developed the idea of chloroplasts as symbionts (Merezkowsky 1905) completely rejected mitochondrial symbosis (Sapp 1994; Sapp, Carrapiço and Zolotonosov 2005). Mereschkowsky (1909) had maintained that nucleus and cytoplasm had also originated as a symbiosis of two kinds of microbes/protoplasms: *mycoplasm* and *ameoboplasm*. In his famed work "La plante considérée comme un complexe symbiotique", Mereschkowsky (Mérejkovsky 1920) denied that plastids originated from mitochondria, and he dismissed the idea that mitochondria arose as symbionts as being unreasonable, and wholly antagonistic to his theory of "the symbiotic nature of plants and of the two plasmas."

## 4.5 Symbionticism and the Origin of Species

In his book *Symbionticism and the Origin of Species* from 1927, Ivan Wallin (1883–1969) at the University of Colorado advanced the theory that the inheritance of acquired bacteria was the source of new genes and the primary mechanism for the origin of species. Wallin had come to symbiosis and mitochondria through studies of tissue differentiation and development in the lamprey *Ammocetes*. The continental research on mitochondria and their proposed roles in development and heredity were introduced to English-speaking researchers by Edmund Cowdry (1918). Wallin had also made a study of Portier's *Les symbiotes*, and he pressed the idea in the context of heredity, development, and evolution.

Wallin emphasized the importance of bacterial symbiosis as a generator of new tissues and new organs. Like others, he maintained that many other organelles and cell structures were products of mitochondria. In his scheme, Golgi bodies, as well as cilia and flagella, were also the products of mitochondrial symbiosis. As Wallin (1927, p. 142) wrote: "The ciliate and flagellate protozoa, apparently, acquired their special locomotor structures through symbiosis with ciliate and flagellate bacteria. We have previously mentioned that some investigators have claimed that cilia formation in the cells of higher organisms is associated with mitochondria." Mitochondria, and repeated symbiosis, he argued were largely responsible for cellular differentiation and morphogenesis in the course of ontogenetic development.

In Wallin's scheme, mitochondria were not organisms that had entered some primitive microbes millions or billions of years ago; they were constantly being added to the germplasm in the course of the origin of species. The mitochondrial population in the egg would represent pleomorphisms of a great number of strains of acquired bacteria. In the chemical environment of the egg cytoplasm, the various strains were resolved into a common morphological type, but during development as new intracellular environments are produced, mitochondria would differentiate hand in hand with the tissue differentiation. Therefore, inasmuch as ontogeny involved mitochondrial differentiation, the different lines that emerged in the course of development were the result of various symbiotic events in the course of evolution. In short, ontogeny recapitulated a symbiotic phylogeny.

Wallin (1924, 1927) also claimed that he had cultured mitochondria outside the living cell to prove their bacterial nature. When he planted bits of the tissue from the liver from embryonic, fetal and newborn rabbits, guinea pigs, and dogs in more commonly used media, the results were negative. He obtain cultures on a human blood medium, but difficulties with procuring sufficient quantities of human blood to make the medium on a large scale and with sterilization of the media made this approach unviable. Wallin devised his own special media, and when bits of liver tissue from the fetal and newborn rabbits were planted in these media, he observed coccoid organisms that could be fixed with a mitochondrial fixative.

Like Portier before him, Wallin (1927) argued that is was merely "a convenient matter" to explain his results away "with the dogmatic cry of 'contamination'". He had assured his readers that all instruments and vessels had been thoroughly sterilized. The fetus or newborn rabbit after decapitation was saturated with 95% alcohol. The instruments used in opening the abdomen and removing the liver were always sterilized in the flame immediately before use. The liver was quickly removed to a sterile petri dish, cut into pieces by inserting a sterile scalpel under the lid. Pieces of the liver tissue were then planted in the media by the usual bacteriological technique.

## 4.6 Against the Current

There were several obstructions to a symbiotic conception of mitochondria that persisted throughout most of the twentieth century:

1. Ever since Pasteur, bacteria had been defined as disease-causing germs, and the notion that bacteria played any beneficial role in the tissues of animals was in virtual conflict with the basic tenets of germ theory. Studies

of microbial natural history and symbiosis were eclipsed by the study of disease. As Wallin (1927, p. 8) wrote: "It is a rather startling proposal that bacteria, the organisms which are popularly associated with disease, may represent the fundamental causative factor in the origin of species. Evidence of the constructive activities of bacteria has been at hand for many years, but popular conceptions of bacteria have been coloured chiefly by their destructive activities as represented in disease." It seemed ridiculous to suggest that bacteria living in tissue could be part of the physiological well-being of animals.

2. Bacteria had no natural history, and they lacked biological definition (Sapp 2005a). Classification of bacteria for genealogical purposes, based on comparative physiology and morphology, was unfruitful. Pathologists were concerned that if mitochondria were bacteria how could one distinguish them from diseases such as that caused by *Rickettsia*. Critics of the symbiotic theory at the Rockefeller Institute thus emphasized what they took to be morphological and physiological differences between mitochondria and bacteria (Cowdry and Olitsky 1922; Wallin 1923).
3. Hereditary symbiosis conflicted with nucleocentric conceptions of the cell. Cell biologists had adopted the Weismannian view of heredity in terms of one germplasm, one organism. This dogma continued into the twentieth century with Mendelian geneticists' insistence that, with the rare exception of plastid inheritance, chromosomal genes were the sole basis of heredity: the concept of one genome, one organism. The leader of the *Drosophila* school, T.H. Morgan (1926, p. 496), put this perspective in a nutshell: "The cytoplasm may be ignored genetically." The same year that Wallin's book appeared, the famed geneticist H.J. Muller (1927) reported that X-rays could dramatically increase the frequency of gene mutations in *Drosophila* by some 1,500 times. Bacterial symbiosis as the source of new genes was zapped by a wave of radiation genetics.
4. Symbiosis conflicted with neo-Darwinian conceptions of evolutionary change, and the perceived incessant struggle for existence underlying nature. The evolutionary synthesis of the 1930s and 1940s was a sterile concept of evolution without microbes. The transfer of genes between species was generally considered to be in violation of nature's laws. Known cases of intracellular symbiosis were typically treated as curiosities, "special aspects of life", of little significance for general biology (Sapp 1994, 2003).
5. Symbiotic theories of cell organelles were speculative. When in the third edition of *The Cell in Development and Heredity* Wilson (1925, p. 739) reviewed ideas that centrioles, plastids, and mitochondria may have risen by symbiosis, he wrote; "To many no doubt, such speculations may appear too fantastic for mention in polite society; nevertheless it is within the range of possibility that they may someday call for some serious consideration." Some of those biologists who pioneered microbial symbiosis in

animals were less open to such a possibility. Paul Buchner (1886–1978) investigated morphological and physiological effects of microbes transmitted through the eggs of many species of insects, during the first half of the twentieth century (Sapp 2002). Yet, Buchner (1965) divorced himself from "the extravagant" claims of Portier and Wallin and from those of Hugo von Schanderl (1947), who claimed to have "regenerated symbiotic bacteriods" from many sterilized plant parts, and to have observed mitochondria transform into free-living bacteria. Buchner saw such far-fetched claims as nothing but a liability to his sound empirical work and to his more basic struggle to change the prevalent view that microbial symbiosis in animals was a rare phenomenon.

## 4.7 Infective Heredity

Hereditary symbiosis captured the attention of geneticists when microbes were domesticated for genetic use (Sapp 1987, 1994). The first generation of bacterial geneticists also showed that bacteria were polygenomic, possessing a main circular genome, bacterial phages, and plasmids. Bacteria also possessed various means for exchanging genes. Plasmids (and fragments of the main genome) could be transferred between different kinds of bacteria by conjugation. Bacteria could also acquire DNA fragments from dead bacteria (transformations), and from viral infections (transductions).

Studies of heredity in bacteria and genetic research on *kappa* in *Paramecium* and *sigma* in *Drosophila* reinforced the idea that genes and genomes acquired by infection could become well integrated into, and form an essential part of, the genetic constitution of host organisms. Tracy Sonneborn (1950) insisted that such infectious genetic particles as *kappa* must be considered as part of heredity, and that mitochondria and chloroplasts which were not infectious today may have originated as symbionts. Biochemists showed that mitochondria were energy-generating organelles of the cell, responsible for oxidative phosphorylation, and contained enzymes of the Krebs cycle (Kennedy and Lehninger 1949; Lehninger 1965). Genetic research programmes on mitochondria and chloroplasts also emerged during the 1950s.

While leading Mendelian geneticists such as Muller (1951) trivialized the significance of cytoplasmic infectious particles for heredity and evolution, others led by Cyril Darlington and Joshua Lederberg (1951) called for a broadening of the term heredity to embrace "infective heredity". Darlington (1951) prophesied that recognition of cytoplasmic genetic entities would enable geneticists "to see the relations of heredity, development and infection and thus be the means of establishing genetic principles as the central framework of biology."

Lederberg (1952) coined the word "plasmid" as a neutral term for any extrachromosomal hereditary determinant regardless of its origin. He argued that hereditary symbiosis was analogous to hybridization. He conceptualized a graded series of symbiosis, from cohabitants of a single chromosome, to plasmids, and to extracellular ecological associations of variable stability and specificity. Symbioses, he reasoned, obscured biological definition of the individual. Lederberg (1952) remained agnostic about the origins of chloroplasts and mitochondria; he was cautious of "uncritical overstatements" that they originated from cyanobacteria and free-living bacteria. The issue, he argued, was irresolvable: "We should not be too explicit in mistaking possibilities for certainties. The general criteria that have been used to decide the historical origin of certain plasmids are unverifiable, and such controversies have tended to be sterile."

The lack of a genealogical classification for bacteria was decried throughout the twentieth century. While lamenting the absence of a definition of bacteria and the impossibility of a phylogenetic classification on physiological and morphological grounds, Roger Stanier and C.B. van Neil (1962) argued, on the basis of electron microscopy, that there was a fundamental dichotomy in nature, a profound structural difference between what Edouard Chatton had referred to decades earlier as the eukaryote and the prokaryote (Sapp 2005a). Eukaryotes had a membrane-bound nucleus, a cytoskeleton, an intricate system of internal membranes, mitochondria that perform respiration, and in the case of plants, chloroplasts. Bacteria (prokaryotes) were smaller and lacked all of these structures; they lacked mitosis:

"The principle distinguishing features of the procaryotic cell are: 1) absence of internal membranes which separate the resting nucleus from the cytoplasm, and isolate the enzymatic machinery of photosynthesis and of respiration in specific organelles; 2) nuclear division by fission, not by mitosis, a character possibly related to the presence of a single structure which carries all the genetic information of the cell; and 3) the presence of a cell wall which contains a specific mucopeptide as its strengthening element. (p. 21 in Stanier and van Neil 1962)".

In *The Microbial World*, Stanier, Michael Douderoff, and Edward Adelberg (1963, p. 85) declared that: "In fact, this basic divergence in cellular structure, which separates the bacteria and blue-green algae from all other cellular organisms, represents the greatest single evolutionary discontinuity to be found in the present-day world." At the same time, they insisted (p. 409 in Stanier et al. 1963) that one could only "discern four principal sub-groups, blue-green algae, myxobacteria, spirochetes, and eubacteria, which seem to be distinct from one another.... Beyond this point, however, any systematic attempt to construct a detailed scheme of natural relationships becomes the purest speculation, completely unsupported by any sort of evidence." Thus they concluded that, "the ultimate scientific goal of biological classification cannot be achieved in the case of bacteria."

## 4.8 The Tipping Point

Symbiotic theories of mitochondria and chloroplasts found a new footing in the context of molecular biology. The nuclear monopoly on genes was broken. DNA was discovered in plastids (Ris 1961; Ris and Plaut 1962) and mitochondria (Nass and Nass 1963; M.N.K Nass 1969; Nass 1971). The second criterion in Stanier and van Neil's definition of the eukaryote had to be modified: all genetic information was not in a single structure. Almost all eukaryotes were polygenomic. In both mitochondria and chloroplasts, the DNA was circular, like that of bacteria. Both organelles possessed ribosomes and a full protein synthesis apparatus. Genetic dissection of their genomes was well under way (Sager 1972; Gilham 1978). The electron microscope (Rasmussen 1997) revealed structural similarities between mitochondria and plastids on the one hand and bacteria on the other.

The paradigm started to turn over rapidly. What was that DNA doing there, if they were not symbionts? Symbiosis was mentioned immediately by those who first reported that those organelles possessed DNA (Ris 1961; Ris and Plaut 1962; Nass and Nass 1963) and the structural and functional similarities between these organelles and bacteria were reviewed by many, including Goksøyr (1967), Sagan (1967), M.M.K Nass (1969, 1971), S. Nass (1969), Wagner (1969), Cohen (1970, 1973), Stanier (1970), Raven (1970), and Margulis (1970). The symbiotic origin of mitochondria and chloroplasts would quickly emerge as the dominant theory against which others would be assessed (Taylor 1974).

The renewed interest in the possible symbiotic origin of mitochondria and chloroplasts did not simply signify a rediscovery of past ideas based on better evidence, and improved techniques. The new symbiosis paradigm differed in fundamental ways. By the 1960s mitochondria were no longer considered the principal basis of cellular differentiation. They were not the source of other organelles such as chloroplasts, and centrioles; they were not held to be responsible for new organs over the course of evolution. Mitochondria were not held to be a diverse population in a cell; one could now speak of the origin of the mitochondrion (John and Whatley 1975).

In the new models of endosymbiosis, mitochondria had not been repeatedly added to organisms, reflecting the course of phylogeny and ontogeny; they were not a heterogenous population as in Wallin's scheme; nor were they organisms that could be cultured outside the cell. Mitochondria and chloroplasts were held to have each originated once from bacteria in the remote past (and in some cases plastids were acquired secondarily by engulfment of a photosynthetic protist, as long had been suggested; Haberlandt 1891). The central rhetorical paradigm was no longer that about microbial symbiosis versus the germ theory of disease. The question for cell biologists was centred on the origin of the eukaryotic cell, and explaining what was held to be the greatest evolutionary discontinuity in the living world: that between a prokaryote and a eukaryote (Margulis 1970; Taylor 1974). The reward of mitochondrial respiratory function assumed paramount importance in all discussions of their host's acquisition billions of years ago.

Jostein Goksøyr (1967) published an outline according to which both organelles were bacterial symbionts which had transferred most of their genes to the chromosomes. The same year Lynn Sagan (1967) (subsequently Margulis), a former student in Hans Ris's class at the University of Wisconsin, also extended the reach of symbiosis to account for the origin of centrioles/kinetosomes, the origin of cell motility, and the microtubular system. Since the nineteenth century, centrioles were thought to divide by fission to spin out spindles and to play a crucial role in cell division by mitosis in many kinds of organisms (Sapp 1998). Margulis's argument also hinged on a morphological analogy with spirochetes, once taken to be cilia, which attach themselves to the protist, *Mixotricha paradoxa*, living in the hind gut of a termite (Cleveland and Grimstone 1964). Both Goksøyr and Margulis (1970) offered a sequence of events in which an aerobic prokaryote (the protomitochondrion) was acquired by and co-evolved in an anaerobic primitive eukaryote at a time in which atmospheric oxygen had increased owing to photosynthetic blue-green algae (cyanobacteria). The ancestor of plastids (cyanobacteria) was added to the now aerobic eukaryote.

The nature of the events leading to the acquisition of mitochondria would become the subject of intense investigation and debate in the 1980s, and alternative scenarios were offered. In the meantime, crucial evidence for their symbiotic origin was lacking. Goksøyr (1967) supposed that his proposed series of events could never be proven. Initially, Margulis (Sagan 1967) imagined that biologists might learn to culture chloroplasts, mitochondria, and centrioles; however, it soon became evident that these organelles were highly integrated into the nuclear genetic system: only a small fraction of the genes needed for mitochondrial and chloroplast functions were actually located in the organelles themselves.

The alternative theory (direct filiation) held that these organelles had arisen by autogenously, that is, they evolved by compartmentalization within the cell (Allsopp 1969; Raff and Mahler 1972, 1975; Uzzel and Spolsky 1974; Bogorad 1975). Such autogeneous theories often entailed the notion that the first eukaryote was already a photosynthetic organism the "Uralga" (Klein and Cronquist 1967; Allsopp 1969; Cavalier-Smith 1975) Blue-green algae provided the missing link, the intermediate form between prokaryote and eukaryote; the latter evolved directly from a blue-green-like prokaryotic stock. Thus, discussions of the origin of mitochondria had to be considered together with the origin of plastids (Taylor 1974; John and Whatley 1975). The origin of those organelles by symbiosis was dubbed the serial endosymbiotic theory, the SET (Taylor 1974). That centriole-like structures and cell motility also arose from endosymbiosis was considered an extension of the theory by Margulis, and was not generally adopted by microbial evolutionists. Centrioles did not possess a double membrane as did the nucleus, mitochondria, and chloroplasts, and electron microscopic studies showed that they did not divide by fission. The evidence for DNA in centrioles/kinetosomes had been on-again, off-again since the 1960s, and was effectively refuted in the 1990s (Hall and Luck 1995; Sapp 1998).

## 4.9 The Birth of Bacterial Phylogenetics

The symbiotic origin of organelles could not be rigorously tested without genealogical methods. Stanier (1970) conveyed this sentiment when he commented that: “Evolutionary speculation constitutes a kind of meta-science, which has the same fascination for some biologists that metaphysical speculation possessed for some medieval scholastics. It can be considered a relatively harmless habit, like eating peanuts, unless it assumes the form of an obsession; then it becomes a vice.” By the end of the decade, the field of molecular evolution had emerged offering a means for discerning microbial phylogenies.

At first, trees depicting a symbiotic origin of mitochondria and chloroplasts were sketchy, and were cobbled together from partial sequences of diverse molecules, 5S RNA, ferredoxin, cytochrome *c*, based on very few organisms (Schwartz and Dayhoff 1978). They were easily criticized (Demoulin 1979). The decisive evidence came from systematic investigations of bacterial phylogeny based on comparisons of the small subunit ribosomal RNA (SSU rRNA) pioneered by the group led by Carl Woese at the University of Illinois.

Woese and George Fox (1977) challenged the fundamental dichotomy of life, arguing on molecular grounds that there were three fundamental lineages, three urkingdoms: Eucaryaea, *Eubacteria*, and *Archaebacteria*. The prokaryote–eukaryote dichotomy based on morphology, they argued, was faulty. Prokaryotes were not a monophyletic group. The argument for “the third form of life”, the archaebacteria, was soon fortified with other evidence of their unique cell-wall chemistry, membrane lipids, and transcription enzymes (Fox et al. 1980). The archaebacterial concept and the comparative analysis of SSU rRNA were immediately applied to the problem of the origin of the eukaryotic organelles (Zablen et al. 1975; Woese and Fox 1977; Fox et al. 1980).

The SSU rRNA technology developed in Urbana was transferred by Linda Bonen, Woese’s former technician, to the laboratories of Ford Doolittle and Michael Gray at Dalhousie University in Canada. Doolittle’s and Woese’s groups sent a collaborative paper to *Journal of Molecular Evolution* in 1975 on the origins of chloroplasts reporting that they were descendants of bacteria: the photosynthetic blue-green bacteria (cyanobacteria) (Doolittle et al. 1975; see also Bonen and Doolittle 1976). The SSU rRNA technology was applied to mitochondrial analysis of wheat embryos in Gray’s laboratory

(Cunningham et al. 1976; Bonen et al. 1977). The mitochondrial ancestor was traced to the α-proteobacteria. There were no comparable data to test whether centrioles/kinetosomes arose as symbionts. Comparing SSU rRNA of chloroplast, mitochondrion, and nucleocytoplasm with each other and with different bacteria was held to provide the rigour and close the main controversy about the origin of mitochondria and chloroplasts (Gray and Doolittle 1982; Gray 1992; Gray et al. 1999).

## 4.10 Just-So Stories

It was generally accepted, by the mid 1980s, that mitochondria (and plastids) originated by symbiosis; but how that symbiosis occurred exactly was far from certain. There were two main aspects of the archetypical plot about how the protomitochondrion was acquired: the biogeological context, and the nature of host–symbiont relationship.

First was the adaptive oxygen context. In most accounts it was assumed that every organelle arose for its present purpose: photosynthesis from photosynthetic bacteria (motility organs from motility symbiosis), and mitochondria for aerobic respiration and ATP. From the outset, the SET proposed an aerobic protomitochondrial symbiont in an anaerobic host, or in a host with a less efficient aerobic respiratory system (Ris 1961; Sagan 1967; Margulis 1970; Stanier 1970; Raff and Mahler 1972; de Duve 1973; Taylor 1974). According to this scenario, as the primitive earth atmosphere began to change from anaerobic to aerobic as a result of photosynthetic oxygen production around 1.5 billion years ago, anaerobic prokaryotes were forced to either adapt to aerobic conditions or become restricted to anaerobic environments. On the grounds that eukaryotes today are restricted to glycolysis for their anaerobic energy, so too the ancestral protoeukaryote utilized glycolysis.

All symbiotic hypotheses assumed that the host possessed this primitive and inefficient mechanism of energy generation until the capacity of aerobic respiration was implanted in it by the acquisition of aerobic protomitochondrial symbionts. Those who had opposed this scenario argued that the modern eukaryote is not an anaerobic cell, containing mitochondria, but that an aerobic pathway of the cytoplasm was a primitive feature of the bacteria that evolved into a prokaryote (Raff and Mahler 1972, 1973; Uzzell and Spolsky 1973). They proposed instead that the mitochondria may have evolved from the enclosure of a bacterial plasmid within a metabolically specialized sac.

The oxygen context for the acquired protomitochondria has remained the principal paradigm for understanding mitochondrial origins to the present day. In the metabolic complementary of symbiont and host, the symbiont was in charge of aerobic respiration and energy (ATP), while the host was responsible for the breakdown of organic substances and protection. Margulis (1992, p. 208) commented that, "the release of ATP to its host would be analogous to throwing cash into the streets." Genes needed only for independent life were lost from the symbiont's genome and some genes needed for respiration were transferred to the host nucleus. The result was the modern mitochondrion with a dramatically reduced genome (John and Whatley 1975; Doolittle 1980, 1998a). While most accounts emphasize the importance of ATP production, some have emphasized that the advantage of the protomitochondria lay first in detoxifying the host of oxygen (Andersson and Kurland 1999).

Various models have been offered to account for why some genes are transferred from symbiont to host nucleus and why others are retained by the mitochondria, from game theory ones, "of paying taxes to a central government" (p. 23 in Maynard Smith and Szathmáry 1995), decreasing mutation load (Kurland 1992), increasing efficiency and regulation, accidental and inevitable subsequent to digesting proto-organelles (Doolittle 1998b; Timmis et al. 2004), to the importance of synthesizing structural proteins of bioenergetic membranes on site (Allen 1993; Race et al. 1999; see also Chap. 3 by Allen et al. in this volume).

Second was the nature of the host. In her book *The Origin of Eukaryotic Cells*, Margulis (1970, p. 183) argued that mitochondria were swallowed, but not digested by an ameoboid prokaryote that is no longer alive today: "If there are primitive protoeukaryotes that lack mitochondria, this theory predicts that they also lack mitosis, the complex flagellum and (unless they have achieved some analogous solution) must be anaerobic." However, there seemed to be a logical problem with this because only eukaryotic cells with a cytoskeleton could accomplish phagocytosis and engulf symbionts (Stanier 1970; Cavalier-Smith 1975). Therefore, the missing host for mitochondria would have been some kind of eukaryote with a system of cortical microtubules which itself could not have arisen by bacterial symbiosis but rather autogenously. Stanier (1970) proposed selective pressure of the eukaryote towards progressive structural evolution using predation as a new biological nutritional means. While the selection pressures on prokaryotes were towards new and diverse modes of energy-yielding metabolism, capturing food by endocytosis, the elaboration of the cytoskeleton and endomembrane systems freed the eukaryotic lineage from the need to evolve the diversity of energy metabolism characteristic of bacteria.

## 4.11 Kingdom Come, Kingdom Go

By the early 1980s it was generally assumed that the protomitochondrion was an aerobic $\alpha$-proteobacterium that had entered a strictly anaerobic amitochondriate eukaryote. Such amitochondriate protists existed. There were more than 1,000 species of them (Fenchel and Finlay 1995). Cavalier-Smith (1983) grouped them into a subkingdom, the Archezoa, a primitive form of life that had emerged before the acquisition of chloroplasts and mitochondria. It included four phyla: Metamonads, Microsporidia, Parabasalia, and Archamoebae, all of which comprised anaerobic amitochondriate protists which satisfied their ATP needs through anaerobic fermentations (Müller 1988, 1993).

Grouping the amitochondriate protists together in the Archezoa subkingdom fortified the notion that the mitochondrial host was indeed a eukaryote. On the basis of comparisons of their transcription machinery (Huet et al. 1983) and other molecular links between eukaryotes and archaebacteria (Iwabe et al. 1989; Gogarten et al. 1989), it seemed that the protoeukaryote's closest relative was archaebacterial.

In the standard narrative the protomitochondrion was acquired by an archezoan for its present-day function of generating ATP in respiration (Cavalier-Smith 1987b; Doolittle 1998b). There were very few deviations of the respiratory ATP adaptive symbiotic scenario of a protomitochondrion and an archezoan (Woese 1977; Searcy 1992). Was this adaptationist paradigm correct? It was possible that the mitochondrion, like many other organismic structures, did not arise for its present purpose. Appreciating this point was also important for understanding the early evolution of complex organs, such as the wings of a bird or insect; before they were used for flying, protowings had other purposes.

Evolutionists sensitive to the fallacy (Gould and Lewontin 1978) that all structures have arisen for their present purpose have invoked Voltaire's Dr. Pangloss: "Things cannot be other than they are, for since everything was made for a purpose, it follows that everything is made for the best purpose. Our noses were made to carry spectacles, so we have spectacles. Legs were clearly intended for breeches, and we wear them" (Voltaire 1947). One could not simply assume that today's function was an explanation for mitochondria's origin.

What is more, the Archezoa subkingdom had been built on a weak foundation. Like the prokaryote–eukaryote dichotomy before it, it was founded on morphological grounds, and was defined negatively in terms of what members lacked. Although Cavalier-Smith (1987b, p. 58) was strongly of the view that "Archezoa are genuinely primitive", it remained possible that the subkingdom was actually polyphyletic: the similarities within and between the four "phyla" an evolutionary illusion arising from convergence. Some or all of its members could once have had mitochondria but lost them. At first, phylogenetic evidence based on SSU rRNA sequencing data was reported (Vossbrinck et al. 1987; Sogin 1989a,b) that corroborated the theory that the Archezoa were indeed an ancient line of eukaryotes that diverged before the advent of the mitochondria-carrying protists. Cavalier-Smith (1987b) favoured the Metamonada as the candidate for the protomitochondrial host; later he favoured the Archamoebae as the ancestral eukaryotes (Cavalier-Smith 1991).

By the mid-1990s severe cracks appeared in the foundational assumptions about the Archezoa and mitochondrial origins. The ancient and monophyletic nature of the Archezoa began to fade away as microbial evolutionists searched for mitochondrial genes (homologues of α-proteobacterial genes) that may have been transferred to the nucleus (Yang et al. 1985). The subkingdom could not withstand molecular scrutiny; it lost constituents rapidly:

The Archamoebae possessed great diversity, and other than being amitochondriate, they possessed few unifying characteristics. There was evidence that the Archamoebae arose from mitochondrial-containing ancestors: mitochondrial-like genes were found in the nucleus, genes that resembled those of α-proteobacteria (Clark and Roger 1995).

The Parabasalia possessed hydrogenosomes (Lindmark and Müller 1973), double membrane bound organelles that produce hydrogen and ATP found in anaerobic amitochondriate protists such as *Trichomonas vaginalis* (Müller 1993). Hydrogenosomes were beginning to be understood, by molecular comparison, to have shared a common ancestor with mitochondria (Embly et al. 1995; Bui et al. 1996: Germot et al. 1996; Roger et al. 1996; Müller and Martin 1999).

Microsporidia had also lost their mitochondria and their ancient status too (Germot et al. 1997). They were found to have had unusual molecular sequences resulting from their parasitic lifestyle that led them to be misplaced in RNA-based phylogenetic trees. Rather than being at the bottom, they were actually at the crown with the fungi (Hirt et al. 1999).

Doubts were also growing about Metamonads, beginning with evidence based on chaperon proteins of *Giardia lamblia* that members of that group may also once have carried mitochondria (Soltys and Gupta 1994; Keeling 1998; Roger 1999).

The Archezoa subkingdom was severely shaken, to some, it seemed, reduced to rubble (Keeling 1998; Roger 1999; Martin and Müller 1998; Martin et al. 2001). Mitochondrial symbiosis seemed to have been established in the ancestor of most, if not all extant eukaryotes.

During the 1990s other molecular anomalies appeared in the standard oxygen–ATP paradigm for mitochondrial origins. It concerned the kinds of genes found in the nucleus. In accordance with the oxygen–ATP account, the eubacterial genes transferred to the primitive eukaryotic nucleus would be those that were involved in mitochondrial biogenesis, maintenance, or respiration. However, many eubacterial-like proteins encoded in the nucleus were found that involved other functions not obviously related to mitochondria today. The eubacterial genes were connected to various functions from carbohydrate metabolism to translation (Roger 1999). Were all these genes derived from the mitochondria? Did mitochondrial symbiosis contribute more to the host than generally assumed by the standard story? Was mitochondrial symbiosis at the very foundation of the eukaryote? Several interpretations were possible.

## 4.12 A Chimeric Paradigm

In 1998, William Martin and Miklós Müller challenged the two core features of the standard archezoan symbiotic scenario:

1. They denied the existence of the Archezoa as a phylogenetic taxon because there was no evidence of protists today that did not once possess mitochondria (Martin and Müller 1998; see also Martin et al. 2001).
2. They rejected the oxygen context for the selective advantage of the protomitochondrial symbionts to their host based on increased efficiency in ATP production through respiratory carbohydrate breakdown.

As they saw it, that scenario was not only purely speculative, but was based on erroneous assumptions. There was no evidence now or then of a bacterium of any kind that did not produce sufficient amounts of ATP, like the host in that narrative. Nor was there any evidence of a bacterium, like the original theorized symbiont, that produced more ATP than it needed to provide to its host. They proposed instead that the nucleated protist evolved after the acquisition of protomitochondria by an archaebacterial host in an anaerobic context.

According to Martin and Muller's "hydrogen hypothesis", the initial advantage of this symbiotic association was not ATP exported from symbiont into the cytosol of a eukaryotic host through respiration, as proposed previously by the traditional oxygen hypothesis for the acquisition of mitochondria, but rather the excretion of molecular H2 produced by the protomitochondrial symbionts in an archaebacterial host. The clue was found in the hydrogen-producing hydrogenosomes postulated to have evolved from the same symbiont as modern-day mitochondria (Embley et al. 1995; Embley and Martin 1998; Martin and Müller 1998; Müller and Martin 1999; Akhmanova et al. 1998).

In short, the hydrogen hypothesis proposed that an anaerobic, autotrophic, and hydrogen-dependent host, living in an environment scarce in hydrogen established a tight relationship with heterotrophic bacteria able to produce molecular H2 through anaerobic fermentation. The engulfed symbiont, an anaerobic $\alpha$-proteobacterium, would initially supply its methanogenic archaebacterial host genes for glycolytic carbohydrate metabolism. Selection to feed the symbiont carbohydrates favoured the transfer of genes from mitochondrial genome to host genome.

Were bacteria incapable of acquiring symbionts? The assumption that only eukaryotes possessed a cytoskeleton and were therefore capable of phagocytosis had been a chief reason for postulating the archezoan host in the first place. As Cavalier-Smith (1987b, p. 56) put it forcefully, "it is only the existence of such fully eukaryotic phagotrophs that makes a symbiotic origin of mitochondria mechanistically plausible; no bacteria, not even predatory ones, can take up or harbor other living cells in their cytoplasm, and to suppose that any ever did is to stray into the realms of science fiction." Martin and Müller (1998) denied that premise, effectively arguing that the primitive archaezoan host was science fiction. But they were uncertain whether their postulated host bacterium possessed a cytoskeleton. Bacteria harbouring bacteria were in fact reported before and after statements about their impossibility (Wujek 1979; von Dohlen et al. 2001).

While heralded as "the first new hypothesis about eukaryotic origins in 30 years to have been really thoroughly articulated at the biochemical, molecular and cellular levels" (p. 15 in Doolittle 1998a) the hydrogen hypothesis has been received with varying degrees of circumspection – from wariness to forceful contradiction (Doolittle 1998a,b; Andersson and Kurland 1999; Roger 1999; Kurland and Andersson 2000; Gray 2005; Gupta 2005). Diverse interpretations of the chimeric nucleus were possible which, in effect, saved the archezoan–mitochondrial host concept. After all, perhaps the primitive amitochondriate protoeukaryote just left no survivors, or perhaps they might still be found hidden among the thousands of amitochondriate protists.

Many microbial phylogeneticists cautioned against the conclusion that the eubacterial genes found in all amitochondriate protists were of protomitochondrial origin. Gray (2005) emphasized that none of the "mitochondrial" genes found in amitochondriate protists have counterparts in contemporary mitochondrial DNA characterized to date. This contrasted with inferences of mitochondrial gene-to-nucleus transfer in which there is evidence that a specific mitochondrial protein is encoded in the nucleus in some organisms but is encoded in mitochondrial DNA in others.

The chimeric nature of the cell nucleus of some amitochondriate protists could have arisen from various mechanisms of lateral gene transfer before and after the emergence of the eukaryote (Fox et al. 1980; Woese 1998). Mitochondria-like genes may have been acquired by repeated symbioses involving α-proteobacteria (Henze et al. 1995) by routine ingestion of food bacteria (Doolittle 1998b) or perhaps the eubacterial genes in the amitochondriate protists resulted from peroxisomes which de Duve (1969, 1996) proposed were of symbiotic ancestry (Roger 1999). Any or all of these scenarios might account for the anomalous eubacterial genes without changing the standard archezoan paradigm for mitochondrial origin in an oxygen context.

There were still other possibilities. A growing number of authors advocated an additional fundamental event forming the protoeukaryote before mitochondria were acquired. Suggestions of such an additional symbiosis arose from diverse perspectives and approaches. For some it resolved the problem that the eukaryote possessed membrane lipids that were like those of eubacteria, yet possessed transcription enzymes that were similar to those of archaebacteria (Zillig et al. 1989; Zillig 1991). To account for other chimeric features of nuclear genomes derived from genome sequencing, James Lake, Maria Rivera, and their collaborators have argued that the nucleus evolved from a fusion of genomes resulting from an engulfed archaebacterial symbiont, an eocyte, and a eubacterial host, a purple sulfur bacterium and an eocyte, linked by hydrogen and sulfur metabolism (Lake and Rivera 1994; Rivera and Lake 2004; Lake et al. 2005). According to this model, informational genes are primarily of archaebacterial origin and operational genes of eubacterial origin (Rivera and Lake 2004).

One of the most detailed nuclear fusion proposals was based on studies of Hsp70 proteins (Gupta and Golding 1996; Gupta 1998, 2005). According to this model, the chimeric genome of the amitochondriate eukaryote resulted from a fusion between a eubacterium (related to proteobacteria) and an archaebacterium. The event would have taken place in an aerobic environment, predominated by antibiotic-producing bacteria. Two selective forces, oxygen and antibiotic sensitivity, would have led to a primitive eukaryote that was antibiotic-resistant and oxygen-tolerant, and which possessed endoplasmic reticulum and a nuclear envelope.

Other nucleus-first models aimed specifically at the origin of the microtubule system in eukaryotes (Hartman and Federov 2002). Mitchell Sogin (1993) suggested that a symbiosis between a now extinct protoeukaryote which had engulfed an archaebacterium resulted in an organism that possessed most protein coding DNA from an archaebacterium and translation apparatus and cytoskeleton from the protoeukaryote. Margulis did not concede to the argument that the protoeukaryotic host had to have possessed a cytoskeleton before acquiring symbionts. In the first edition of *Symbiosis in Cell Evolution* (following Searcy et al. 1978), Margulis (1981) suggested that mitochondria originated from predatory bacteria (e.g. *Bdellovibrio*) that invaded an archaebacterium that was similar to *Thermoplasma*. Later Margulis (Margulis 1992, 1996; Margulis et al. 2000; Melnitsky et al. 2005) argued that the microtubule system originated from motility symbiosis between a spirochete and a *Thermoplasma*-like archaebacterium before mitochondria and plastids.

Few developed mitochondria-first models. Denis Searcy (Searcy et al. 1978; Searcy 1992, 2002) argued that mitochondria (and perhaps plastids) evolved from sulfur-based associations: the host cell, in Searcy's scheme was not an Archezoan, but rather an archaebacterium; *Thermoplasma acidophilus* reduced sulfur and the symbionts oxidized it. Vellai et al. (2001) postulated that mitochondria arose from acquired α-purple photosynthetic Gram-negative eubacteria that reorganized the metabolism of their archaebacterial-like host. On the basis of energetic aspects of nuclear genome organization, they argue that the emergence of the eukaryote was promoted by the establishment of an efficient energy-converting protomitochondrion.

Martin et al. (2001, p. 1526) have emphasized that there is "currently no evidence of any type to indicate that the cell that acquired the mitochondrion was in fact a eukaryote." Still, disproving the existence of anything is well nigh impossible. As Keeling (1998, p. 93) commented: "Even if none of the eukaryotes we know today evolved before the acquisition of the mitochondrion, we might still find an archezoan somewhere in the long branch between the archaebacteria and eukaryotes." Summarizing the evidence for and against the archezoan host concept, Gray's (2005, p. 226) judgment was that "we cannot yet answer with compelling conviction the question of whether the mitochondrion originated under anaerobic or aerobic conditions, and whether it emerged simultaneously with or subsequent to other defining elements of the eukaryotic cell, such as the nucleus." The Archezoa as a primitive subkingdom fell but the conceptual foundation persists.

## 4.13 Recapitulation

The history of ideas about mitochondrial origins is punctuated by different ways of conceiving mitochondrial symbiosis, and assessing theories about them. During the early twentieth century, mitochondria were often conceived of playing crucial roles in tissue differentiation and development, and as the source of diverse structures, including chloroplasts and cilia. Evidence of their symbiotic nature relied on comparative physiology and morphology at a time when bacteriologists had abandoned a phylogenetic approach to bacterial classification on that basis. Ultimately, proof of their symbiotic nature entailed culturing them outside the cell, and distinguishing them from contaminating germs. The rhetorical paradigm was in the context of the germ theory of disease. And the notion that cells were chimeric confronted nucleocentric genetics and cell biology. A concept of infectious heredity was developed in the 1950s with the rise of microbial genetics, but the question of organellar symbiosis remained irresolvable.

A new paradigm of organellar origins by symbiosis did not come about by resolving anomalies in the old. It emerged from wholly new molecular techniques, wholly new concepts about the nature of mitochondria, and new questions about evolution. The weight of opinion about the origins of mitochondria (and chloroplasts) shifted during the 1960s and 1970s when they were understood in molecular biological terms as each possessing their own organellar DNA and complete protein-synthesizing machinery. Mitochondria were no longer conceived of as organ-forming organelles, but rather as energy-generating respiratory organelles. They no longer represented a diverse community of symbionts that had been acquired repeatedly and could be cultured independently, they were at most vestiges of bacteria; the bulk of their genes had been transferred to the nuclear genome. The question was no longer about their role in the origin of species, but about the origin of the eukaryotic cell. Rhetoric about the germ theory of disease faded away. A natural classification of bacteria was now deemed possible with the development of the field of molecular evolution. Comparisons of SSU rRNA sequences of nuclear, organellar, and bacterial genomes closed the main controversy over their symbiotic origin.

A subsequent phase beginning in the 1980s is characterized by research and discussion about the nature of the mitochondrial host. Stories about the nature of the protomitochondrial symbiont–host relationship emerged from a fusion of knowledge about bacteria and protist structure and function, paleobiology, natural history, as well from molecular evolution. The general assumption that the mitochondrial host was an anaerobic eukaryotic organism, that later acquired and domesticated aerobic bacteria, was fortified by the proposal of the Archezoa to unify extant amitochondriate protists. That subkingdom had been built on shaky morphological grounds and, by the end of the century, proved not to be able to withstand molecular scrutiny. The Archezoa was not a phylogenetically coherent group. Vestiges of "mitochondrial-like" genes were found, and so were related organelles, the hydrogenosomes in some archezoa. There was no clear evidence of the existence of any amitochondriate eukaryotes. Arguments over the protomitochondrial host swing today between two extremes – that a bacterial host of bacterial symbionts is fiction, but primitive amitochondriate protists may exist – and the other – that the Archezoa is fiction, and bacterial hosts do exist.

The Archezoa kingdom was devastated, but the archezoan concept survived, supported by the argument since 1970 that components enabling phagocytosis preceded symbiosis. It is not certain if the eubacterial genes found in all amitochondriate protists were of mitochondrial ancestry. Various schemes involving engulfment or fusion, and lateral gene transfer have been offered to account for the chimeric nature of the primitive nuclear genome. The issue today is not a matter of distinguishing mitochondria from germs in the air; it now involves distinguishing mitochondrial genes from genes from other exogenous possibilities.

## References

Akhmanova A, Voncken F, van Alen T, van Hoek A, Boxma B, Vogels G, Veenhuis M, Hackstein HP (1998) A hydrogenosome with a genome. Nature 396:527–528

Allen JF (1993) Control of gene expression by redox potential and the requirement for chloroplast and mitochondrial genomes. J Theor Biol 165:609–631

Allsopp A (1969) Phylogenetic relationships of the procaryota and the origin of the eucaryotic cell. New Phytol 68:591–612

Altmann R (1890) Die Elementarorganismen. Veit, Leipzig

Andersson SGE, Kurland CG (1999) Origins of mitochondria and hydrogenosomes. Curr Opin Microbiol 2:535–541

Benda C (1898) Weitere Mitteilungen über die Mitochondria. Verh Physiol Ges Berlin 376–383

Bogorad L (1975) Evolution of organelles and eukaryotic genomes. Science 188:891–898

Bonen L, Doolittle WF (1976) Partial sequences of 16S rRNA and the phylogeny of blue-green algae and chloroplasts. Nature 261:669–673

Bonen L, Cunningham RS, Gray MW, Doolittle WF (1977) Wheat embryo mitochondrial 18S ribosomal RNA:Evidence for its prokaryotic nature. Nucleic Acids Res 4:663–667

Buchner P (1965) Endosymbiosis of animals with plant microorganisms. Interscience, New York

Bui ET, Bradley PJ, Johnson PJ (1996) A common evolutionary origin for mitochondria and hydrogenosomes. Proc Natl Acad Sci USA 93:9651–9695
Cavalier-Smith T (1975) The origin of nuclei and of eukaryotic cells. Nature 256:463–468
Cavalier-Smith T (1983) Endosymbiotic origin of the mitochondrial envelope. In: Schwemmler W, Schenk HEA (eds) Endocytobiology II. Intracellular space as an oligogenetic ecosystem. De Gutyer, Berlin, pp 265–279
Cavalier-Smith T (1987a) Eukaryotes with no mitochondria. Nature 326:332–333
Cavalier-Smith T (1987b) The simultaneous symbiotic origin of mitochondria, chloroplasts and microbodies. Ann N Y Acad Sci 503:55–71
Cavalier-Smith T (1991) Archamoebae: the ancestral eukaryotes? Biosystems 25:25–38
Clark CG, Roger AJ (1995) Diect evidence for secondary loss of mitochondria in *Entamoeba histolytica*. Proc Natl Acad Sci USA 92:6518–6521
Cleveland LR, Grimstone AV (1964) The fine structure of the flagellate *Mixotricha paradoxa* and its associated micro-organisms. Proc R Soc Lond B Biol Sci 159:668–686
Cohen SS (1970) Were mitochondria and chloroplasts microorganisms? Am Sci 58:282–289
Cohen SS (1973) Mitochondria and chloroplasts revisited. Am Sci 61:437–445
Cowdry EV (1916) The General functional significance of mitochondria. Am J Anat 19:423–446
Cowdry EV (1918) The mitochondrial constituents of protoplasm. Carnegie Inst Wash Publ Contrib Embryol 8:(25):41–144
Cowdry EV, Olitsky PK (1922) Differences between mitochondria and bacteria. J Exp Med 36:521–536
Cunningham RS, Bonen L, Doolittle WF, Gray MW (1976) Unique species of 5S, 18S and 26S ribosomal RNA in wheat mitochondria. FEBS Lett 69:116–122
Darlington CD (1951) Mendel and the determinants. In; Dunn LC (ed) Genetics in the twentieth century. Macmillan, New York, pp 315–332
de Duve C (1969) Evolution of the peroxisome. Ann N Y Acad Sci 168:369–381
de Duve C (1973) Origin of mitochondria. Science 182:85
de Duve C (1982) Peroxisomes and related particles in historical perspective. Ann N Y Acad Sci 1982 386:1–4
de Duve C (1996) The birth of complex cells. Sci Am April 50–57
Demoulin V (1979) Protein and nucleic acid sequence data and phylogeny. Science 205:1036–1038
Doolittle WF (1980) Revolutionary concepts in evolutionary biology. Trends Biochem Sci 5:146–149
Doolittle WF (1998a) A paradigm gets shifty. Nature 392:15–16
Doolittle WF (1998b) You are what you eat: a gene transfer ratchet could account for bacterial genes in eukaryotic nuclear genomes. Trends Genet 14:307–311
Doolittle WF, Woese CR, Sogin ML, Bonen L, Stahl D (1975) Sequence studies on 16S ribosomal RNA from a blue-green alga. J Mol Evol 4:307–315
Duesberg J (1913) Plastosomes et 'organ-forming substances' dans l'œuf des Ascidiens. Bull Acad R Belg 5:463–475
Duesberg J (1919) On the present status of the chondriosome problem. Biol Bull 36:71–81
Embley MT, Martin W (1998) A hydrogen-producing mitochondrion. Nature 396:517–519
Embley TM, Finlay BJ, Dyal PL, Hirt RP, Wilkinson M, Williams AG (1995) Multiple origins of anaerobic ciliates with hydrogenosomes within the radiation of aerobic ciliates. Proc R Soc Lond B Biol Sci 262:87–93
Fenchel T, Finlay BJ (1995) Ecology and evolution in anoxic worlds. Oxford University Press, Oxford
Fox G et al. (1980) The phylogeny of prokaryotes. Science 209:457–463
Germot A, Philiipe H, Le Guyader H (1996) Presence of a mitochondrial-type HSP70 in *Trichomonas* suggests a very early mitochondrial endosymbiosis in eukaryotes. Proc Natl Acad Sci USA 93:14614–14617
Germot A, Philippe H, Le Guyader H (1997) Evidence for loss of mitochondria in Microsporidia from a mitochondrial-type HSP70 in *Nosema locustae*. Mol Biochem Parasitol 87:159–168

Gillham N (1978) Organelle heredity. Raven, New York
Gogarten JP, Kiblak H, Dittrich P, Taiz L, Bowman EJ, Bowman BJ, Manolson NF, Poole RJ, Date T, Oshima T, Konishi J, Deda K, Yoshida M (1989) Evolution of the vacuolar H+-ATPase: implications for the origin of eukarotes. Proc Natl Acad Sci USA 86:6661–6665
Goksøyr J (1967) Evolution of eukaryotic cells. Nature 214:1167
Gould S, Lewontin R (1978) The spandrels of San Marco and the Panglossian paradigm: a critique of the adaptationist program. Proc R Soc Lond B Biol Sci 205:581–598
Gray M, Gertraud B, Lang BF (1999) Mitochondrial evolution. Science 283:1476–1481
Gray MW (1992) The endosymbiont hypothesis revisited. Int Rev Cytol 141:233–257
Gray MW (2005) Contemporary issues in mitochondrial origins and evolution. In: Sapp J (ed) Microbial phylogeny and evolution: concepts and controversies. Oxford University Press, New York, pp 224–237
Gray MW, Doolittle WF (1982) Has the endosymbiont hypothesis been proven? Microb Rev 46:1–42
Guilliermond A (1914) Etat actuel de la question de l'évolution et du rôle physiologique des mitochondries d'après les travaux récents de cytologie végétale. Rev Gen Bot 26:129–143
Gupta R (1998) Protein phylogenies and signature sequences: a reappraisal of evolutionary relationships among archaebacteria, eubacteria and eukaryotes. Microbiol Mol Biol Rev 62:1435–1491
Gupta R (2005) Molecular sequences and the early history of life. In: Sapp J (ed) Microbial phylogeny and evolution: concepts and controversies. Oxford University Press, New York, pp 160–183
Gupta R, Golding GB (1996) The origin of the eukaryotic cell. Trends Biochem Sci 21:166–170
Haberlandt G (1891) Über den Bau und die Bedeutung der Chlorophyll-zellen von Convoluta roscoffsensis. In: von Graff L (ed) Die Organization der Turbellaria Acoela. Engelmann, Leipzig, pp 75–90
Hall JL, Luck DJ, (1995) Basal body-associated DNA: in situ studies in *Chlamydomonas reinhardtii*. Proc Natl Acad Sci USA 92:5129–5133
Hartman H, Federov A (2002) The origin of the eukaryotic cell: a genomic investigation. Proc Natl Acad Sci USA 99:1420–1425
Henze K, Badr M, Wettern R, Cerff R, Martin W (1995) A nuclear gene of eubacterial origin in *Euglena gracilis* reflects cryptic endosymbiosis during protist evolution. Proc Natl Acad Sci USA 92:9122–9126
Hirt RP, Logsdon JM Jr, Healy B, Dorey MW Doolittle WF, Embley TM (1999) Microsporidia are related to fungi: evidence from the largest subunit of RNA polymerase II and other proteins. Proc Natl Acad Sci USA 96:580–585
Huet J, Schanbel R, Sentenac A, Zillig W (1983) Archaebacteria and eukaryotes posess DNA-dependent RNA polymerases of a common type. EMBO J 2:1291–1294
Iwabe N, Kuma KI, Hasegawa M, Osawa S, Miyara T (1989) Evolutioanry relationship of archaebacteria, eubacteria, and eukaryotes inferred from phylogenetic trees of duplicated genes. Proc Natl Acad Sci USA 86:9355–9359
John P, Whatley FR (1975) *Paracoccus denitrificans* and the evolutionary origin of the mitochondrion. Nature 254:495–498
Keeble F (1910) *Plant-animals*. A study in symbiosis. Cambridge University Press, Cambridge
Keeling PJ (1998) A Kingdom's progress: Archezoa and the origin of eukaryotes. BioEssays 20:87–95
Kennedy EP, Lehninger AL (1949) Oxidation of fatty acids and tricarboxylic acid cycle intermediates by isolated rat liver mitochondria. J Biol Chem 179:957–972
Klein R, Conquist A (1967) A consideration of the evolutionary and taxonomic significance of some biochemical, micromorphological and physiological characteristicsm in the Thallophytes. Q Rev Biol 42:105–296
Kurland CG (1992) Evolution of mitochondrial genomes and the genetic code. BioEssays 14:709–714
Kurland CG, Andersson SGE (2000) Origin and evolution of the mitochondrial proteome. Microbiol Mol Biol Rev 64:786–820

Lake J, Rivera M (1994) Was the nucleus the first symbiont? Proc Natl Acad Sci USA 91:2880–2881
Lake J, Moore JE, Simonson A, Rivera MC (2005) In: Sapp J (ed) Microbial phylogeny and evolution: concepts and controversies. Oxford University Press, New York, pp 184–206
Lederberg J (1951) Genetic studies in bacteria. In: Dunn LC (ed) Genetics in the twentieth century. Macmillan, New York, pp 263–289
Lederberg J (1952) Cell genetics and hereditary symbiosis. Physiol Rev 32:403–430
Lehninger AL (1965) The mitochondrion. Benjamin, New York
Lindmark DG, Müller M (1973) Hydrogenosome, a cytoplasmic organelle of the anaerobic flagellate *Trichomonas foetus*, and its role in pyruvate metabolism. J Biol Chem 248:7724–7728
Margulis L (1970) Origin of eukaryotic cells. Yale University Press, New Haven
Margulis L (1981) Symbiosis in cell evolution. Freeman, San Francisco
Margulis L (1992) Symbiosis in cell evolution. 2nd edn. Freeman, San Francisco
Margulis L (1996) Archaeal-eubacterial mergers in the origin of Eukarya: phylogenetic classification of life. Proc Nat Acad Sci USA 93:1071–1076
Margulis L, Dolan MF, Guerrero R (2000) The chimeric eukaryote: origin of the nucleus from the karyomastigont in amitochondriate protists. Proc Natl Acad Sci 97:6954–6959
Martin W, Müller M (1998) The hydrogen hypothesis for the first eukaryote. Nature 392:37–41
Martin W, Hoffmeister M, Rotte C, Henze K (2001) An overview of endosymbiotic models for the origins of eukaryotes, their ATP-producing organelles (mitochondria and hydrogenosomes) and their heterotropic lifestyle. Biol Chem 382:1521–1539
Maynard Smith J, Szathmáry E (1995) The major transitions in life. Freeman, San Francisco
Mazzerello P, Bentivoglio M (1998) The centenarian Golgi apparatus. Nature 392:543–544
Melnitsky H, Rainey FA, Margulis L (2005) The karyomastigont model of eukaryosis. In: Sapp J (ed) Microbial phylogeny and evolution: concepts and controversies. Oxford University Press, New York, pp 261–280
Mérejkovsky C (1920) La plante considérée comme un complexe symbiotique. Bull Soc Nat 6:17–98
Mereschkowsky C (1909) Theorie der zwei Plasmaarten als Grundlage der Symbiogenesis, einer neuen Lehre von der Entstehung der Organismen. Biol Zentralbl 30:277–303, 321–347, 353–367
Merezkowsky C (1905) Über Natur und Ursprung der Chromatophoren im Pflanzenreiche. Biol Zentralbl 25:593–604
Meves F (1918) Die Plastosomentheroie der Verebung. Eine Antwort auf verschiedene Einwände. Arch Mikro Anat 92:42–136
Morgan TH (1926) Genetics and the physiology of development Am Nat 60:489–515
Muller HJ (1927) Artificial transmutation of the gene. Science 46:84–87
Muller HJ (1951) The development of the gene theory. In: Dunn LC (ed) Genetics in the twentieth century. Macmillan, New York, pp 77–100
Müller M (1988) Energy metabolism of protozoa without mitochondria. Annu Rev Microbiol 42:465–488
Müller M (1993) The hydogenosome. J Gen Microbiol 139:2879–2889
Müller M, Martin W (1999) The genome of *Rickettsia prowazekii* and some thoughts on the origins of mitochondria and hydrogenosomes. BioEssays 21:377–381
Nass MMK (1969) Mitochondrial DNA: advances, problems and goals. Science 165:25–35
Nass MMK (1971) Properties and biological significance of mitochndrial DNA. In: Harris PJ (ed) Biological ultrastructure: the origin of cell organelles. Oregon State University Press, Corvalis, pp 41–63
Nass S (1969) The significance of the structural and functional similarities of bacteria and mitochondria. Int Rev Cytol 25:55–129
Nass S, Nass M (1963) Intramitochondrial fibers with DNA characteristics. J Cell Biol 19:613–628
Portier P (1918) Les symbiotes. Masson, Paris
Race HL, Hermann, RG, Martin W (1999) Why have organelles retained genomes? Trends Genet 15:364–370

Raff RA, Mahler HR (1972) The non-symbiotic origin of mitochondria. Science 177:575–582
Raff RA, Mahler HR (1973) Origin of mitochondria. Science 180:517–518
Rasmussen N (1997) Picture control: the electron microscope and the transformation of biology in America, 1940-1960. Stanford University Press, Stanford
Raven PH (1970) A multiple origin for plastids and mitochondria Science 169:641–646
Regaud E (1909) Attribution aux 'formation mitochondriales' de la fonction générale d'extraction et de fixation électives, exercée par les cellules vivantes sur les substances dissoutes dans le milieu ambiant. C R Soc Biol 66:919–912
Ris H (1961) Ultrastructure and molecular organization of genetic systems. Can J Genet Cytol 3:95–120
Ris H, Plaut W (1962) Ultrastructure of DNA-containing areas in the chloroplasts of *Chlamydomonas*. J Cell Biol 12:383-391
Rivera MC, Lake JA (2004) The ring of life provides evidence for a genome fusion origin of eukaryotes. Nature 432:152–155
Roger AJ (1999) Reconstructing early events in eukaryotic evolution. Am Nat 154:S146–S163
Roger AJ, Clark CG, Doolittle WF (1996) A possible mitochondrial gene in the early-branching amitochondriate protist *Trichomonas vaginalis*. Proc Natl Acad Sci USA 93:14618–14622
Sagan L (1967) On the origin of mitosing cells. J Theor Biol 14:225–274
Sager R (1972) Cytoplasmic genes and organelles. Academic, New York
Sapp J (1987) Beyond the gene. cytoplasmic inheritance and the struggle for authority in genetics. Oxford University Press, New York
Sapp J (1998) Freewheeling centrioles. Hist Philos Life Sci 20:255–290
Sapp J (1994) Evolution by association: a history of symbiosis. Oxford University Press, New York
Sapp J (2002) Paul Buchner and hereditary symbiosis in insects. Int Microbiol 5:145–160
Sapp J (2003) Genesis: the evolution of biology. Oxford University Press, New York
Sapp J (2005a) The bacterium's place in nature. In: Sapp J (ed) Microbial phylogeny and evolution: concepts and controversies. Oxford University Press, New York, pp 1–52
Sapp J (2005b) The Prokaryote-eukaryote dichotomy: meanings and mythology. Microbiol Mol Biol Rev 69:292–305
Sapp J, Carrapiço F, Zolotonosov M (2002) Symbiogenesis: the hidden face of Constantin Merezhkowsky. Hist Philos Life Sci 24:413–440
Schanderl H (1947) Botanische Bakteriologie und Stickstoffhaushalt der Pflanzen auf neuer Grunglage. Ulmer, Stuttgart
Schimper AFW (1883) Ueber die Entwickelung der Chlorophyllkörner und Farbkörper. Bot Zeit 41:105–114, 112–113
Schimper AFW (1885) Untersuchungen über die Chlorophyllkörner und die ihnen homologen Gebilde. Jahrb Wiss Bot 16:1–247
Schrader F (1953) Mitosis: the movements of chromosomes in cell division. Columbia University Press, New York
Schwartz R M, Dayhoff MO (1978) Origins of prokaryotes, eukaryotes, mitochondria, and chloroplasts. Science 199:395–403
Searcy DG (1992) Origins of mitochiondria and chloroplasts from sulphur-based symbioses. In: Hartman H, Matsuno K (eds) The origin and evolution of the cell. World Scientific, Singapore, pp 47–78
Searcy DG (2002) Syntrophic models for mitochondrial origin. In: Seckbach J (ed) Symbiosis: mechanisms and model systems. Kluwer, Dordrecht, pp 163–183
Searcy DG, Stein DB, Green GR (1978) Phylogenetic affinities between eukaryotic cells and a thermoplasmic mycoplasm. BioSystems 10:19–28
Sogin ML (1989a) Evolution of eukaryotic microorganisms and their small subunit ribosomal RNA from *Giardia lamblia* Science 243:75–77
Sogin ML (1989b) Evolution of eukaryotic microorganisms and their small subunit ribosomal RNAs. Am Zool 29:487–499
Sogin ML (1993) The origin of eukaryotes and evolution into major kingdoms. In: Bengtson S (ed) Early life on earth: Nobel symposium 84. Columbia University Press, New York

Soltys BJ, Gupta RS (1994) Presence and cellular distribution of a 60-kDa protein related to mitochondria Hsp60 in *Giardia lamblia.* J Protisol 80:580–588
Sonneborn TM (1950) The cytoplasm in heredity. Heredity 4:11–36
Stanier R (1970) Some aspects of the biology of cells and their possible evolutionary significance. In: Charles HP, Knight BC (eds) Organization and control in prokaryotic cells. 20th symposium of the Society for General Microbiology. Cambridge University Press, Cambridge, pp 1–38
Stanier R, van Niel CB (1962) The concept of a bacterium. Arch Mikrobiol 42:17–35
Stanier R, Douderoff M, Adleberg E (1963) The microbial world, 2nd edn. Prentice-Hall, Engelwood Cliffs
Taylor FJR (1974) Implications and extensions of the serial endosymbiosis theory of the origin of eukaryotes. Taxon 23:229–258
Timmis JN, Ayliffe MA, Huang CY, Martin W (2004) Endosymbiotic gene transfer: organelle genomes forge eukaryotic chromosomes. Nat Rev Genet 5:123–136
Uzzel T, Spolsky C (1974) Mitochondria and plastids as endosymbionts: a revival of special creation? Am Sci 62:335–343
Uzzell T, Spolsky C (1973) Origin of mitochondria. Science 180:516–517
Vellai T, Takács K, Vida G (2001). A new aspect to the origin and evolution of eukaryotes. J Mol Evol 46:499–507
Voltaire (1947) Candide or Optimism, translated by Butt J. Penguin, London
von Dohlen CD, Kohler S, Alsop ST, McManus WR (2001) Mealybug β-proteobacterial ensymbionts contain γ-proteobacterial symbionts. Nature 412:433–435
Vossbrinck CR, Maddox JV, Friedman S, Debrunner-Vossbrinck BA, Woese CR (1987) Ribosomal RNA sequence suggests microsporidia are extremely ancient eukaryotes. Nature 326:411–414
Wagner RP (1969) Genetics and phenogenetics of mitochondria Science 180:516–517
Wallin IE (1923) The mitochondria problem. Am Nat 57:255–261
Wallin IE (1924) On the nature of mitochondria, VII. The independent growth of mitochondria in culture media. Am J Anat 33:147–173
Wallin IE (1927) Symbionticism and the origin of species. Williams and Wilkins, Baltimore
Watasé S (1893) On the nature of cell organization. Woods Hole Biol Lect 83–103
Wilson EB (1896) The cell in development and inheritance. Macmillan, New York
Wilson EB (1925) The cell in development and heredity, 3rd edn. Macmillan, New York
Woese CR (1977) Endosymbionts and mitochondrial origins. J Mol Evol 10:93–96
Woese CR (1998) The universal ancestor. Proc Natl Acad Sci 95:6854–6859
Woese CR, Fox GE (1977) Phylogenetic structure of the prokaryotic domain: the primary kingdoms. Proc Natl Acad Sci USA 74:5088–5090
Wujek DE (1979) Intracellular bacteria in the blue-green-algae *Pluerocapsa minor.* Trans Am Microsc Soc 98:143–145
Yang D, Oyaizu Y, Oyaizu H, Olsen GJ, Woese CR (1985) Mitochondrial origins Proc Natl Acad Sci USA 82:4443–4447
Zablen LB, Kissle MS, Woese CR, Buetow DE (1975) The phylogenetic origin of the chloroplast and the prokaryotic nature of its ribosomal RNA. Proc Natl Acad Sci USA 72:2418–2422
Zillig W (1991) Comparative biochemistry of archaea and bacteria. Curr Opin Genet Dev 1:544–551
Zillig W, Klenk HP, Palm P, Leffers H, Puhler G, Gropp F, Garrett RA (1989) Did eukaryotes originate by a fusion event? Endocyt Cell Res 6:1–25

# 5 Anaerobic Mitochondria: Properties and Origins

Aloysius G.M. Tielens, Jaap J. van Hellemond

## Summary

Mitochondria are usually considered to be the powerhouses of the cell and to be responsible for the aerobic production of ATP. However, it is now known that not all mitochondria are the same, and that not all mitochondria comply with the classic textbook image of an oxygen-dependent organelle. Several other types of mitochondria exist that do not use oxygen for ATP production. Anaerobic energy-generating organelles exist that produce ATP with the help of proton-pumping electron transport (anaerobic mitochondria with oxidative phosphorylation), whereas other types produce hydrogen and lack an electron-transport chain (hydrogenosomes). All ATP-producing organelles supposedly originated from one single endosymbiotic ancestor, which raises the question whether those anaerobically functioning mitochondria evolved directly from this early endosymbiont? Possible evolutionary origins of these mitochondria will be discussed, and it will be argued that anaerobically functioning mitochondria most likely did not originate directly from the ancestral endosymbiont, but arose later in evolution from aerobic types of mitochondria. In our opinion, the anaerobically functioning mitochondria are the result of adaptations of aerobic ones to hypoxic environments. This implies that these anaerobic mitochondria are in fact a further evolution and not a more primitive form. Mitochondria are rather diverse and adapted to distinct conditions and this diversity has evolved via multiple adaptations that occurred independently in various unrelated eukaryotic lineages.

## 5.1 Introduction

Life on earth started anaerobically and anaerobic organisms flourished for more than 500 million years before oxygen became available and started to play a role in the further evolution of life (Fenchel and Finlay 1994). The evolution of photosynthesis in cyanobacteria resulted in a steady increase in the amount of oxygen in the atmosphere. This presence of oxygen opened new ways for the degradation of substrates and resulted in the evolution of aerobic energy metabolism, i.e. production of ATP. Many prokaryotes use oxygen

Origin of Mitochondria and Hydrogenosomes (ed. by William F. Martin and Miklós Müller)

in their energy metabolism, but it is the eukaryotes and especially the metazoa that gained most from this aerobic production of ATP and many became oxygen-dependent organisms. These eukaryotes contain mitochondria, ATP-producing organelles, which use an electron-transport chain to deliver the electrons resulting from substrate oxidation to the final electron acceptor, oxygen. In these mitochondria, most of the ATP is produced via oxidative phosphorylation: electrons are transported via NADH or $FADH_2$ to oxygen by the proton-pumping protein complexes of the electron-transport chain, and the backflow of the pumped protons via the mitochondrial ATP-synthase results in ATP formation (Saraste 1999). Oxygen deprivation in these classical mitochondria rapidly results in halted mitochondrial ATP synthesis and often in cell death. Yet many eukaryotes exist that have specific adaptations which enable them to survive hypoxia, either for relatively short periods of time or sometimes even continuously.

## 5.2 Possible Variants in Anaerobic Metabolism

When organisms have to function (temporarily) without oxygen as terminal electron acceptors, they have to maintain redox balance without aerobic respiration; hence, the reduced cofactors produced by the catabolic pathways have to be oxidized by an alternative process. Organisms with an anaerobic energy metabolism can be broadly divided into two different classes: those that perform anaerobic respiration using an alternative electron acceptor present in the environment, such as nitrate, and those that perform fermentation using an endogenously produced organic electron acceptor, such as pyruvate or fumarate. In contrast to bacteria, eukaryotes using alternative electron acceptors present in the environment are rare and they will not be discussed extensively here. This chapter is devoted to adaptations to hypoxia involving fermentation pathways where an endogenously produced electron acceptor is used. Organisms using such an endogenous electron acceptor can, again, be divided into two classes: those that use cytosolic processes only to produce ATP, and those that use specific energy-generating organelles for this purpose. In fermentation pathways for the degradation of organic compounds and generation of ATP, oxidation–reduction reactions occur in the absence of any external electron acceptors. Fermentation of substrates avoids the net production of reduced cofactors, such as NADH, because both oxidation and reduction of the substrate occur. In some cases (e.g., fumarate reduction) this is linked to the electron-transport chain. Carbohydrates are suitable substrates for fermentation. Lipids, however, are too reduced to be fermented as both oxidation and reduction of the substrate must occur.

It turns out that organisms that usually function aerobically, producing ATP via oxidative phosphorylation in their "classical" mitochondria, can in the absence of oxygen only use cytosolic pathways for the generation of ATP. These mitochondria apparently cannot produce ATP in the absence of oxygen. On the other hand, eukaryotes exist that have specially adapted mitochondria. These facultative anaerobes possess mitochondria that can produce ATP via oxidative phosphorylation even in the absence of oxygen. Next to all these organisms that use oxygen in their production of ATP when this final electron acceptor is available in the environment (or tissue in question) organisms also exist that never use oxygen for the production of ATP. These eukaryotes that do not use oxygen as a terminal electron acceptor are always unicellular and live in anaerobic environments such as marine or freshwater sediments, or the rumen and intestinal tracts of their host. These protists produce via anaerobic reactions ATP exclusively by substrate-level phosphorylations, and they were divided by Müller (1998) into two groups, designated type I and type II, depending on the absence or presence of hydrogenosomes, respectively. Type I amitochondriate protists do not possess ATP-producing organelles and are fully dependent on substrate-level phosphorylations for ATP production, which occur exclusively in the cytosol. *Giardia* and *Entamoeba*, human intestinal parasites, are the best studied representatives of these type I anaerobic protists. Type II anaerobic organisms, on the other hand, do contain specialized organelles, which produce ATP and hydrogen and are therefore called hydrogenosomes. Hydrogenosomes are double membrane bound organelles, clearly related to mitochondria but they lack oxidative phosphorylation and produce ATP by substrate-level phosphorylation only. Several types of hydrogenosomes exist (Fig. 5.1) and their characteristics are further discussed in the chapters by Tachezy and Doležal (Chap. 6), Hackstein et al. (Chap. 7) and Tovar (Chap. 11) of this volume. This chapter will focus on anaerobic mitochondria.

From the classification described already, it is clear that most eukaryotes contain specific organelles involved in ATP production, such as mitochondria and hydrogenosomes. This could indicate that ATP-producing organelles are characteristic for eukaryotes; however, as discussed before, also several groups of unicellular eukaryotes exist that lack ATP-producing organelles. On the other hand, it was recently discovered that these organisms, such as *Giardia* and *Entamoeba* species, possess mitochondrial remnants, called mitosomes, which contain enzymes for iron–sulphur cluster biosynthesis but lack enzymes involved in energy metabolism (Tovar et al. 1999, 2003). The presence of mitochondrion-like remnants indicates that these organisms once possessed mitochondria, but that these organelles lost their ATP-generating capacity during further evolution; therefore, it is now suggested that there are no truly amitochondriate eukaryotes and that all eukaryotes once possessed organelles specialized in the production of ATP, among other things (Müller 2000; Martin et al. 2001; Roger and Silberman 2002; Williams et al. 2002; Henze and Martin 2003; Embley et al. 2003a, Gabaldón and Huynen 2004).

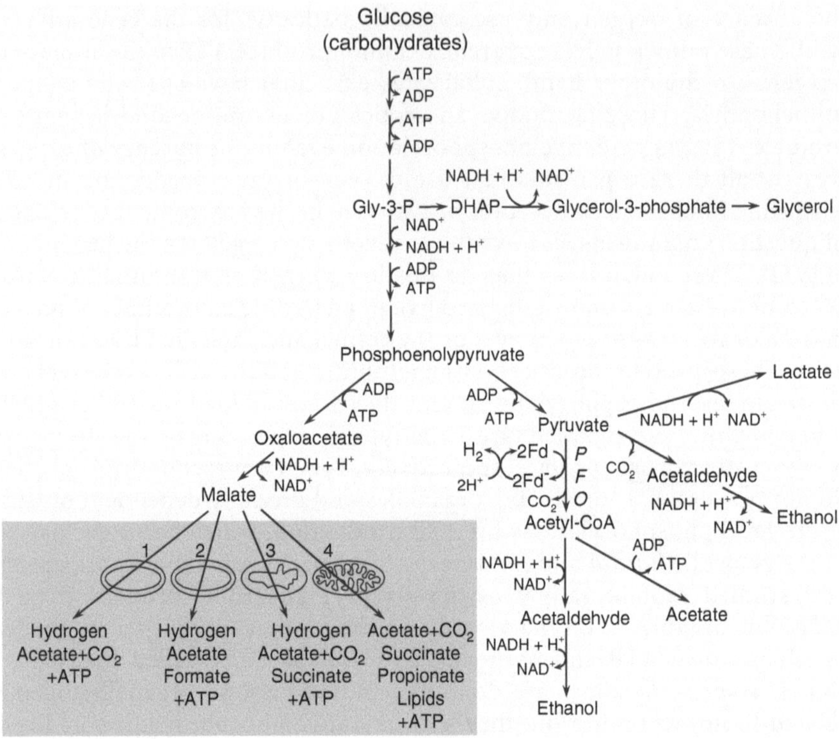


**Fig. 5.1.** Fermentation pathways present in facultative anaerobic eukaryotes. Examples of fermentation pathways present in the cytosol and in subcellular compartments. Fermentation processes localized in hydrogenosomes (1–3) and mitochondria (4) are indicated by the *shaded box*. Examples of the anaerobic ATP-producing organelles shown can be found in trichomonads (1), chytridiomycete fungi (2), *Nyctotherus ovalis* (3), and parasitic helminths, bivalves and *Euglena gracilis* (4). *CoA* coenzyme A, *DHAP* dihydroxyacetone phosphate, *Fd* ferredoxin, *Gly-3-P*, glyceraldehyde-3-phosphate, *PFO* pyruvate:ferredoxin oxidoreductase

## 5.3 Cytosolic Adaptations to an Anaerobic Energy Metabolism

Some eukaryotes can survive hypoxia by using simple fermentations in which the electrons from glycolysis are transferred to pyruvate or a derivative of it. Many variations of this type of fermentation exist, resulting in end products such as lactate or ethanol (Fig. 5.1). The formation of lactate produces 2 mol of ATP per glucose degraded, is found in all phyla, and it is the sole anaerobic pathway of evolutionarily more advanced species like arthropoda and vertebrates (Livingstone 1991).

Conversion of pyruvate into ethanol is a variant of cytosolic fermentation, which occurs, for instance, in yeast and hypoxia-tolerant fish, such as crucian carp and goldfish (van Waarde et al. 1993; Lutz and Nilsson 1997). Ethanol can be formed from pyruvate via two distinct pathways. The most common pathway for ethanol production, which is present in yeast, consists of a decarboxylation of pyruvate to acetaldehyde by pyruvate decarboxylase followed by reduction to ethanol by alcohol dehydrogenase (Fig. 5.1). However, ethanol can also be formed from pyruvate via acetylcoenzyme A (acetyl-CoA), a pathway in which pyruvate is first oxidatively decarboxylated to acetyl-CoA, which is then reduced in two subsequent reactions by acetaldehyde dehydrogenase and alcohol dehydrogenase. This pathway occurs, for example, in the human intestinal parasites *Giardia lamblia* and *Entamoeba histolytica*, where pyruvate is oxidized by pyruvate:ferredoxin oxidoreductase (PFO), which yields acetyl-CoA and carbon dioxide, like the mitochondrial reaction catalysed by pyruvate dehydrogenase (PDH) (Horner et al. 1999). However, in the case of *Giardia* and *Entamoeba*, ferredoxin acts as electron acceptor and not $NAD^+$, as is the case in the reaction catalysed by PDH. The reduced ferredoxin donates the electrons to $H^+$, resulting in the formation of hydrogen. In *G. lamblia* and *E. histolytica*, which lack energy-generating organelles, the acetyl-CoA formed is further degraded in the cytosol into a mixture of ethanol and acetate (Fig. 5.1). The ethanol is produced by an alcohol dehydrogenase E (ADH-E), a bifunctional enzyme that combines aldehyde dehydrogenase and alcohol dehydrogenase activities (Bruchhaus and Tannich 1994; Sanchez 1998). Acetate is produced by acetyl-CoA synthase with the concomitant production of ATP. The extra ATP produced by this further degradation of pyruvate fluctuates between 0 and 2 mol of ATP per glucose degraded, as the relative amounts of ethanol and acetate produced depend on environmental conditions (Lloyd 1996; Müller 1998; Martin 2000).

Another cytosolic fermentation variant is found in several trichomonad and yeast species, which increase their glycerol production during anoxia (Steinbüchel and Müller 1986; Valadi et al. 2004). In this case, the glycolytic intermediate dihydroxyacetone phosphate (DHAP) is converted to glycerol-3-phosphate by an NADH-dependent glycerol-3-phosphate dehydrogenase and subsequently to glycerol by glycerol kinase (Fig. 5.1). Glycerol production from glucose thus results in net consumption of NADH, and it is essential in respiratory-incompetent yeast cells (Valadi et al. 2004).

## 5.4 Anaerobically Functioning ATP-Generating Organelles

Next to the previously described organisms, where only cytosolic adaptations to hypoxic functioning occur, organisms exist that possess organelles which can produce ATP anaerobically. These organelles, hydrogenosomes as well as anaerobically functioning mitochondria, are specifically adapted to generate ATP without the use of oxygen as the final electron acceptor. Mitochondria and hydrogenosomes share many characteristics, like the surrounding double membrane, protein import machinery and the production of ATP (Embley et al. 2003b). Hydrogenosomes are characterized by their capacity to produce hydrogen, whereas mitochondria contain an electron-transport chain in their inner membrane and perform oxidative phosphorylation.

It is important to note that in those organisms that contain anaerobically functioning mitochondria or hydrogenosomes, adaptations to the hypoxic way of living are not exclusively restricted to these organelles. Next to the adaptations to hypoxia inside the anaerobic-functioning ATP-producing organelles, cytosolic adaptations to anaerobic functioning can be present in the same organism as well. For instance, in the anaerobic chytrids *Neocallimastix* and *Piromyces*, which lack mitochondria but contain hydrogenosomes, most of the pyruvate is degraded inside the hydrogenosomes, but approximately one third of it is degraded in the cytosol, mainly to the end products ethanol and formate, with minor amounts of lactate and succinate (Boxma et al. 2004). In these two fungal symbionts in the gastrointestinal tract of many herbivorous animals, pyruvate is in the cytosol as well as inside the hydrogenosomes degraded to acetyl-CoA and formate by pyruvate:formate lyase (PFL). In the cytosol, this acetyl-CoA is then converted into ethanol by ADH-E, whereas in the hydrogenosome, acetyl-CoA is supposed to be converted into acetate (Boxma et al. 2004).

Organisms that contain anaerobically functioning mitochondria or hydrogenosomes often do not (solely) produce pyruvate as an end product of their cytosolic metabolism. In these organisms, carbohydrates can be degraded to phosphoenolpyruvate (PEP) via the usual glycolytic pathway. PEP is then mainly carboxylated by phosphoenolpyruvate carboxykinase (PEPCK) to form oxaloacetate, which is reduced to malate (Fig. 5.1). This malate, produced in the cytosol, is transported into the anaerobic mitochondria for further degradation (see later). Cytosolic malate formation is comparable to lactate formation in that it generates 2 mol of ATP, while redox balance in the cytosol is maintained.

## 5.5 Energy Metabolism in Anaerobically Functioning Mitochondria

Organisms with anaerobic mitochondria can be divided into two different types: those which perform anaerobic respiration and use an alternative electron acceptor present in the environment, such as nitrate or nitrite, and those which perform fermentation reactions using an endogenously produced, organic electron acceptor, such as fumarate (Martin et al. 2001; Tielens et al. 2002). An example of the first type is the nitrate respiration that occurs in several ciliates (Finlay et al. 1983), and fungi (Kobayashi et al. 1996; Takaya et al. 2003), which use nitrate and/or nitrite as the terminal electron acceptor of their mitochondrial electron-transport chain, producing nitrous oxide as the reduced end product. These specific electron-transport chains contain special terminal oxidoreductases that donate electrons to an acceptor present in the environment, in this case nitrite or nitrate instead of oxygen. Little is known of these mitochondria and they will not be discussed further in this chapter.

The other class of anaerobic mitochondria, those which use fermentation reactions, is present in organisms that are highly adapted for either prolonged survival or continuous functioning in the absence of oxygen, such as parasitic helminths, or in organisms that are adapted to alternating periods in the presence and absence of oxygen, such as mussels, oysters and lugworms, which are intermittently dependent on this process when the tides of the sea force them to function anaerobically (De Zwaan 1991; Grieshaber et al. 1994). The energy metabolism in these anaerobic mitochondria differs principally and significantly from that in aerobic mitochondria, as no external final electron acceptors are used. Therefore, this mitochondrial metabolism has to be truly fermentative, in other words the number of NADH-producing reactions has to equal the number of NADH-consuming reactions without the use of oxygen or other external electron acceptors. Many mitochondrial catabolic pathways produce NADH, and anaerobically functioning mitochondria are adapted in that they also comprise catabolic pathways that consume NADH, processes that can be used as an electron sink. So far, in anaerobic mitochondria two distinct processes have been detected that are used as an electron sink to reoxidize the NADH produced by the oxidative catabolic pathways in these mitochondria. The reduction of fumarate to succinate during the fermentative malate dismutation pathway is the most commonly used electron sink in anaerobic mitochondria. Some organisms, however, do not only use this reduction of fumarate, but also use distinct reactions involved in lipid biosynthesis as an electron-sink during anoxic conditions (see later).

Malate dismutation is a fermentation pathway, which involves the use of an especially adapted electron-transport chain and the reduction of endogenously produced fumarate as an electron sink (Tielens and Van Hellemond 1998; Fig. 5.2). In organisms that are adapted to anoxic functioning using malate dismutation, carbohydrates are degraded by the usual glycolytic pathway to PEP, which is then converted to malate (as described before). This malate, produced in the cytosol, is transported into the mitochondria for further degradation (Fig. 5.2). In a split pathway, one portion of this malate is oxidized via pyruvate and acetyl-CoA to acetate and another portion is reduced to succinate, which is often further metabolized to propionate (Tielens 1994). Although several variations of malate dismutation with various end products occur, the use of succinate production from fumarate as an electron sink is universal. The reduction of malate to succinate occurs in two reactions that reverse part of the Krebs cycle, and the reduction of fumarate is the essential NADH-consuming reaction used to maintain redox balance. Fumarate reduction is linked to electron transport via electron-transferring

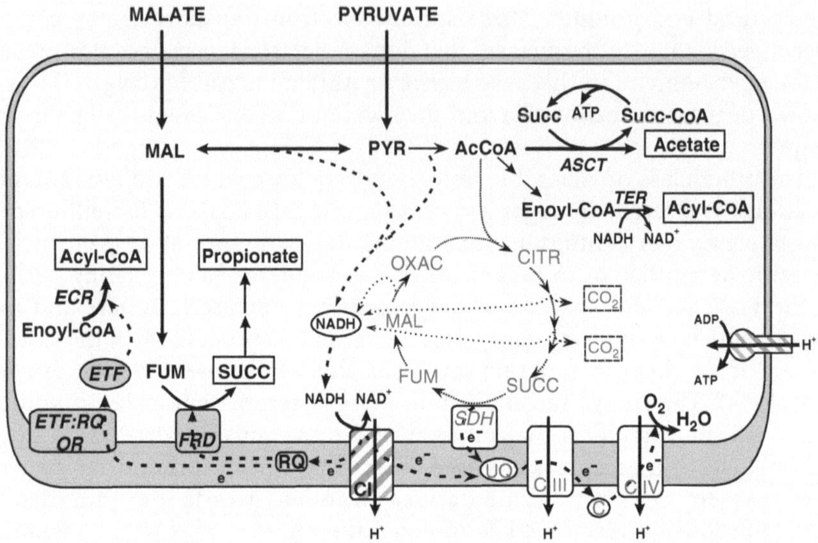


**Fig. 5.2.** Possible metabolic pathways in facultative anaerobic mitochondria. *Shaded boxes* show components of the electron-transport chain used during hypoxia, *open boxes* are components used during aerobiosis, and the *hatched boxes* (complex I and ATP-synthase) are components used under aerobic as well as anaerobic conditions. *ASCT* acetate:succinate CoA-transferase, *C* cytochrome *c*, *CI*, *CIII* and *CIV* complexes I, III and IV of the respiratory chain, *CITR* citrate, *ECR* enoyl-CoA reductase (such as present in *Ascaris suum*), *ETF* electron-transfer flavoprotein, *ETF:RQ OR* electron-transfer flavoprotein:rhodoquinone oxidoreductase, *FRD* fumarate reductase, *FUM* fumarate, *MAL* malate, *OXAC* oxaloacetate, *PYR* pyruvate, *RQ* rhodoquinone, *SDH* succinate dehydrogenase, *SUCC* succinate, *Succ-CoA* succinyl-CoA, *TER* *trans*-2-enoyl-CoA reductase (such as present in *E. gracilis*), *UQ* ubiquinone

enzyme complexes in an anaerobically functioning electron-transport chain (Fig. 5.2). Characteristic components of these anaerobically functioning mitochondria are fumarate reductase (FRD), the enzyme catalysing the reduction of fumarate to succinate, and rhodoquinone (RQ), the quinone shuttling the electrons from NADH dehydrogenase (complex I) towards FRD for the reduction of fumarate.

Next to fumarate reduction, some organisms use specific reactions in lipid biosynthesis as an electron sink to maintain redox balance in anaerobically functioning mitochondria. In anaerobic mitochondria two variants are known: the production of branched-chain fatty acids and the production of wax esters. The parasitic nematode *Ascaris suum* reduces fumarate in its anaerobic mitochondria, but instead of only producing acetate and succinate or propionate, like most other parasitic helminths, this organism also use the intermediates acetyl-CoA and propionyl-CoA to form branched-chain fatty acids (Komuniecki et al. 1989). This pathway is similar to reversal of β-oxidation and a complex mixture of the end products acetate, propionate, succinate and branched-chain fatty acids is excreted. In this pathway, the condensation of an acetyl-CoA and a propionyl-CoA, or of two propionyl-CoA molecules, ultimately results in the formation of the excreted end products 2-methylbutanoate and 2-methylpentanoate, respectively (Figs. 5.2, 5.3). After the condensation, the resulting intermediates are first reduced by NADH and are then hydrated in reactions comparable to those occurring in mammalian mitochondrial β-oxidation. In the final reaction of the pathway, the NADH-dependent reduction of 2-methyl branched-chain enoyl-CoAs, both membrane-bound and soluble components are involved. Complex I, rhodoquinone and electron-transport flavoprotein reductase comprise the membrane-bound components (Fig. 5.2, 5.3). The soluble components consist of two flavoproteins: electron-transport flavoprotein reductase and

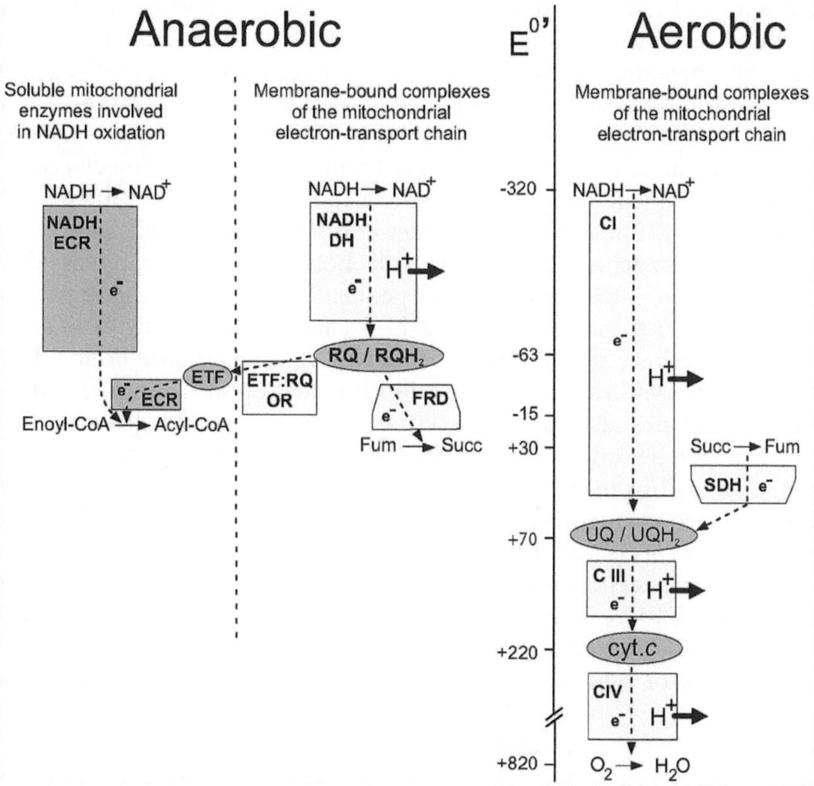


**Fig. 5.3.** The major components involved in mitochondrial NADH oxidation in facultative anaerobic mitochondria. In anaerobically functioning mitochondria, NADH is oxidized either by soluble enzymes (*left*) or by membrane-bound complexes of the electron-transport chain (*middle*). Under aerobic conditions, a classic respiratory chain is used to oxidize NADH (*right*). Proton translocation is indicated by *$H^+$ with arrows*. *Ovals* represent the electron transporters RQ, UQ and cytochrome *c* (*cyt. c*), and electron transport is indicated by *dashed arrows*. The *vertical bar* represents a scale for the standard redox potentials in millivolts. *Fum* fumarate, *NADH-DH* NADH dehydrogenase, *NADH-ECR* soluble NADH:enoyl-CoA reductase, *$RQH_2$* rhodoquinol, *Succ* succinate, *$UQH_2$* ubiquinol

2-methyl branched-chain enoyl-CoA reductase (Komuniecki and Harris 1995; Fig. 5.2). This branched-chain fatty acid formation provides an additional pathway for the oxidation of NADH, and could provide increased flexibility in regulating intramitochondrial NADH to $NAD^+$ ratios under the reducing conditions in the gut of the host, by serving as an important sink for excess reducing power (Komuniecki and Harris 1995).

The photosynthetic flagellate *Euglena gracilis* utilizes another variant of mitochondrial lipid biosynthesis as alternative electron sink to survive hypoxia. This adaptation enables the mitochondrion of this organism to produce ATP also in the absence of oxygen. Under aerobic conditions, *Euglena* performs a more or less typical oxidative phosphorylation in association with a modified Krebs cycle and respiratory chain. Pyruvate from glycolysis enters the mitochondrion and undergoes oxidative decarboxylation; the resulting acetyl-CoA enters a modified Krebs cycle with a succinate-semialdehyde shunt, circumventing the step catalysed by $\alpha$-ketoglutarate dehydrogenase. Electrons from glucose breakdown are transferred to oxygen as the terminal electron acceptor, and oxidative phosphorylation generates most of the ATP. During hypoxia, mitochondrial fatty acid synthesis serves in *Euglena* cells as an electron sink when wax esters are formed from its reserve glucose polymer, paramylon (Inui et al. 1984). Waxes are esters of long-chain saturated or unsaturated fatty acids with long-chain alcohols. In the single mitochondrion of *Euglena* cells, a malonyl-CoA-independent fatty acid synthetic pathway exists, which has the capability to synthesize fatty acids directly from acetyl-CoA as both primer and $C_2$ donor, using NADH as an electron donor. These fatty acids are synthesized by reversal of $\beta$-oxidation in that it proceeds via CoA intermediates instead of via acyl carrier protein (ACP). However, a key difference with $\beta$-oxidation is that enoyl-CoA reductase functions instead of acyl-CoA dehydrogenase to reduce the double bond. A portion of the fatty acids produced are reduced to alcohols, esterified with another fatty acid and the wax esters formed are deposited in the cytosol. Upon return to aerobiosis, these wax esters are degraded rapidly and paramylon is resynthesized then (Inui et al. 1982). This mitochondrial system for fatty acid synthesis produces fatty acids with chain lengths of eight to 16 carbons, with a majority of $C_{14}$. Synthesis of odd-numbered fatty acids also occurs and starts from propionyl-CoA. Formation of the propionate necessary for this pathway functions in itself as an extra electron sink during hypoxia as this propionate is most likely produced via fumarate reduction in the same pathway that occurs also in the anaerobically functioning mitochondria of many parasitic helminths (Fig. 5.2). The acetyl-CoA used in wax ester formation in *Euglena* cells stems from pyruvate via an unusual oxygen-sensitive enzyme pyruvate:$NADP^+$ oxidoreductase (PNO) (Inui et al. 1987). The core catalytic component of this PNO is PFO, an enzyme also present in amitochondriate protists and in the hydrogenosomes of trichomonads. PNO exists in *Euglena* in the mitochondrion alongside a classical mitochondrial PDH, with messenger RNA expression patterns converse to that of PNO in response to hypoxia (Hoffmeister et al. 2004).

The two variants of the use of mitochondrial lipid biosynthesis as an electron sink in the absence of oxygen, as described before, have a very comparable crucial reaction in common. In the production of the branched-chain fatty acids in *Ascaris* mitochondria as well as in the malonyl-CoA-independent fatty acid biosynthesis in *Euglena* mitochondria, the final reaction is catalysed by a 2-methyl branched-chain enoyl-CoA reductase. However, the *Ascaris* enzyme is clearly distinct from the *Euglena* one (Hoffmeister et al. 2005). As described before, the *Ascaris* enzyme accepts electrons from NADH via complex I, rhodoquinone and an electron-transporting flavoprotein (Komuniecki and Harris 1995; Figs. 5.2, 5.3), whereas it was recently shown that the *Euglena* enzyme accepts electrons from NADH directly (Hoffmeister et al. 2005). The gene of the *Ascaris* enzyme shows sequence similarities to mitochondrial acyl-CoA dehydrogenases found in mitochondria of most eukaryotes (Duran et al. 1993), but for the gene of the *Euglena* enzyme, no homologues were found among other eukaryotes (Hoffmeister et al. 2005).

## 5.6 Adaptations in Electron-Transport Chains in Anaerobic Mitochondria

In classical aerobic mitochondria, the respiratory chain is used for the reoxidation of reduced cofactors, such as NADH and $FADH_2$, which are produced in large amounts by Krebs cycle activity. In these aerobically functioning mitochondria, electrons are transferred from NADH and succinate to ubiquinone via complexes I and II of the respiratory chain, respectively. Subsequently, electrons are transferred from ubiquinol to oxygen via complexes III and IV of the respiratory chain (Fig. 5.3). Oxidation of NADH in most anaerobically functioning mitochondria is also linked to electron transport and oxidative phosphorylation; however, these anaerobic mitochondria do not use oxygen as a terminal electron acceptor but use endogenously produced organic molecules instead (see before). These anaerobic mitochondria possess specific terminal oxidoreductase complexes that are able to transfer electrons to these endogenously produced electron acceptors (for recent reviews see Tielens and Van Hellemond 1998; Tielens et al. 2002; Kita and Takamiya 2002).

Most anaerobically functioning mitochondria use endogenously produced fumarate as a terminal electron-acceptor (see before) and thus contain a FRD as the final respiratory chain complex (Behm 1991). The reduction of fumarate is the reversal of succinate oxidation, a Krebs cycle reaction catalysed by succinate dehydrogenase (SDH), also known as complex II of the electron-transport chain (Fig. 5.3). The interconversion of succinate and fumarate is readily reversible by FRD and SDH complexes in vitro. However, under standard conditions in the cell, oxidation and reduction reactions preferentially occur when electrons are transferred to an acceptor with a higher standard redox potential; therefore, electrons derived from the oxidation of succinate to fumarate ($E^{0'} = +30$ mV) are transferred by SDH to ubiquinone, which is then reduced to ubiquinol ($E^{0'}$= + 100 mV). FRD, on the other hand, catalyses the opposite reaction, in which an electron donor is oxidized and electrons are transferred to fumarate, which is then reduced to succinate. Therefore, the electron donor used by FRD is expected to possess a lower standard electron potential than the $E^{0'}$= + 30mV of the fumarate/succinate couple (Fig. 5.3). Accordingly, these anaerobic mitochondria do not use ubiquinone ($E^{0'}$= + 100 mV), but use rhodoquinone ($E^{0'}$= – 73 mV) for transport of electrons from NADH to fumarate instead (Allen 1973; Van Hellemond et al. 1995). Although succinate oxidation and fumarate reduction are reversed and are similar reactions, distinct enzyme complexes coexist for these reactions in facultative anaerobic organisms (Ackrell et al. 1992; Hederstedt and Ohnishi 1992). The fact that, in one and the same organism (for instance *Escherichia coli*), SDH and FRD complexes coexist and are differentially expressed demonstrates that in this case it is advantageous to use a different enzyme for the two opposite directions of the reaction in question. This correlates with their binding of distinct quinones and with the transfer of electrons in opposite directions through the enzyme complex, which requires differences in the affinity for electrons (standard redox potential) of the distinct electron-binding domains within these enzyme complexes (Ackrell et al. 1992; Van Hellemond and Tielens 1994; Tielens and Van Hellemond 1998). Therefore, the transition from succinate oxidation during aerobic conditions to fumarate reduction during anaerobic conditions has major consequences for the composition and use of the mitochondrial electron-transport chain, because (1) ubiquinone is replaced by rhodoquinone (Van Hellemond et al. 1995), (2) SDH is replaced by FRD (Roos and Tielens 1994; Saruta et al. 1995; Tielens and Van Hellemond 1998) and (3) the expression of complexes III and IV of the classic aerobic respiratory chain is strongly reduced (Hoffmeister et al. 2004).

As described before, also the formation of branched-chain fatty acids by enoyl-CoA reductase activity is coupled to electron transport (Komuniecki and Harris 1995). In this case electrons are transported from NADH to rhodoquinone via complex I and subsequently to the electron-transfer flavoprotein (ETF) via ETF-reductase (Fig. 5.3). The soluble, non-membrane-bound ETF then transfers electrons to enoyl-CoA reductase, which uses the electrons for the condensation of two short-chain ($C_2$–$C_3$) acyl-CoA moieties for the formation of branched-chain fatty acids.

## 5.7 Structural Aspects of Anaerobically Functioning Electron-Transport Chains

Fumarate reduction is the most commonly used electron-sink for the reoxidation of reduced cofactors in anaerobically functioning mitochondria (see before). Fumarate reduction is catalysed in these mitochondria by a FRD complex, which is structurally very similar to the SDH complexes present in aerobically functioning mitochondria. Both complexes usually consist of four nonidentical subunits: a flavin-containing A subunit (Fp subunit), a B subunit that contains three iron–sulphur clusters (Ip subunit) and two hydrophobic, cytochrome *b* containing subunits C and D that are essential for the attachment of the catalytic subunits A and B to the membrane and for the interaction of the catalytic subunits with the quinones (Ackrell et al. 1992; Hederstedt and Ohnishi 1992). The Fp and Ip subunits of SDH are highly conserved in different species and are also closely related to the Fp and Ip subunits of FRD. Analyses of enzyme kinetics, as well as the known differences in primary structures of prokaryotic and eukaryotic complexes that reduce fumarate, led to the suggestion that fumarate-reducing eukaryotes possess an enzyme complex for the reduction of fumarate that is structurally related to SDH-type complex II, but that has the functional characteristics of the FRD complexes of prokaryotes (Van Hellemond et al. 1995). Despite the limited amount of data available so far, the sequences of eukaryotic FRDs are clearly more closely related to those of SDHs than to those of bacterial FRDs, indicating that during the evolution of anaerobic mitochondria, their SDH is modified to preferentially catalyse the reverse reaction and use lower-potential quinones (Tielens and Van Hellemond 1998; Tielens et al. 2002).

Most prokaryotes are known to use menaquinone for electron transport to FRD, whereas the anaerobically functioning mitochondria of eukaryotes use rhodoquinone. Although menaquinone and rhodoquinone are functional equivalents, because they both have a low standard electron potential, their structure differs significantly. The prokaryotic electron transporter menaquinone is a naphthoquinone, whereas rhodoquinone and ubiquinone are benzoquinones; therefore, the quinone involved in electron transport to fumarate in eukaryotes (rhodoquinone) is structurally more similar to the quinone involved in aerobic mitochondrial electron transport (ubiquinone) than to its functional equivalent in prokaryotes, menaquinone. The quinone structure thus also suggests that electron transport in anaerobically functioning mitochondria is more closely related to that in aerobically functioning mitochondria than to its functional homologue in prokaryotes (Van Hellemond et al. 1995; Tielens and Van Hellemond 1998). Furthermore, the specific enzyme enoyl-CoA reductase, involved in the biosynthesis of branched-chain fatty acids in anaerobically functioning mitochondria of *Ascaris*, shows sequence similarities to mitochondrial acyl-CoA dehydrogenases found in many aerobic functioning eukaryotes (Duran et al. 1993). This is further evidence that the mechanisms to reoxidize reduced cofactors in the absence of oxygen evolved from components already present in aerobically functioning mitochondria.

## 5.8 Evolutionary Origin of Anaerobic Mitochondria

It is now generally accepted that the classical aerobic mitochondria evolved by a single endosymbiotic event between an anaerobic host and an $\alpha$-proteobacterium. Endosymbiotic theories for the origin of cellular organelles had already been postulated at the end of the nineteenth and early in the twentieth century, but the publications of Margulis in the second half of the twentieth century renewed interest in the theory that mitochondria originate from an endosymbiotic event (Margulis 2005; see also Chap. 4 by Sapp in this volume). However, that theory needed refinement as various organelles were recently shown to exist that, although different, are clearly related to mitochondria (Martin and Müller 1998; Gray et al. 1999). Next to the classical aerobic mitochondria, which contain a respiratory chain and use oxygen as a final electron acceptor, anaerobically functioning mitochondria were identified that also contain a respiratory chain and perform oxidative phosphorylation, but do not use oxygen as a terminal electron acceptor (see before). In addition, several eukaryotes were identified that contained hydrogenosomes, double membrane bound, ATP-producing organelles that are related to mitochondria (Embley et al. 2003b; Dyall et al. 2004). Hydrogenosomes, however, differ from mitochondria as they lack a respiratory chain and produce hydrogen (see Chap. 6 by Tachezy and Doležal, Chap. 7 by Hackstein et al. and Chap. 11 by Tovar for details). Several types of hydrogenosomes exist (Fig. 5.1) and they have evolved independently in unrelated lineages of eukaryotic microorganisms (Yarlett et al. 1984; Embley et al. 1995; Martin et al. 2001; Hackstein et al. 2001; Van der Giezen et al. 2005). Furthermore, several eukaryotes were shown to lack compartmentation of energy metabolism, but these organisms were nevertheless shown to contain organelles related to mitochondria (mitosomes) (see before and Chap. 10 by Barberà et al. for details). Although it is generally believed that all these organelles related to mitochondria share a common ancestor, the ancestral endosymbiont, evolution of the distinct types is hotly debated. It is conceivable that the distinct variants of anaerobically ATP-generating organelles all evolved directly from the ancestral endosymbiont. Alternatively, the different anaerobic types of organelles could have evolved from each other or from aerobic ones in a more dynamic fashion, where adaptations to the absence of oxygen turned aerobic mitochondria into anaerobically functioning ones (or where the energy-generating function of the organelle was even lost all together). It should be noted, however, that the distinct types of organelles related to mitochondria did not necessarily develope via the same evolutionary pathway.

Several observations indicate that anaerobically functioning mitochondria evolved from the classical aerobically functioning mitochondria, and did not originate directly (by adaptation to an anaerobic environment) from the facultative anaerobic ancestral cell that was the result of the endosymbiosis of an $\alpha$-proteobacterial symbiont and the host cell. First, as discussed before, all sequence data available up to now on the fumarate-reducing enzymes (FRDs) of these organisms demonstrate that these enzymes are closely related to the SDHs of classical aerobic mitochondria. Second, the same argument holds true for the quinone used for this anaerobic fumarate reduction. All anaerobically functioning, fumarate-reducing eukaryotes investigated so far use rhodoquinone for the transport of electrons from complex I to the fumarate-reducing enzyme, while the more ancient prokaryotic systems use menaquinone. Again, the anaerobically functioning mitochondria use a molecule (rhodoquinone) which, as it is a benzoquinone, is structurally more related to the molecule of the classical mitochondria, ubiquinone (also a benzoquinone) than to the molecule used by the anaerobic prokaryotes, menaquinone (a naphthoquinone). As the synthesis of rhodoquinone probably differs only in one of the last steps from the synthesis of ubiquinone, this indicates that also the electron transporter was adapted to anaerobically functioning after the synthesis of the originally present transporter (menaquinone) had been lost by adaptation to an aerobic environment. Third, many organisms containing anaerobically functioning mitochondria evolved from aerobic ancestors, which argues for adaptation of classical aerobic mitochondria to an anaerobic environment instead of an earlier adaptation of the ancestral endosymbiont to an anaerobic environment. An example of such organisms is parasitic helminths, which are supposed to have evolved from free-living worms, which most likely functioned aerobically, like free-living worms nowadays. Fourth, in distinct lineages distinct mechanisms for mitochondrial reoxidation of reduced cofactors are used, such as fumarate reduction, wax-ester formation or synthesis of branched-chain fatty acids (see before). The existence of these different mechanisms in separate lineages can more easily be explained by the evolution of different adaptations rather than by differential loss. The extensive similarity observed between components involved in anaerobic and aerobic metabolism in mitochondria (see before) suggests that anaerobic mitochondria evolved by adaptation of classical aerobic mitochondria to hypoxic conditions (Tielens and Van Hellemond 1998; Tielens et al. 2002).

The recent characterization of the hydrogenosomes of the anaerobic ciliate *Nyctotherus ovalis*, which thrives in the hindgut of cockroaches, supports the hypothesis that hydrogenosomes and anaerobic mitochondria can evolve from aerobic mitochondria (Boxma et al. 2005). The *N. ovalis* organelle was identified as a missing link between aerobic mitochondria and hydrogenosomes, because it comprises hydrogen production together with the presence of a genome encoding active electron transport chain components (Boxma et al. 2005). In addition, phylogenetic analyses revealed that the proteins of this electron-transport chain, and all nuclear genes encoding mitochondrial proteins, cluster with their homologues from aerobic ciliates, which demonstrated that this organelle is closely related to aerobic mitochondria. Furthermore, this organelle contains biochemical features characteristic of anaerobic mitochondria, such as the presence of rhodoquinone and the ability to produce succinate. These results strongly suggest that this organelle evolved from a mitochondrion of aerobic ciliates, that adapted to anaerobic functioning. Subsequently, the mitochondrial hallmarks, electron transport chain complexes that are partially encoded by a mitochondrial genome, started to degenerate after the acquisition of alternative electron sinks, a hydrogenase and fumarate reduction.

The mitochondrion of the photosynthetic flagellate *E. gracilis* was also recently discovered as an intermediate between mitochondria and hydrogenosomes that unites biochemical properties of hydrogenosomes and of both aerobic and anaerobic mitochondria. The mitochondrion of *Euglena* contains an electron-transport chain that uses ubiquinone, SDH and cytochrome-containing complexes in the presence of oxygen, and rhodoquinone, FRD and endogenously produced fumarate during hypoxia. Furthermore, this mitochondrion comprises not only PDH, the mitochondrial enzyme for conversion of pyruvate to acetyl-CoA (Hoffmeister et al. 2004), but also a PNO, which consists of a C-terminal NADPH-cytochrome P450 reductase domain fused to an N-terminal PFO domain (Rotte et al. 2001). PFO has so far only been found in those eukaryotes that lack mitochondria and lack PDH; therefore, the mitochondrion of *E. gracilis* is an intermediate that unites biochemical properties of aerobic and anaerobic mitochondria and hydrogenosomes, as it comprises characteristic components of these organelles.

## 5.9 Conclusion

Anaerobic metabolism is usually considered to be an old and rather primitive way of life, but the scenario described here would imply that anaerobically functioning mitochondria are an adaptation of the traditional type of mitochondria to anaerobic environments; hence, these anaerobic mitochondria are in fact a further evolution of aerobic mitochondria.

Similar to the observed diversity in hydrogenosomes (Fig. 5.1), it is now known that not all mitochondria are the same (Scheffler 1999; Tielens et al 2002, Burger et al. 2003; Gabaldón and Huynen 2004). Over the past few years many different variants have been identified, and we might be only at the beginning of the discovery of many more interesting variants of ATP-producing organelles. Mitochondria are no longer the once-thought uniform ATP-producing organelles, identical in all organisms, but are rather diverse and adapted to distinct conditions. Recent analyses suggest that all mitochondrial variants, including anaerobically functioning ones, have evolved from a single endosymbiotic ancestor. The mitochondrial diversity has evolved by multiple mitochondrial adaptations that occurred independently in multiple nonrelated lineages.

## References

Ackrell BAC, Johnson MK, Gunsalus RP, Cecchini G (1992) Structure and function of succinate dehydrogenase and fumarate reductase. In: Muller F (ed) Chemistry and biochemistry of flavoenzymes, vol III. CRC, Boca Raton, pp 229–297

Allen PC (1973) Helminths: comparison of their rhodoquinone. Exp Parasitol 34:211–219

Behm CA (1991) Fumarate reductase and the evolution of electron transport systems. In: Bryant C (ed) Metazoan life without oxygen. Chapman & Hall, London, pp 88–108

Boxma B, Voncken F, Jannink S, van Alen T, Akhmanova A, Van Weelden SWH, Van Hellemond JJ, Ricard G, Huynen M, Tielens AGM, Hackstein JH (2004) The anaerobic chytridiomycete fungus *Piromyces* sp. E2 produces ethanol via pyruvate:formate lyase and an alcohol dehydrogenase E. Mol Microbiol 51:1389–1399

Boxma B, de Graaf RM, van der Staay GW, van Alen TA, Ricard G, Gabaldon T, van Hoek AH, Moon-van der Staay SY, Koopman WJ, Van Hellemond JJ, Tielens AGM, Friedrich T, Veenhuis M, Huynen MA, Hackstein JH (2005) An anaerobic mitochondrion that produces hydrogen. Nature 434:74–79

Bruchhaus I, Tannich E (1994) Purification and molecular characterization of the NAD(+)-dependent acetaldehyde/alcohol dehydrogenase from *Entamoeba histolytica*. Biochem J 303(Pt 3):743–748

Burger G, Gray MW, Lang BF (2003) Mitochondrial genomes: anything goes. Trends Genet 19:709–716

De Zwaan A (1991) Molluscs. In: Bryant C (ed) Metazoan life without oxygen. Chapman & Hall, London, pp 186–217

Duran E, Komuniecki RW, Komuniecki PR, Wheelock MJ, Klingbeil MM, Ma YC, Johnson KR (1993) Characterization of cDNA clones for the 2-methyl branched-chain enoyl-CoA reductase. An enzyme involved in branched-chain fatty acid synthesis in anaerobic mitochondria of the parasitic nematode *Ascaris suum*. J Biol Chem 268:22391–22396

Dyall SD, Brown MT, Johnson PJ (2004) Ancient invasions: from endosymbionts to organelles. Science 304:253–257

Embley TM, Finlay BJ, Dyal PL, Hirt RP, Wilkinson M, Williams AG (1995) Multiple origins of anaerobic ciliates with hydrogenosomes within the radiation of aerobic ciliates. Proc R Soc Lond B Biol Sci 262:87–93

Embley TM, van der GM, Horner DS, Dyal PL, Bell S, Foster PG (2003a) Hydrogenosomes, mitochondria and early eukaryotic evolution. IUBMB Life 55:387–395

Embley TM, van der GM, Horner DS, Dyal PL, Foster P (2003b) Mitochondria and hydrogenosomes are two forms of the same fundamental organelle. Philos Trans R Soc Lond B Biol Sci 358:191–201

Fenchel T, Finlay BJ (1994) The Evolution of Life Without Oxygen. Am Sci 82:22–29

Finlay BJ, Span ASW, Harman JM (1983) Nitrate respiration in primitive eukaryotes. Nature 303:333–336

Gabaldón T, Huynen MA (2004) Shaping the mitochondrial proteome. Biochim Biophys Acta 1659:212–220

Gray MW, Burger G, Lang BF (1999) Mitochondrial evolution. Science 283:1476–1481

Grieshaber MK, Hardewig I, Kreutzer U, Pörtner HO (1994) Physiological and metabolic responses to hypoxia in invertebrates. Rev Physiol Biochem Pharmacol 125:43–147

Hackstein JHP, Akhmanova A, Voncken F, van Hoek AH, van Alen T, Boxma B, Moon-van der Staay S, van der Staay GW, Leunissen J, Huynen M, Rosenberg J, Veenhuis M (2001) Hydrogenosomes: convergent adaptations of mitochondria to anaerobic environments. Zoology 104:290–302

Hederstedt L, Ohnishi T (1992) Progress in succinate:quinone oxidoreductase research. In: Ernster L (ed) Molecular mechanisms in bioenergetics. Elsevier, Amsterdam, pp 163–198

Henze K, Martin W (2003) Evolutionary biology: essence of mitochondria. Nature 426:127–128

Hoffmeister M, van der Klei A, Rotte C, van Grinsven KW, Van Hellemond JJ, Henze K, Tielens AGM, Martin W (2004) *Euglena gracilis* rhodoquinone:ubiquinone ratio and mitochondrial proteome differ under aerobic and anaerobic conditions. J Biol Chem 279:22422–22429

Hoffmeister M, Piotrowski M, Nowitzki U, Martin W (2005) Mitochondrial trans-2-enoyl-CoA reductase of wax ester fermentation from *Euglena gracilis* defines a new family of enzymes involved in lipid synthesis. J Biol Chem 280:4329–4338

Horner DS, Hirt RP, Embley TM (1999) A single eubacterial origin of eukaryotic pyruvate:ferredoxin oxidoreductase genes: implications for the evolution of anaerobic eukaryotes. Mol Biol Evol 16:1280–1291
Inui H, Miyatake K, Nakano Y, Kitaoka S (1982) Wax ester fermentation in *Euglena gracilis.* FEBS Lett 150:89–93
Inui H, Miyatake K, Nakano Y, Kitaoka S (1984) Fatty acid synthesis in mitochondria of *Euglena gracilis.* Eur J Biochem 142:121–126
Inui H, Ono K, Miyatake K, Nakano Y, Kitaoka S (1987) Purification and characterization of pyruvate:NADP+ oxidoreductase in *Euglena gracilis.* J Biol Chem 262:9130– 9135
Kita K, Takamiya S (2002) Electron-transfer complexes in Ascaris mitochondria. Adv Parasitol 51:95–131
Kobayashi M, Matsuo Y, Takimoto A, Suzuki S, Maruo F, Shoun H (1996) Denitrification, a novel type of respiratory metabolism in fungal mitochondrion. J Biol Chem 271:16263–16267
Komuniecki R, McCrury J, Thissen J, Rubin N (1989) Electron-transfer flavoprotein from anaerobic *Ascaris suum* mitochondria and its role in NADH-dependent 2-methyl branched-chain enoyl-CoA reduction. Biochim Biophys Acta 975:127–131
Komuniecki R, Harris BG (1995) Carbohydrate and energy metabolism in helminths. In: Marr JJ, Müller M (eds) Biochemistry and molecular biology of parasites. Academic, London, pp 49–66
Livingstone DR (1991) Origins and evolution of pathways of anaerobic metabolism in the animal kingdom. Am Zool 31:522–534
Lloyd D (1996) Obligate anaerobe or not. Nature 381:121
Lutz PL, Nilsson GE (1997) Contrasting strategies for anoxic brain survival - glycolysis up or down. J Exp Biol 200:411–419
Margulis L (2005) Origin of eukaryotic cells. Yale University Press, New Haven
Martin W (2000) Primitive anaerobic protozoa: the wrong host for mitochondria and hydrogenosomes? Microbiology 146:1021–1022
Martin W, Müller M (1998) The hydrogen hypothesis for the first eukaryote. Nature 392:37–41
Martin W, Hoffmeister M, Rotte C, Henze K (2001) An overview of endosymbiotic models for the origins of eukaryotes, their ATP-producing organelles (mitochondria and hydrogenosomes), and their heterotrophic lifestyle. Biol Chem 382:1521–1539
Müller M (1998) Enzymes and compartmentation of core energy metabolism of anaerobic protist – a special case in eukaryotic evolution? In: Coombs GH, Vickerman K, Sleigh MA, Warren A (eds) Evolutionary relationships among protozoa. Kluwer, Dordrecht, pp 109–131
Müller M (2000) A mitochondrion in *Entamoeba histolytica*? Parasitol Today 16:368–369
Roger AJ, Silberman JD (2002) Cell evolution: mitochondria in hiding. Nature 418:827–829
Roos MH, Tielens AGM (1994) Differential expression of two succinate dehydrogenase subunit-B genes and a transition in energy metabolism during the development of the parasitic nematode *Haemonchus contortus.* Mol Biochem Parasitol 66:273–281
Rotte C, Stejskal F, Zhu G, Keithly JS, Martin W (2001) Pyruvate:NADP+ oxidoreductase from the mitochondrion of Euglena gracilis and from the apicomplexan Cryptosporidium parvum: a biochemical relic linking pyruvate metabolism in mitochondriate and amitochondriate protists. Mol Biol Evol 18:710–720
Sanchez LB (1998) Aldehyde dehydrogenase (CoA-acetylating) and the mechanism of ethanol formation in the amitochondriate protist, *Giardia lamblia.* Arch Biochem Biophys 354:57–64
Saraste M (1999) Oxidative phosphorylation at the fin de siecle. Science 283:1488–1493
Saruta F, Kuramochi T, Nakamura K, Takamiya S, Yu Y, Aoki T, Sekimizu K, Kojima S, Kita K (1995) Stage-specific isoforms of complex II (succinate-ubiquinone oxidoreductase) in mitochondria from the parasitic nematode, *Ascaris suum.* J Biol Chem 270:928–932
Scheffler I (1999) Mitochondria. Wiley-Liss, New York
Steinbüchel A, Müller M (1986) Glycerol, a metabolic end product of *Trichomonas vaginalis* and *Tritrichomonas foetus.* Mol Biochem Parasitol 20:45–55

Takaya N, Kuwazaki S, Adachi Y, Suzuki S, Kikuchi T, Nakamura H, Shiro Y, Shoun H (2003) Hybrid respiration in the denitrifying mitochondria of *Fusarium oxysporum*. J Biochem (Tokyo) 133:461–465

Tielens AGM (1994) Energy generation in parasitic helminths. Parasitol Today 10:346–352

Tielens AGM, Rotte C, Van Hellemond JJ, Martin W (2002) Mitochondria as we don't know them. Trends Biochem Sci 27:564– 572

Tielens AGM, Van Hellemond JJ (1998) The electron transport chain in anaerobically functioning eukaryotes. Biochim Biophys Acta 1365:71–78

Tovar J, Fischer A, Clark CG (1999) The mitosome, a novel organelle related to mitochondria in the amitochondrial parasite *Entamoeba histolytica*. Mol Microbiol 32:1013–1021

Tovar J, Leon-Avila G, Sanchez LB, Sutak R, Tachezy J, van der Giezen M, Hernandez M, Müller M, Lucocq JM (2003) Mitochondrial remnant organelles of *Giardia* function in iron-sulphur protein maturation. Nature 426:172–176

Valadi A, Granath K, Gustafsson L, Adler L (2004) Distinct intracellular localization of Gpd1p and Gpd2p, the two yeast isoforms of $NAD^+$-dependent glycerol-3-phosphate dehydrogenase, explains their different contributions to redox-driven glycerol production. J Biol Chem 279:39677–39685

Van der Giezen M, Tovar J and Clark CG (2005) Mitochondrion-derived organelles in protists and fungi. Int Rev Cytol 244:175–225

Van Hellemond JJ, Tielens AGM (1994) Expression and functional properties of fumarate reductase. Biochem J 304:321–331

Van Hellemond JJ, Klockiewicz M, Gaasenbeek CPH, Roos MH, Tielens AGM (1995) Rhodoquinone and complex II of the electron transport chain in anaerobically functioning eukaryotes. J Biol Chem 270:31065–31070

van Waarde A, Van den Thillart G, Verhagen M (1993) Ethanol formation and pH-regulation in fish. In: Hochachka PW, Lutz PL, Sick T, Rosenthal M, Van den Thillart G (eds) Surviving hypoxia, mechanisms of control and adaptation. CRC, Boca Raton, pp 157–170

Williams BA, Hirt RP, Lucocq JM, Embley TM (2002) A mitochondrial remnant in the microsporidian *Trachipleistophora hominis*. Nature 418:865–869

Yarlett N, Coleman GS, Williams AG, Lloyd D (1984) Hydrogenosomes in known species of rumen entodiniomorphid protozoa. FEMS Microbiol Lett 21:15–19

# 6 Iron–Sulfur Proteins and Iron–Sulfur Cluster Assembly in Organisms with Hydrogenosomes and Mitosomes

JAN TACHEZY, PAVEL DOLEŽAL

## 6.1 Introduction

Mitochondrion is the name reflecting a typical morphology of organelles originally observed as granules in muscle cells (*mitos* meaning "thread" and *chondros* meaning "grain"). This organelle is usually presented as a cellular power plant producing ATP by oxidative phosphorylation. The process is dependent on consumption of oxygen as the terminal electron acceptor, which is coupled with the citric acid cycle generating reducing equivalents and the cytochrome-dependent respiratory chain. Additionally, mitochondria are involved in a number of metabolic pathways, such as the fatty acid catabolism by β-oxidation, amino acid and biotin synthesis, and the urea cycle. As a hallmark of eukaryotic cells, mitochondria are unequivocally present in all multicellular organisms. However, structural studies of unicellular eukaryotes (protists) reveal that a number of free-living as well as parasitic protists, which inhabit oxygen-poor environments, do not possess organelles with characters of typical mitochondria. These organisms were long designated as "amitochondriate."

The formation of iron–sulfur (FeS) clusters is a novel fundamental function of mitochondria, which is required for maturation of FeS proteins (Lill et al. 1999; Lill and Kispal 2000; Lill and Mühlenhoff 2005; Rouault and Tong 2005). Studies of eukaryotic diversity showed that a number of "textbook" mitochondrial pathways are highly modified or absent under specific environmental conditions or at certain developmental stages of various organisms (Tielens et al. 2002). For example, a number of aquatic invertebrate species living in sulfidic habitats oxidize sulfide in mitochondria, which is coupled with oxygen consumption and ATP synthesis (Grieshaber and Volkel 1998). The lumen-dwelling helminths, which inhabit niches where oxygen tension is low, utilize the unique mitochondrial respiratory chain that enables them to employ fumarate instead of oxygen as a terminal electron acceptor (Kita et al. 1997). The cytrochrome-dependent respiratory chain as well as citric acid cycle are repeatedly activated and inactivated during the life cycle of African trypanosomes (Besteiro et al. 2005; van Weelden et al. 2005). Nevertheless, both organisms with fully active mitochondria and those with altered mitochondrial metabolism require functions of FeS proteins; thus,

Origin of Mitochondria and Hydrogenosomes (ed. by William F. Martin and Miklós Müller)

FeS cluster assembly is likely the only essential function of mitochondria, inactivation of which is lethal (Lill et al. 1999; Lill and Kispal 2000; Mühlenhoff and Lill 2000). The intriguing question is how and in which cellular compartment this function takes place in "amitochondriate" organisms.

## 6.2 Mitochondrion-Related Organelles in "Amitochondriate" Eukaryotes

It is generally accepted that mitochondria evolved from endosymbiotic bacteria most likely of proteobacterial origin (Esser et al. 2004). A debated question of interest is whether the ancestral host was an amitochondriate protoeukaryote, which gained mitochondria at a certain stage of eukaryogenesis, or whether the appearance of the mitochondria was directly associated with the origin of the eukaryotic cell. The latter idea was elaborated in two symbiosis hypotheses: the hydrogen hypothesis (Martin and Müller 1998) and the syntrophy hypothesis (Moreira and Lopez-Garcia 1998). Both proposed a symbiotic metabolic association in anaerobic environments between an archaeal methanogen and a proteobacterion, although they provide different views concerning the nature of the eubacterial partners (discussed in Lopez-Garcia and Moreira 1999). Accordingly, the absence of mitochondria in "amitochondriates" was considered to be a result of two possible scenarios: contemporary "amitochondriates" are either (1) descendants of primarily amitochondrial protoeukaryotes, which separated from the main eukaryotic trunk before the mitochondrial endosymbiosis, or (2) they lost their mitochondria in a secondary event owing to their specific adaptation to anaerobic or oxygen-restricted conditions. The first scenario was suggested for organisms long considered most ancient among eukaryotes, such as diplomonads, parabasalids, archamoebae, and microsporidia. The absence of apparent mitochondria was postulated as a "primitive" character, and the amitochondriate organisms were grouped into the taxon named Archezoa to indicate their ancient origin (Cavalier-Smith 1987a, b). Although later phylogenetic analysis eroded the Archezoa hypothesis (Roger et al. 1999), diplomonads and parabasalids still remain among the candidates that might represent the deepest diverging lineages of eukaryotes (Adam 2001; Best et al. 2004). The second scenario may explain amitochondriate status in organisms that belong to the monophyletic taxons together with organisms possessing typical mitochondria such as the apicomplexan *Cryptosporidium* or microsporidia, which are currently placed among fungi. However, if the appearance of eukaryotes was interrelated with the invention of mitochondria, a secondary loss of mitochondria can be expected even in "early branching" eukaryotes. Indeed, several genes regarded as mitochondrial in origin, including mitochondrial-type heat shock protein 70 kDa (Hsp70) (Arisue et al. 2002; Bui et al. 1996; Germot et al. 1996), chaperonin 60 kDa (Cpn60) (Horner et al. 1996; Roger et al. 1998), valyl-transfer RNA synthetase (Hashimoto 1998), and cysteine desulfurase IscS (Tachezy et al. 2001) were identified in genomes of some diplomonads and parabasalids, supporting secondary absence of mitochondria in these organisms. However, do they really lack mitochondria?

### 6.2.1 Hydrogenosomes

Although organelles corresponding to the textbook definition of mitochondria are not present in amitochondriates, parabasalids and some other protists possess double membrane bounded organelles called hydrogenosomes. The hydrogenosomes are ATP-generating organelles employing substrate-level phosphorylation and protons as an electron sink. They represent a rather heterogenous group of organelles described in organisms of considerably distant phylogenetic positions. Hydrogenosomes were found in trichomonads and other parabasalids (Čerkasovová et al. 1973; Lindmark et al. 1975; Lindmark and Müller 1973), chytrid fungi (*Neocalimastix* sp.) (Yarlett et al. 1986a), amoeboflagellates (*Psalteriomonas lanterna*) (Broers 1992), and some free-living (Embley et al. 1995; Fenchel and Finlay 1994) as well as symbiotic (Paul et al. 1990; Snyers et al. 1982; Yarlett et al. 1981, 1983) ciliates. These organisms belong to separate lineages of protists, some of them forming monophyletic groups with organisms harboring typical mitochondria (Fig. 6.1). Therefore, the hydrogen-producing organelles have been derived repeatedly in various unicellular eukaryotes (Embley et al. 1995; Yarlett and Hackstein 2005).

The biogenesis and core metabolism of hydrogenosomes have been studied mainly in trichomonads and fungi, while information about the hydrogenosomes in other organisms is scanty (Dyall and Johnson 2000; Hackstein et al. 1999; Müller 2003). The main similarities between hydrogenosomes and mitochondria are (1) the presence of a double membrane surrounding the organelles, (2) the generation of ATP, (3) catalytic components of respiratory complex I (NADH dehydrogenase), (4) calcium storage, (5) mechanism of organelle division, (6) mode of protein targeting and maturation, and (7) mechanism of FeS cluster assembly. The main differences are (1) the absence of a hydrogenosomal genome with the exception of hydrogenosomes in some ciliates, (2) the absence of oxidative phosphorylation, citric acid cycle, and β-oxidation of fatty acids, and (3) the production of hydrogen under anaerobic conditions.

On the basis of ultrastructural studies, comparative biochemistry, and molecular phologeny, it is now generally accepted that hydrogenosomes of ciliates and fungi evolved from aerobic mitochondria, and thus represent mitochondrial adaptation to oxygen-poor environments (Hackstein et al. 1999; Yarlett and Hackstein 2005). A rudimentary genome in hydrogenosomes of *Nyctotherus ovalis* coding for several components of a mitochondrial-type electron transport chain strongly suggests that these organelles represent a very recent adaptation (Akhmanova et al. 1998; Boxma

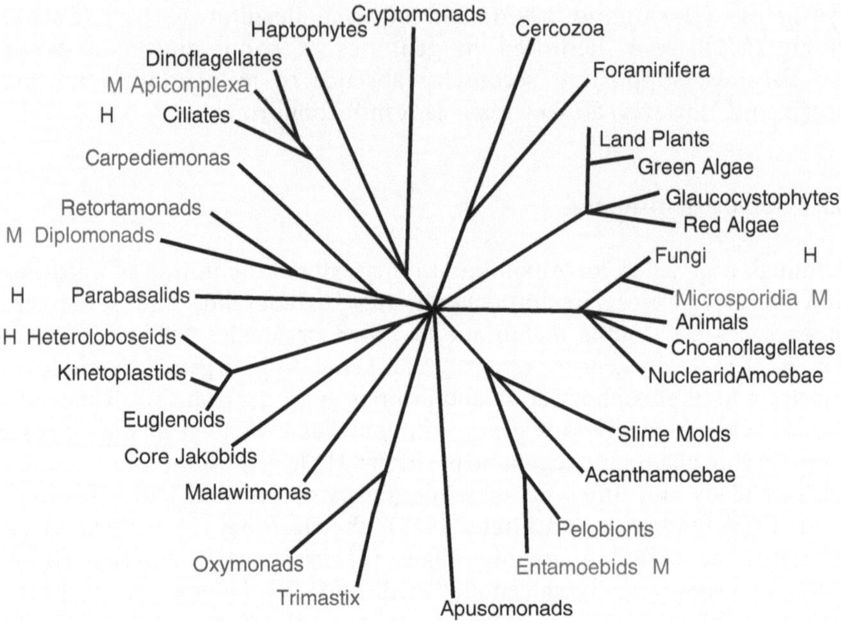


**Fig. 6.1.** Distribution of hydrogenosomes (*H*) and mitosomes (*M*) in eukaryotes. The organisms harboring putative hydrogenosome- or mitosome-like organelles which have not been characterized are in *green*. (The eukaryotic tree was adapted according to Dacks et al. 2003)

et al. 2005). The origin of the trichomonad hydrogenosome is less apparent and has been a matter of debate since their discovery in 1973 (Čerkasovová et al. 1973; Lindmark and Müller 1973). The most frequently considered scenario is that they evolved together with mitochondria from a common eubacterial ancestor (Dyall et al. 2004a; Dyall and Johnson 2000; Embley et al. 2003a, 2003b; Yarlett and Hackstein 2005)

### 6.2.2 Mitosomes

Mitosomes (synonym crypton) is the name given to the double membrane bounded organelles considered to be highly reduced mitochondria. Unlike mitochondria and hydrogenosomes, mitosomes are not involved in ATP synthesis. The mitosomes were first recognized in *Entamoeba histolytica* as the organelles possessing a mitochondrial marker Cpn60 (Mai et al. 1999; Tovar et al. 1999). Although mostly one mitosome per amoeba cell was originally observed, over 200 mitosomes per cell were reported in a later study (Leon-Avila and Tovar 2004). In addition to Cpn60, the entamoebid mitosomes likely possess Hsp10 (van der Giezen et al. 2005) and a single ADP/ATP carrier (Chan et al. 2005). The function of these organelles, however, remains enigmatic. On the basis of genome analysis, the presence of mitosomes was also predicted in the microsporidian *Encephalitozoon cuniculi* (Katinka et al. 2001), and was then demonstrated experimentally in another microsporidian *Trachipleistophora hominis* (Williams et al. 2002) as an organelle in which a Hsp70 protein homologue was localized. Organelles of similar characteristics have been found in *Cryptosporidium parvum*, originally described as remnant mitochondria (Keithly et al. 2005; Riordan et al. 1999; Šlapeta and Keithly 2004), and finally in *Giardia intestinalis* (Tovar et al. 2003). Like the hydrogenosomes, the mitosomes appear to have lost their own genomes, and genes coding for mitosomal proteins were transferred into the nucleus. The general mode of the protein synthesis, targeting, translocation, and processing seems to operate on a similar molecular basis in mitochondria, hydrogenosomes, and mitosomes (Doležal et al. 2005; Regoes et al. 2005).

## 6.3 Iron–Sulfur Cluster, an Ancient Indispensable Prosthetic Group

FeS clusters represent one of the chemically simplest and functionally diverse prosthetic groups of proteins. According to the chemoautotropic theory of the origin of life, FeS clusters may represent the most ancient catalysts mediating the abiological formation of C–C bonds (Huber and Wächtershäuser 1997). While the early anaerobic and high-temperature conditions on Earth changed dramatically, FeS clusters remained ubiquitous cofactors of a number of reactions that occur nowadays (Johnson et al. 2005).

The most common FeS clusters found in nature are the rhombic [2Fe2S] and the cubane [4Fe4S] clusters. More complex clusters have been described and include the so-called P-cluster and the FeMo cofactor of nitrogenase and the H-cluster of hydrogenase, which consists of a [4Fe4S] cluster linked to another [2Fe2S] cluster. The iron ions are usually coordinated in proteins by four cysteinyl ligands, although they can be replaced by histinidyl and CO/CN ligands in Rieske protein and hydrogenase, respectively. The particular character of the ligands and the protein domains involved in the cluster coordination determine the redox potential, which ranges between –700 and 400 mV. Such versatility allows FeS proteins to serve in a wide variety of processes: as electron carriers, enzymes, structural components, and sensors of oxygen or cellular iron levels (Johnson et al. 2005).

## 6.4 Iron–Sulfur Proteins in Mitochondria and Other Cell Compartments

The FeS proteins of eukaryotes were recently reviewed (Lill and Mühlenhoff 2005; Rouault and Tong 2005). Typically, the majority of cellular FeS proteins are present in mitochondria and participate in pathways leading to ATP production by oxidative phosphorylation. These include about eight FeS protein components of NADH dehydrogenase (complex I), succinate dehydrogenase (complex II), the Rieske protein of ubiquinone cytochrome *c* oxidoreductase (complex III) of the respiratory chain, and the citric acid cycle component aconitase. Mitochondria also contain [2Fe2S] ferredoxin (also called andrenodoxin), which is involved in biosynthesis of steroids, heme and FeS cluster formation, ferrochelatase, which is involved in heme synthesis, and biotin synthase. An increasing number of FeS proteins are also found in the cytosol. They are involved in amino acid metabolism (glutamate dehydrogenase, isopropylmalate isomerase, sulfite reductase), regulation of iron homeostasis (cytosolic aconitase homologue iron-regulatory proteins-1, IRP-1), ribosome formation and transcription regulation (ABC protein Rli1) (Dong et al. 2004; Kispal et al. 2005; Yarunin et al. 2005), and FeS cluster assembly (P-loop NTPase Cfd1p/Nbp35, hydrogenase-like protein Narf/Nar1) (Balk et al. 2005; Hausmann et al. 2005). Finally, FeS clusters are present in nuclear enzymes such as DNA glyosylase involved in DNA repair (McGoldrick et al. 1995). In photosynthetic eukaryotes, FeS proteins are present in plastids to sustain function of the photosynthetic electron transport chain (Kapazoglou et al. 2000). [2Fe2S] ferredoxin has been found also in apicoplast, a remnant nonphotosynthetic plastid of apicomplexan parasitic protists such as *Plasmodium falciparum* (Vollmer et al. 2001).

## 6.5 Iron–Sulfur Proteins in Organisms Harboring Hydrogenosomes and Mitosomes

The common feature of organisms with hydrogenosomes or mitosomes is that they inhabit oxygen-poor environments. Their energy metabolism is fermentative, producing pyruvate via a classic Embden–Mayerhof pathway; however, the further metabolism of pyruvate, a key intermediate product, which is linked to ATP production by substrate-level phosphorylation only, is significantly different. Consequently, a different set of FeS proteins is involved in this process. These proteins are compartmentalized either into hydrogenosemes (*Trichomonas*, *Neocalimastix*) or into the cytosol (*Entamoeba*, *Giardia*, *Cryptosporidium*, microsporidia) (Hackstein et al. 1999; Katinka et al. 2001; Müller 1988, 2003; Xu et al. 2004; Table 6.1).

In trichomonad hydrogenosomes, pyruvate is oxidatively decarboxylated by [4Fe4S] pyruvate:ferredoxin oxidoreductase (PFO). Released electrons are transferred via [2Fe2S] ferredoxin to [Fe]-hydrogenase, producing $H_2$. In contrast, in fungal hydrogenosomes, pyruvate is cleaved by pyruvate: formate lyase, in which ferredoxin is not involved. Malate is another hydrogenosomal substrate, which is oxidatively decarboxylated to pyruvate by NAD(P)-dependent malic enzyme. In trichomonads, the transfer of electrons released during this reaction from NADH to ferredoxin can be monitored as NADH:ferredoxin or NADH: methylviologen oxidoreductase

**Table 6.1.** Distrubution of FeS proteins in organisms with hydrogenosomes or mitosomes

*PFO* puruvate: ferredoxin oxidoreductase, *HCP* hybrid cluster protein, *PNO*

| | Trichomonas vaginalis | Giardia intestinalis | Entamoeba histolytica | Cryptosporidium parvum | Encephalitozoon cuniculi |
|---|---|---|---|---|---|
| Hydrogenosomes/ mitosomes | PFO | | | | |
| | Hydrogenase | | | | |
| | [2Fe-2S] ferredoxin | [2Fe-2S] ferredoxin | | [2Fe-2S] ferredoxin | [2Fe-2S] ferredoxin |
| | Tvh47(complex I) | | | | |
| | Tvh22(complex I) | | | | |
| | IscU | IscU | | IscU | IscU |
| | IscA | IscA | | | |
| | FeS flavoprotein | | | | |
| Cytosol | | PFO | FeS flavoprotein | PNO | |
| | | Hydrogenase? | Hydrogenase? | | |
| | | [4Fe-4S] ferredoxin | [4Fe-4S] ferredoxin | | |
| | | [3Fe-4S] ferredoxin | [3Fe-4S] ferredoxin | | |
| | | | NifU | | |
| | HCP | HCP | | | |
| | Rli | Rli | Rli | Rli | Rli |
| | Nbp35 | Nbp35 | Nbp35 | Nbp35 | Nbp35 |
| | Nar/Narf | Nar/Narf | Nar/Narf | Nar/Narf | Nar/Narf |
| Nucleus | DNA glycosylase | DNA glycosylase | DNA glycosylase | | DNA glycosylase |

*PFO* puruvate:ferredoxin oxidoreductase, *HCP* hybrid cluster protein, *PNO* pyruvate:NADH oxidoreductase, *Nbp35* P-loop NTPase, *Nar/Narf* hydrogenase-like protein, *Rli* ATP-binding cassette protein activity (Steinbühel and Müller 1986; Thong and Coombs 1987). Recently it was shown that this step of electron transport is catalyzed by heterodimeric NADH dehydrogenase with homology to the promotory module of mitochondrial respiratory complex I (Hrdý et al. 2004). In mitochondria, NADH dehydrogenase is a membrane-bound multi-sub-unit protein complex, which is composed of up to 43 subunits (Yano 2002). It catalyzes transfer of electrons from NADH to the lipid-soluble electron carrier ubiquinone, which is coupled to the proton translocation. It is organized into two parts: a hydrophilic (promotory) part, containing most of the redox cofactors, and a hydrophobic (proton-pumping) part that is anchored in the membrane. The promotory part contains most of the FeS proteins. The 51-kDa subunit binds NADH and contains one flavin mononucleotide and one [4Fe4S] cluster as the redox cofactors. The 24-kDa subunit is probably involved in NADH binding and possesses one [2Fe2S] cluster. These two subunits were isolated from *Trichomonas vaginalis* hydrogenosomes as a heterodimer (Tvh47 and Tvh24) and their ability to transfer electrons from NADH to [2Fe2S] ferredoxin was demonstrated (Hrdý et al. 2004). In addition to the 51- and 24-kDa subunits, the gene coding for the 75-kDa subunit of complex I was found in the genome of *N. ovalis* (Boxma et al. 2005). Neither ubiqinone nor rodoquinone has been detected in hydrogenosomes (Dyall et al. 2004b); thus, ferredoxin is likely the principal electron acceptor for the hydrogenosomal NADH dehydrogenase. No other FeS proteins of the promotory part were found in the complete genome of *T. vaginalis*; however, both *T. vaginalis* and *N. ovalis* contain genes coding for some subunits of the hydrophobic part (Boxma et al. 2005; Tachezy, unpublished results). Interestingly, the *N. ovalis* genome also contains two genes encoding mitochondrial complex II subunits including the FeS protein SDH-β (Boxma et al. 2005).

Hydrogenase is the canonical enzyme of the hydrogenosomes. Although hydrogenosomes have evolved several times in different eukaryotic lineages, they all contain hydrogenases, which belong to the iron-only [Fe] hydrogenase class (Horner et al. 2000). These hydrogenases are characterized by the presence of an oxygen-sensitive catalytic site, named the H-cluster. It consists of a [4Fe4S] cluster bridged by a cysteinyl residue to a binuclear [2Fe] center. The H-cluster, which is located in the interior of the proteins, is linked to additional [2Fe2S] and [4Fe4S] clusters, which are required for electron transport from soluble mediators (Vignais et al. 2001). The presence of a phylogenetically distinct hydrogenase of the iron–nickel–selenium class was reported only in hydrogenosomes of the anaerobic fungus *Neocallimastix frontalis* L2 (Marvin Sikkema et al. 1993); however, the identity of the putative [NiFeSe]-hydrogenase was not confirmed by microsequencing and attempts to provide evidence for the presence of corresponding genes in chytrids were not successful (Voncken et al. 2002). In contrast, [Fe]-hydrogenase was later identified in *N. frontalis* L2 hydrogenosomes by two independent groups (Davidson et al. 2002; Voncken et al. 2002).

In *T. vaginalis*, there are at least four types of [Fe] hydrogenases, which possess different arrangements of N-terminal electron-transporting clusters and other functional domains (Fig. 6.2):

1. The 50-kDa [Fe] hydrogenase contains two electron-transporting [4Fe4S] clusters preceding the H-cluster. This "short" [Fe] hydrogenase has also been found in some organisms that do not harbor hydrogenosomes, including the parasitic protists *Giardia*, *Spironucleus* (Horner et al. 2000), and *Entamoeba* (Nixon et al. 2003), and hydrogen photoproducing green algae (Vignais et al. 2001).
2. Genes coding a putative hydrogenase with an additional N-terminal [2Fe2S] cluster of the ferredoxin-type were revealed by sequencing of the *T. vaginalis* genome.
3. The "long" 64-kDa [Fe] hydrogenase was partially purified from *T. vaginalis* by Payne et al. (1993) and the corresponding gene was later isolated and analysed by Horner et al. (2000). This metalloenzyme possesses an N-terminal [2Fe2S] cluster followed by three [4Fe4S] clusters. It is

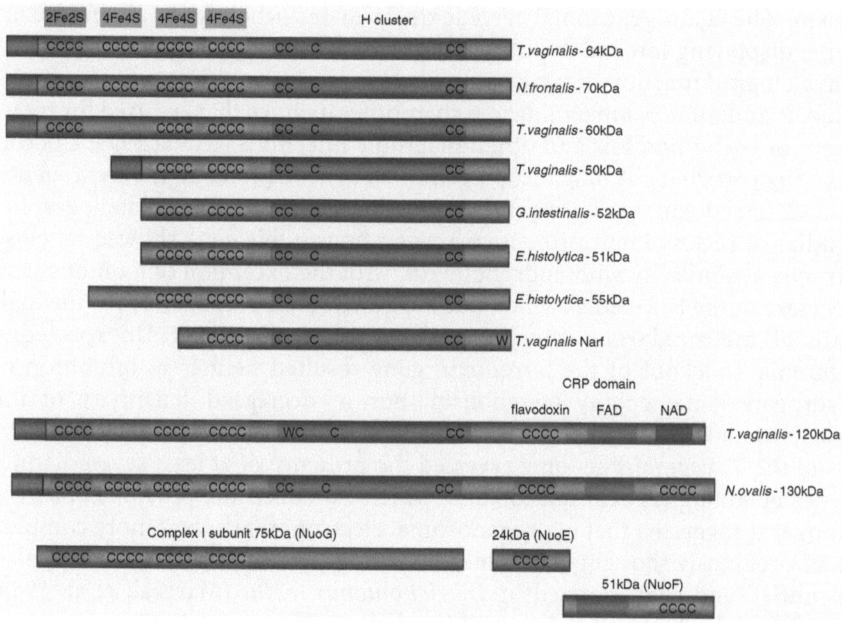


**Fig. 6.2.** [Fe] hydrogenases in protists harboring hydrogenosomes and mitosomes. Putative signals targeting the proteins into the hydrogenosomes are in *dark blue*. Conserved cysteines (*C*) involved in coordination of [2Fe2S], [4Fe4S], and the H-cluster are in *pink boxes*, *yellow boxes*, and *red boxes*, respectively. *Trichomonas vaginalis* fusion hydrogenase contains a carboxyl-terminal diflavin domain with similarities to the NADPH–cytochrome P450 oxidoreductase (*CRP*), while the fusion hydrogenase of *Nyctotherus ovalis* contain at its carboxyl terminus two domains with homology to the 24- and 51-kDa subunits of mitochondrial complex I. *W* typical tryptophan at the carboxyl terminus of *T. vaginalis* Narf homologue

noteworthy that the same arrangement of FeS clusters is present in the 75-kDa (NuoG) subunit of mitochondrial complex I. The evolutionary relationship between some components of complex I and hydrogenase has been suggested (Finel 1998; Pilkington et al. 1991).
4. Finally, *T. vaginalis* possesses genes encoding "fused" proteins which consist of an N-terminal long-type [Fe] hydrogenase domain and a C-terminal diflavin domain with similarities to the NADPH–cytochrome P450 oxidoreductase (CPR) domain of pyruvate:NADP oxidoreductase (PNO) of *Euglena gracilis* and *C. parvum* (Rotte et al. 2001). The "fused" hydrogenase in *T. vaginalis* is thus distinct from the hydrogenase found in *N. ovalis*, which contains a C-terminal domain with homology to the 51-kDa (NuoF) and the 24-kDa (NuoE) subunits of complex I (Akhmanova et al. 1998). It can be inferred that *T. vaginalis* "fused" hydrogenases may catalyze NAD(P)-dependent formation of hydrogen; however, their physiological role has yet to be established.

[2Fe2S] ferredoxin is a principal electron carrier in trichomonad hydrogenosomes that is required for both pyruvate-as well as malate-dependent catabolism. The hydrogenosomal ferredoxin is of mitochondrial (adrenodoxin) type, displaying low redox potential (Yarlett et al. 1986b). In addition to its physiological function, it is a key molecule for reducing activation of metronidazole and other 5-nitroimidazole chemotherapeutics that are used for treatment of trichomoniasis and other anaerobic infections (Kulda 1999; Upcroft and Upcroft 2001). A single copy gene was believed to code the hydrogenosomal ferredoxin in *T. vaginalis* (Johnson et al. 1990). Crystallographic studies of recombinant protein corresponding to this gene showed its close structural similarity with andrenodoxin, with the exception of a unique cavity close to the FeS center, which was suggested to be responsible for the high rate of metronidazole activation (Crossnoe et al. 2002). Unexpectedly, genomic knockout of the ferredoxin gene resulted neither in inhibition of hydrogenosomal energy metabolism, nor in decreased sensitivity of the trichomonads to metronidazole (Land et al. 2004). More recently, the analysis of the *T. vaginalis* genome revealed the presence of at least seven distinct genes encoding [2Fe2S] ferredoxins, which explained the previous observations and suggested that hydrogenosomal electron transport is more complex than previously thought. The genes encoding [2Fe2S] ferredoxins were also identified and characterized in *Tritrichomonas foetus* (Marczak et al. 1983; Suchan et al. 2003) and *P. lanterna* (Brul et al. 1994).

In mitosome-harboring organisms such as *G. intestinalis* and *E. histolytica*, the pyruvate-dependent ATP synthesis as well as transport of electrons generated during carbohydrate catabolism take place in the cytosol; hence, the FeS proteins PFO and ferredoxin are localized within this cellular compartment (Müller 1988, 2003). The cytosolic ferredoxins of *Giardia* and *Entamoeba* are of the bacterial type, containing either two cubane [4Fe4S] clusters or one [4Fe4S] and one [3Fe4S] cluster (Nixon et al. 2002a). A [2Fe2S] ferredoxin of the adrenodoxin type with an N-terminal targeting sequence was found in *Giardia* and *Cryptosporidium*, but not in *E. histolytica*. These proteins are most likely targeted to mitosomes in order to function as a component of the iron cluster assembly machinery.

Recently, [Fe] hydrogenases have been shown to be expressed in *Giardia* and *Entamoeba* (Lloyd and Harris 2002; Nixon et al. 2003). In *Giardia*, a single gene coding "short"-type hydrogenase was found to be transcribed (Nixon et al. 2003) and hydrogenase activity, as well as hydrogen production by the cell, was detected (Lloyd et al. 2002). *E. histolytica* possesses two different genes. One encodes a "short"-type hydrogenase, similar to that in trichomonads and *Giardia*, and the other encodes a putative 55-kDa hydrogenase, which appears to be more related to eubacterial [Fe] hydrogenases. Although a recombinant short-type hydrogenase of *E. histolytica* displays hydrogenase activity, this activity was not found in *Entamoeba* cells. The cellular localization of giardial and entamoebal hydrogenases has not yet been clarified. The presence of an N-terminal sequence similar to that required for organellar targeting has been suggested; however, no experimental evidence is available (Nixon et al. 2003).

An unusual and most likely cytosolic FeS enzyme of *C. parvum* is PNO, which consists of an N-terminal PFO domain followed by a C-terminal CPR (Rotte et al. 2001). Unlike the PNO of *E. gracilis*, which is targeted into mitochondria, the PNO of *Cryptosporidium* does not possess an N-terminal targeting extension and its presence in a "relict mitochondrion" is therefore unlikely.

A novel essential FeS protein found in the eukaryotic cytosol, is the ATP-binding cassette (ABC) protein Rli1p from *Saccharomyces cerevisiae* (Decottignies and Goffeau 1997). Unlike ABC transporters, this soluble protein is not a transporter and instead carries two N-terminal cysteine motifs predicted to coordinate two [4Fe4S] clusters. Although the exact role of these clusters is not known, Rli1p is essential for cell viability and is required for maturation and nuclear export of ribosome subunits and for efficient initiation of translation (Dong et al. 2004; Kispal et al. 2005; Yarunin et al. 2005). Homologues of Rli1p are present in all eukaryotes, including hydrogenosome- and mitosome-bearing protists, and in archaebacteria.

Cytosolic aconitase, also named iron-regulatory protein (IRP-1), plays a pivotal role in iron homeostasis of animal cells; yet, this protein has not been found in hydrogenosome- or mitosome-harboring organisms. Interestingly, unlike other eukaryotes, *T. vaginalis* possesses genes coding for the hybrid cluster protein (HCP) (originally called "prismane"), which coordinates a cubane [4Fe4S] center and a hybrid cluster comprising both O and S bridges between iron atoms (van den Berg et al. 2000). The protein is probably localized in the trichomonad cytosol (Tachezy, unpublished results). Previously, HCP was found in obligate and facultative anaerobic bacteria, for which its involvement in anaerobic nitrate and/or nitrite respiration was suggested. In eukaryotes, HCP-coding genes have only been identified in the genomes of anaerobic aerotolerant protists (*G. intestinalis*, *Spironucleus barkhanus* and *E. histolytica*), providing an example of lateral gene transfer from prokaryotes (Andersson et al. 2003; Han et al. 2004).

## 6.6 Iron–Sulfur Cluster Assembly Machineries

The study of FeS cluster formation is an emerging field, with FeS cluster assembly first characterized in the nitrogen-fixing bacterium *Azotobacter vinelandii* in 1992 (Kennedy et al. 1992). In bacteria there are at least three different systems involved in this process: (1) the NIF (nitrogen-fixing) system, which is primarily involved in the FeS cluster formation of nitrogenase (Kennedy et al. 1992); (2) the ISC (FeS cluster) assembly machinery, dedicated to forming FeS clusters in proteins involved in various "housekeeping" functions (Zheng et al. 1998); and (3) the SUF (sulfur-mobilization) system, which has been implicated in the maintenance and repair of FeS clusters under oxidative stress and iron-restricted conditions (Takahashi and Tokumoto 2002). In addition, the HYD system, required for maturation of the H-cluster of [Fe] hydrogenases has been found in bacteria as well as in the plastids of some green algae (Posewitz et al. 2004). In eukaryotes, FeS cluster assembly is a fundamental function of mitochondria. The disruption of FeS cluster assembly leads to mitochondrial iron overload and cell death. The machinery mediating this process is of the ISC type and evolutionary studies indicate that the eukaryotic ISC machinery was inherited from the proteobacterial ancestor, which is consistent with the endosymbiotic origin of mitochondria. Similarly, components of the SUF system have been found in plastids and apicoplasts, and these *suf* genes were most likely inherited from cyanobacteria, the plastid ancestors (Tachezy et al. 2001). Surprisingly, the function of mitochondrial FeS cluster machinery was shown to be indispensable not only for maturation of mitochondrial FeS proteins, but also for FeS cluster formation in the cytosol and nucleus. The central role of mitochondria in cellular FeS cluster assembly then led to the proposal that formation of FeS clusters represents the only essential function of mitochondria common to all eukaryotes (Lill and Mühlenhoff 2005).

### 6.6.1 Iron–Sulfur Cluster Assembly in *Saccharomyces cerevisiae*

The current model for FeS cluster formation is based mainly on studies of *S. cerevisiae*, for which methods of genetic manipulation are well established (reviewed in Lill and Mühlenhoff 2005) (Fig. 6.3). The central pair of components of Fe cluster formation consist of mitochondrial IscS (Nfs), a pyridoxal phosphate-dependent desulfurase, and IscU (Isu), which serves as a scaffold for the formation of a transient FeS cluster. Sulfur released by IscS from cysteine is transferred to IscU and combined with iron to form a labile FeS

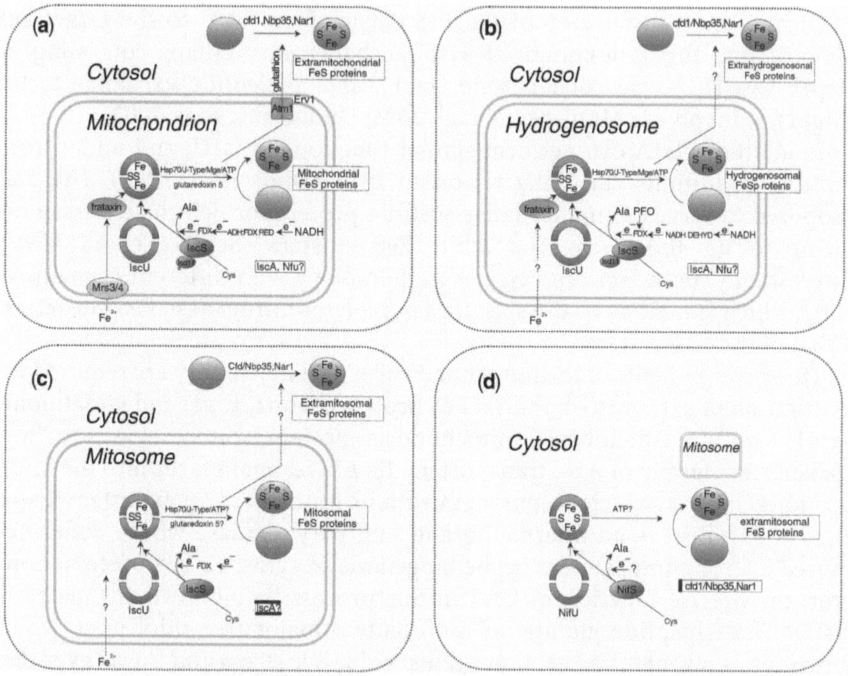


**Fig. 6.3.** Comparison of FeS cluster assembly machinery in *Saccharomyces cerevisiae* (**a**), *T. vaginalis* (**b**), *Giardia intestinalis* (**c**), and *Entamoeba histolytica* (**d**). *T. vaginalis* and *G. intestinalis* contain a complete set of the main organellar and cytosolic components required for biogenesis of FeS clusters observed in yeast. However, components of the mitochondrial export machinery (Atm1, Erv1, and glutathione) are not present in these anaerobic protists. Reducing power and ATP required for FeS cluster assembly are generated in trichomonad hydrogenosomes, while in *Giardia*, these factors are unlikely to be generated within mitosomes. Unlike other eukaryotes, *E. histolytica* possesses a bacterial NIF-like system in its cytosol. *FDX* ferredoxin, *NADH:FDX RED* NADH:ferredoxin reductase, *PFO* pyruvate:ferredoxin oxidoreductase, *NADH DEHYD NADH* dehydrogenase activity (complex I) that reduces ferredoxin in *T. vaginalis*

cluster, which is subsequently transferred from IscU to mitochondrial or extramitochondrial apoproteins (Smith et al. 2001). The candidate iron donor is frataxin, a mitochondrial protein whose impaired expression is associated with Friedriech ataxia in humans (Campuzano et al. 1996). The formation of FeS clusters requires reduction of sulfur $S^0$ which is provided by IscS to sulfide $S^{2-}$ present in the cluster (Lill and Mühlenhoff 2005). While direct evidence is still missing, two components, [2Fe2S] ferredoxin (Yah1) and ferredoxin reductase (Arh1), are proposed to provide the reducing power. The early step of FeS cluster formation also requires ISC biogenesis desulfurase interacting protein (Isd11), which seems to stabilize IscS in mitochondria (Adam et al. 2005), and glutaredoxin 5, which was suggested to regulate the redox state of important cysteine residues in proteins involved in ISC biogenesis (Alves et al. 2004).

During the transfer step of the FeS cluster from IscU to the target FeS apoproteins, the mitochondrial Hsp70 chaperone system, consisting of Hsp70 (Ssq1), J-type cochaperone (Jac1), and nucleotide exchange factor (Mge1), is involved (Mühlenhoff et al. 2003; Dutkiewicz et al. 2003).

In addition, IscA can perform similar functions to IscU, and under iron-limiting conditions can deliver iron to IscU (Ding et al. 2004). Nfu was proposed to serve as an alternative scaffold protein for FeS cluster assembly owing to its ability to bind labile FeS clusters (Schilke et al. 1999). Interestingly, both IscU and Nfu share homology with different domains of NifU, which functions in the specific assembly of nitrogenase FeS clusters as a part of the NIF system.

Three components of the mitochondrial export machinery are required for maturation of extramitochondrial FeS proteins: Atm1, Erv1, and glutathione. Atm1 is an inner-membrane mitochondrial protein, which belongs to "half transporter" family of ABC transporters. Its ABC domain faces into the mitochondrial matrix, which is consistent with its function as an exporter (Kispal et al. 1999). Erv1 is an intermembrane sulfhydryl oxidase whose deficiency caused a strong impairment in the biogenesis of cytosolic FeS protein; however, the precise function of Erv1 in this process is unknown (Lange et al. 2001). The tripeptide glutathion represents a major free thiol pool, which functions as a general protection against oxidative stress and it was hypothesized that transporting compounds may be chelated by glutathion, and thus stabilized during their transport from the mitochondria to the cytosol (Sipos et al. 2002). However, the nature of the compound exported from the mitochondria to the cytosol is not known.

Three cytosolic factors were recently characterized that contribute specifically to the maturation of cytosolic and nuclear FeS proteins. Cfd1 (Roy et al. 2003) and Nbp35 (Hausmann et al. 2005) are essential soluble P-loop ATPases. Except for a short N-terminal extension in Nbp35 that itself carries an FeS cluster, these two proteins are structurally very similar. Together with the third component, yeast Nar1 (or human Narf, nuclear prelamin A recognition factor), these proteins have dual nuclear and cytosolic localization. It is not clear what the molecular role these cytosolic factors play in the maturation of FeS proteins is. An attractive hypothesis is that the FeS clusters or their precursors are transferred into the target apoproteins with the assistance of cytosolic factors after their export from mitochondria. Alternatively, the cytosolic proteins may facilitate de novo synthesis of FeS clusters in the cytosol using some compounds which are produced by mitochondrial FeS proteins (Lill and Mühlenhoff 2005).

### 6.6.2 *Trichomonas vaginalis*

The identification of IscS homologues in the genome of *T. vaginalis* and *G. intestinalis* indicates that systems analogous to that operating in animals and fungi are conserved in these anaerobic organisms (Tachezy et al. 2001).

Phylogenetic analysis of eukaryotic and bacterial IscS genes showed the affinity of the protist homologues to the clade of eukaryotic mitochondrial enzymes, while the IscS of proteobacteria formed a sister group (Emelyanov 2003; Tachezy et al. 2001). This analysis thus suggested that *isc*-related genes were present in the proteobacterial ancestor of mitochondria and hydrogenosomes. Subsequently, the hydrogenosomal localization of IscS was demonstrated, as well as the ability of isolated hydrogenosomes to catalyze assembly and insertion of the FeS cluster into apoproteins (Šut'ák et al. 2004). In contrast to the key hydrogenosomal proteins involved in energy metabolism (PFO, hydrogenase, ferredoxin, malic enzyme), the transcription of *T. vaginalis iscS* is markedly upregulated under iron deficiency, which might reflect increased demand for the synthesis of new FeS clusters (Šut'ák et al. 2004). In addition to IscS, other components of the mitochondrial-type ISC assembly machinery can be identified in the *T. vaginalis* genome (Fig. 6.3). Although, the function of a single IscU and that of multiple IscA homologues have not been experimentally tested, all sequences contain putative hydrogenosomal targeting signals, suggesting their colocalization inside the organelle. Frataxin is another important component of FeS cluster biosynthesis which was recently found in hydrogenosomes (Doležal, unpublished results). Consistent with IscS, transcription of the frataxin gene is upregulated under iron deficiency, indicating functional connection between these FeS cluster assembly components. Interestingly, this is the opposite of what is found with the yeast mitochondrial frataxin (Yfh1), expression of which is strongly stimulated by iron (Santos et al. 2004). As frataxin expression in *T. vaginalis* is upregulated under conditions of iron deficiency, it is unlikely that this protein serves as an iron-storage molecule in hydrogenosomes as proposed for the mitochondrial homologue (Gakh et al. 2002). More likely, it donates iron for the formation of hydrogenosomal FeS clusters. This assumption is supported by the ability of *T. vaginalis* frataxin to replace mitochondrial frataxin in yeast, restoring defects in FeS cluster synthesis of the *Δyfh1* mutant (Doležal, unpublished results). Interestingly, recombinant *T. vaginalis* frataxin interacted with yeast ferrochelatase (Doležal, unpublished results), which completes heme synthesis via the insertion of iron into the protoporphyrin ring (Dailey et al. 2000). Additionally, expression of *T. vaginalis* frataxin restored heme synthesis in *Δyfh1* yeast mutants. This finding is unexpected, as neither heme-containing proteins nor components involved in heme synthesis have yet been identified in *T. vaginalis*.

Sequencing of the complete *T. vaginalis* genome revealed multiple copies of [2Fe2S] ferredoxin genes with the gene products predicted to localize to the hydrogenosomes. It is possible that some of these gene products serve specifically as electron donors for FeS cluster biogenesis, while others are required for electron transport associated with the energy metabolism of hydrogenosomes. The functional redundancy or specialization of these particular ferredoxins has yet to be established. Similarly, multiple copies of hydrogenosomal Hsp70 are present in *T. vaginalis*; however, their particular involvement in various functions, including FeS cluster assembly, is not known. The genes coding for homologues of cochaperons Jac1(DnaJ) and Mge (GrpE) as well as Isd11 are also present in the *T. vaginalis* genome.

Besides the formation of simple rhombic or cubane FeS clusters, *T. vaginalis* also needs to assemble more complex H-clusters, required for activity of the [Fe] hydrogenase in hydrogenosomes. As found in *Chlamydomonas reinhardtii*, [Fe] hydrogenases require specific accessory proteins, called HydE, HydF, and HydG, which are necessary for H-cluster maturation and probably for the mobilization of iron (Posewitz et al. 2004). Homologues of these proteins are exclusively distributed in [Fe] hydrogenase containing bacteria, and were recently also found in the hydrogenosomes of *T. vaginalis* (Pütz et al. 2006). Two independent machineries therefore might function side by side in *T. vaginalis* hydrogenosomes. While the mitochondrial-type ISC machinery might be responsible for general FeS cluster maturation, including [2Fe2S] and [4Fe4S] clusters at the N-terminal regions of hydrogenases, the HYD machinery specifically ensures the proper maturation of H hydrogenase clusters.

In *T. vaginalis,* no functional information on the assembly of extrahydrogenosomal FeS cluster is available to date; however, searches in the *T. vaginalis* genome database for homologues involved in extramitochondrial FeS cluster assembly revealed the presence of putative Narf and Nbp35/Cfd1 homologues. It is noteworthy that candidates for two known membrane components, Erv1 and the ABC transporter Atm1p, were not found. Although at least 20 sequences related to Atm1 were found in the *T. vaginalis* genome, none of them possess a putative N-terminal targeting presequence and a membrane-spanning region (Lill, personal communication). Thus, whether hydrogenosomes are involved in the biogenesis of cytosolic FeS proteins is still a challenging question.

### 6.6.3 *Giardia intestinalis*

The first conclusions from the identification and phylogenetic reconstruction of IscS in *G. intestinalis* were analogous to the situation of the mitochondrial-type Hsp60 found in this protist (Roger et al. 1998). The *G. intestinalis* IscS is closely related to mitochondrial homologues (Emelyanov 2003; Tachezy et al. 2001), but there is no putative targeting signal at the N-terminus of the protein which would argue for its mitochondrial localization. Because such a targeting signal is typical for all other eukaryotic IscS proteins, secondary loss of the mitochondrion in *G. intestinalis* was suggested (Tachezy et al. 2001). However, a putative N-terminal targeting sequence was later predicted in the *G. intestinalis* IscU homologue (Tovar et al. 2003). Cellular localization of IscU and IscS showed their common distribution in tiny discrete vesicles limited by two membranes. Afterwards, it was demonstrated that while IscU contains a cleavable targeting signal, IscS utilizes multiple internal targeting signals that ensure its proper localization (Doležal et al. 2005). Both proteins were present in a sedimentable fraction of *G. intestinalis* lysate, which was shown to catalyze the formation of new FeS clusters in vitro (Tovar et al. 2003). Taken together, these data suggest that *G. intestinalis* not only retained proteins of mitochondrial origin but still possesses mitochondrion-related compartments called mitosomes, to which the FeS cluster assembly machinery is localized (Tovar et al. 2003). It was shown recently that [2Fe-2S] ferredoxin also contains an N-terminal targeting signal, which facilitates targeting and translocation of ferredoxin into the mitosomes (Doležal et al. 2005; Regoes et al. 2005). In addition, a gene homologous to IscA with a mitochondrial targeting signal predicted in its product was identified in the *G. intestinalis* genome and glutaredoxin 5 was found in the proteome of *Giardia* mitosome-rich fraction (Tachezy, unpublished results). Thus, it is likely that at least five components, IscS, IscU, IscA, ferredoxin, and glutaredoxin, assist FeS cluster synthesis within the same mitosomal compartment, thus displaying general similarity to the mitochondrial scheme. However, some of the factors mediating important steps in mitochondrial FeS cluster synthesis were not found in *G. intestinalis*. Genome-mining did not reveal any giardial frataxin homologue nor any other possible iron donor. Moreover, a ferredoxin-specific reductase with mitosomal localization is also missing. The most intriguing question is what is the source of ATP and reducing power, which are required for FeS cluster assembly, as no ATP-generating pathway that would supply the ISC assembly machinery with energy equivalents has been found in the mitosomes thus far. Whether this is because of our incomplete information on mitosomal metabolism or whether these commodities are specifically transported into the organelles from the cytosol remains to be elucidated.

### 6.6.4 *Cryptosporidium parvum*

Limited experimental data are available concerning FeS cluster formation in *C. parvum*; however, since it is affiliated with the mitochondriate group of Apicomplexa, it is highly probable that *C. parvum* possesses a functional set of components of mitochondrial FeS cluster assembly. Indeed, localization of homologues of mitochondrial heat shock proteins Hsp70 and Hsp60 demonstrated the presence of a relic mitochondrial compartment in *C. parvum* (Riordan et al. 2003; Putignani et al. 2004; Šlapeta and Keithly 2004).

Moreover, IscS and IscU homologues found in the *C. parvum* genome were shown to contain targeting signals efficient for translocation of proteins into yeast mitochondria, indicating that the mitochondrial relict is the site of FeS cluster assembly (LaGier et al. 2003). Several other components, such as frataxin, [2Fe2S] ferredoxin (LaGier et al. 2003), and Narf (Stejskal et al. 2003) were shown to be expressed in *C. parvum*. The genome of *C. parvum* also contains a homologue of Nbp35.

### 6.6.5 Microsporidia

*E. cuniculi* belongs to the group of Microsporidia, unicellular eukaryotes which were recently affiliated to Fungi. The microsporidial lifestyle as obligatory intracellular parasites is proposed as the reason for the dramatic overall reduction of cellular metabolism, including the mitochondrial compartment. For this reason Microsporidia were for some time considered amitochondriate eukaryotes until a homologue of mitochondrial Hsp70 was localized in the double membrane bounded mitosomes of the human parasite *T. hominis* (Williams et al. 2002). So far, this is the only protein known to be expressed within the microsporidial mitosome and other information comes solely from the bioinformatic analysis of the *E. cuniculi* genome (Katinka et al. 2001). The genes coding for homologues of IscU, IscS, ferredoxin, frataxin, and also for Atm1 and Erv1 were identified, suggesting that the functional pathway is present in *E. cuniculi.* Interestingly, none of these conceptual proteins contain a mitochondrial targeting sequence, which would argue for their mitosomal localization. However, since neither mitosome-localized Hsp70 isoform contains such a targeting sequence this may rather reflect the peculiarity of the microsporidial protein targeting system, rather than indicate the extramitochondrial localization of FeS cluster formation.

Similarly to the situation in *G. intestinalis*, the ATP production pathway is predicted to be absent in the mitochondrial remnants of *E. cuniculi* and *C. parvum* (see before), and FeS cluster formation is currently the only known pathway of mitochondrial origin which likely functions in these organelles.

### 6.6.6 *Entamoeba histolytica*

Unlike other eukaryotes, the identification of the components involved in FeS cluster assembly in *E. histolytica* did not indicate a role for the amoeba mitosome in this pathway. Instead, the structure and the origin of the *E. histolytica* FeS cluster assembly appears to be unique amongst all eukaryotes. A genome-wide search revealed that instead of the ISC system, *E. histolytica* possesses two components of a nonredundant NIF system (Ali et al. 2004). As deduced from the phylogenetic reconstruction of the *E. histolytica* NifS and NifU genes, both were apparently acquired by horizontal gene transfer from the group of ε-proteobacteria, with the closest relationship to genes from the *Campylobacter* and *Helicobacter* species (Putignani et al. 2004). These bacteria are incapable of nitrogen fixation but were shown to contain the NIF system, as their sole mechanism for FeS assembly. Therefore, in the absence of both ISC and SUF systems it was suggested that, in this case, the NIF system is able to mediate the general formation of all type of FeS clusters (Tokumoto et al. 2004). Analogously, the absence of the ISC or the SUF system postulates an exclusive role for the NIF machinery in *E. histolytica.* While the structure of NifS is very similar to that of IscS in mitochondria and bacteria, NifU is a modular protein with which IscU shares similarity only at the N-terminal third that binds the labile FeS cluster. The central part of NifU binds a stable [2Fe2S] cluster, in a manner analogous to mitochondrial [2Fe2S] ferredoxin, and the C-terminal part shares homology with a mitochondrial Nfu protein whose role in FeS cluster assembly was shown but not specified in detail (Schilke et al. 1999). When expressed in a strain of *Escherichia coli* with lethal absence of both ISC and SUF systems, NifU and NifS of *E. histolytica* were able to complement the growth of the mutant bacteria, though only under anaerobic conditions (Ali et al. 2004). This result indicated the ability of the NIF components to catalyze general FeS cluster assembly, but likely with the inability to overcome the deleterious effects of iron exposed to oxygen. Moreover, analysis of both proteins did not reveal the presence of a putative mitochondrial targeting signal that would predict their localization within the mitosomal compartment and instead suggested a cytosolic localization (van der Giezen et al. 2004). Such a localization, if verified experimentally, raises important questions: How does *E. histolytica* keep the synthesis of FeS clusters nontoxic for other cellular processes? And if the NifS and NifU components were acquired by horizontal gene transfer from an ε-proteobacterium, what was the preceding arrangement of the FeS cluster assembly machinery before this event, and what was the selective advantage of replacing the original mechanism with a heterologous NIF system?

Interestingly, NIF components are found in other species of the *Entamoeba* genus, indicating that this is not a unique deviation in the intestinal human parasite *E. histolytica* but a common pattern for related amoebas. In addition, a more distant relative from the group Conosa, the aerobic organism *Dictyostelium discoideum*, also contains homologues of the mitochondrial ISC system. Investigation of another anaerobic representative of Conosa, *Mastigamoeba balamuthi*, may provide more details on the distribution and the history of this eukaryotic curiosity.

## 6.7 Iron–Sulfur Cluster Biosynthesis and the Evolution of Mitochondria

Eukaryotic cells have several internal compartments, each of which has a discrete metabolic function. There are different evolutionary scenarios attempting to explain the history of these organelles and the selection pressure that propelled their genesis. To further this aim, an important point is to define the essential functions of the organelles conserved in organisms amongst all branches of the eukaryotic tree.

Mitochondria were derived some two billion years ago from endosymbiotic bacteria, most likely of α-proteobacterial origin (Gray et al. 1999; Kurland and Andersson 2000). Modern mitochondria have evolved into a compartment harboring dozens of metabolic pathways. Up to 1,000 distinct yeast proteins are involved in the many functions of mitochodria, for instance, mitochondrial genome maintenance, energy metabolism, protein translocation, and lipid metabolism (Sickmann et al. 2003). Interestingly, while energy metabolism, a typical mitochondrial function, involves only 14% of the proteins identified, 25% of the proteins still have unknown function. Almost half of the yeast mitochondrial proteins are eukaryote-specific, created during the evolution of the eukaryotic cell, for which corresponding bacterial homologues are not found. In relation to their representation in the mitochondrial proteome, iron metabolism (including synthesis of FeS clusters) represents one of the marginal metabolic pathways. Up to 11 proteins mediate formation of FeS clusters for the target apoproteins that fulfill a variety of functions in various cellular compartments. These mitochondrial proteins are highly conserved throughout Eukarya in spite of various functional modifications of mitochondria in different organisms and their independent phylogenetic position.

Recent investigations of "amitochondrial" organisms focused on phylogenetic analysis of components of their FeS cluster assembly machineries. Studies of their intracellular localization, as well as the physiological function, provided strong evidence that "amitochondriates" in fact possess a mitochondrion-related organelle. Namely, it revealed that a mitochondrial type of FeS cluster assembly machinery operates in trichomonad hydrogenosomes (Šut'ák et al. 2004; Tachezy et al. 2001), and it led to the discovery of highly reduced mitochondria (mitosomes) in *Giardia*, the organism considered in earlier studies among descendants of primitive eukaryotes that diverged prior to the endosymbiotic origin of mitochondria (Doležal et al. 2005; Tachezy et al. 2001; Tovar et al. 2003). So far, FeS cluster assembly is the only known function of the *Giardia* mitosome and it most likely operates also in mitosomes of microsporidia (Katinka et al. 2001) and *Cryptosporidium* (LaGier et al. 2003). The only exception is represented by *Entamoeba* mitosomes, which appear not to harbor any FeS cluster assembly machinery and the function of these mitosomes is unclear. It was suggested previously that the essential role of FeS assembly machinery for cells might be the evolutionary reason why highly reduced organelles such as mitosomes are retained in some eukaryotes. The advantage in preserving such an organelle might be compartmentalization of toxic ferrous ions and sulfide, which are required for FeS cluster assembly. In addition, anaerobic or microaerophilic protists have a considerably higher nutritional requirement for iron than aerobes, although they do not employ ferritin to store iron in nontoxic form. Interestingly, hydrogenosomes contain about a 10–100-fold higher concentration of iron than mitochondria (Suchan et al. 2003). Thus, the compartmentalization of iron metabolism could be the essential step for these organisms.

The origin of eukaryotic cells and particularly mitochondria is an old but still hotly debated topic. The endosymbiotic model proposed the acquisition of a proteobacterial endosymbiont by a primitive amitochondrial eukaryotic host (Yang et al. 1985). In this view, protists such as trichomonads and *Giardia*, which lack readily distinguishable mitochondria, could represent descendants of premitochondrial eukaryotes. In contrast, the hydrogen/syntrophic hypotheses propose that the appearance of the endosymbiotic organelle was directly associated with eukaryogenesis, assuming a common origin of mitochondria and hydrogenosomes (Lopez-Garcia and Moreira 1999; Martin and Müller 1998). Studies of FeS cluster assembly machinery have contributed significantly to the view that mitochondria, hydrogenosomes, and mitosomes are different forms of the same fundamental organelle. Hydrogenosomes and mitosomes are found predominantly in parasitic protists, which are adapted for life in specific niches in oxygen-poor environments of their hosts. However, hydrogenosomes were also observed in the free-living amoeboflagellate *P. lanterna* (Broers 1992) and structures reminiscent of such organelles were observed in other free-living organisms, such as flagellate *Postgaardi mariagerensis* (Simpson et al. 1997), pelobiont *M. balamuthi* (our unpublished results), and the euglenozoan *Carpediomonas membranifera* (Simpson and Patterson 1999). The lack of unequivocally amitochondriate organisms strengthens the hydrogen/syntrophic hypotheses, although we cannot rule out that the primitive genuine amitochondriate eukaryotes (archezoa) did not survive to the current day.

Another major difference between the serial endosymbiotic and hydrogen/syntrophic hypotheses is the view regarding the selection pressure that confined the endosymbiont within the host. The serial endosymbiotic model proposes that the endosymbiont (organelle) served as an oxygen scavenger, thereby providing protection to the anaerobic host. The hydrogen hypothesis was based on a symbiosis between a methanogenic archaebacterium (the host) and a facultatively anaerobic proteobacterium (the endosymbiont), which provided its metabolic waste products such as hydrogen, carbon dioxide, and acetate for the host. Interestingly, Nar/Narf proteins are homologues of [Fe] hydrogenases (Balk et al. 2005; Horner et al. 2000, 2002), lacking the complete set of cysteines required to coordinate the H-cluster of hydrogenase and with a typical tryptophan residue at the end of the protein (Fig. 6.2). These proteins are highly conserved in eukaryotes from trichomonads to humans. The common ancestry of Nar-/Narf-like proteins and [Fe] hydrogenase, together with their retention, provides further support for the hydrogen/syntrophic hypotheses.

The essential character of the FeS cluster assembly machinery and its conservation in all eukaryotes suggests that FeS cluster assembly might have been among the original and essential functions of the ancestral endosymbiont, which gave rise to mitochondria and related organelles. If so, eubacterial FeS cluster assembly should provide a functional advantage in comparison with the archaebacterial machinery. At the same time, the archaebacterial host should have been dependent on the function of the target FeS proteins, making the FeS cluster assembly machinery essential. Interestingly, eubacteria possess two independent systems involved in the general maturation of FeS cluster assembly, ISC and SUF, with partially overlapping roles. While inactivation of *isc* genes causes severe effects on FeS cluster assembly in *E. coli*, disruption of the *suf* operon does not cause any major defect. However, synthetic lethality was found when both the *isc* and *suf* operones were disrupted. Notably, the SUF machinery appears to be the exclusive [FeS] cluster biosynthetic system available for some archeabacteria, including methanogenic species. If we presuppose that the ancient archebacterial host possessed only a simple SUF machinery, the acquisition of an endosymbiont with an ISC system (1) would allow the host compartmentalization of FeS cluster biogenesis, avoiding the toxicity of reduced iron and sulfur, (2) could provide a backup system for this vital function under specific conditions, and (3) hypothetically would increase both the ability to synthesize a wider and more complex variety of FeS clusters, as well as the efficiency of the cellular FeS cluster assembly itself. To test this hypothesis, further comparative studies of the ISC and SUF systems in various organisms are required.

The second question of interest is: What are the key FeS proteins and their cellular tasks for which the function of mitochondrial FeS cluster assembly is indispensable? In addition to the obvious importance of FeS proteins in intermediary metabolism, a recently characterized cytosolic FeS protein Rli1 connects the biogenesis of FeS clusters with two fundamental cellular functions: maturation of ribosome subunits, and the initiation of translation (Kispal et al. 2005; Yarunin et al. 2005). Rli1 is highly conserved in all eukaryotes, including those bearing hydrogenosomes and mitosomes. Interestingly, Rli1 homologues were found in archeabacteria, but were not identified in eubacteria, indicating that the eukaryotic Rli homologue is of archaebacterial origin as are other components of cytosolic translation machinery. Thus, it seems that in contemporary eukaryotes, the eubacterial type of FeS cluster assembly machinery, operating in mitochondria, hydrogenosomes, or mitosomes, is required for biogenesis of the archaebacterial Rli1 protein, thus providing a functional link between an ancient key cellular task, FeS proteins, and biogenesis of FeS clusters. Conversely, during the same period of evolution the *suf* genes were possibly lost, as no genes coding for SUF homologues have been found in eukaryotic genomes so far, with the exception of those containing plastids.

The discovery of FeS cluster assembly in *A. vinelandii* was a remarkable achievement accomplished in the late 1980s by Jacobson et al. (1989) with broad impact on cell biology. Thanks to progress in the analysis of genomes and genomics, in a few years, homologues of bacterial FeS cluster assembly components were found in all eukaryotic organisms, including humans. An essential function of FeS cluster assembly for eukaryotic mitochondria has been established, and provided new insights into diseases associated with impaired FeS cluster biogenesis and mitochondrial iron overload.

High conservation of FeS cluster assembly machinery in prokaryotes, as well as in eukaryotes, reflects its ancient and indispensable role and provides a powerful tool for tracing endosymbiotic evolutionary history. In particular, the functional and phylogenetic analysis of the FeS cluster assembly machinery changes our view of "amitochondriates" from organisms that lack mitochondria, to organisms harboring organelles that evolved either from aerobic mitochondria or together with mitochondria from a common endosymbiotic ancestor under differing selection pressures.

## References

Adam AC, Bornhövd C, Prokisch H, Neupert W, Hell K (2005) The Nfs1 interacting protein Isd11 has an essential role in Fe/S cluster biogenesis in mitochondria. EMBO J 25:174–183

Adam RD (2001) Biology of *Giardia lamblia*. Clin Microbiol Rev 14:447–475

Akhmanova A, Voncken F, van Alen T, van Hoek A, Boxma B, Vogels G, Veenhuiss M, Hackstein JHP (1998) A hydrogenosome with a genome. Nature 396:527–528

Ali V, Shigeta Y, Tokumoto U, Takahashi Y, Nozaki T (2004) An intestinal parasitic protist, *Entamoeba histolytica*, possesses a non-redundant nitrogen fixation-like system for iron-sulfur cluster assembly under anaerobic conditions. J Biol Chem 279:16863–16874

Alves R, Herrero E, Sorribas A (2004) Predictive reconstruction of the mitochondrial iron-sulfur cluster assembly metabolism. II. Role of glutaredoxin Grx5. Proteins Struct Funct Bioinform 57:481–492

Andersson JO, Sjogren AM, Davis LAM, Embley TM, Roger AJ (2003) Phylogenetic analyses of diplomonad genes reveal frequent lateral gene transfers affecting eukaryotes. Curr Biol 13:94–104

Arisue N, Sáchez LB, Weiss LM, Müller M, Hashimoto T (2002) Mitochondrial-type hsp70 genes of the amitochondriate protists, *Giardia intestinalis*, *Entamoeba histolytica* and two microsporidians. Parasitol Int 51:9–16

Balk J, Pierik AJ, Guilar Netz DJ, Mühlenhoff U, Lill R (2005) Nar1p, a conserved eukaryotic protein with similarity to Fe-only hydrogenases, functions in cytosolic iron-sulphur protein biogenesis. Biochem Soc Trans 33:86–89

Best AA, Morrison HG, McArthur AG, Sogin ML, Olsen GJ (2004) Evolution of eukaryotic transcription: insights from the genome of *Giardia lamblia*. Genome Res 14:1537–1547

Besteiro S, Barrett MP, Riviere L, Bringaud F (2005) Energy generation in insect stages of *Trypanosoma brucei*: metabolism in flux. Trends Parasitol 21:185–191

Boxma B, de Graaf RM, van der Staay GWM, van Alen TA, Ricard G, Gabaldon T, van Hoek AHAM, Moon-van der Staay S, Koopman WJH, van Hellemond JJ, Tielens AGM, Friedrich T, Veenhuis M, Huynen MA, Hackstein JHP (2005) An anaerobic mitochondrion that produces hydrogen. Nature 434:74–79

Broers CAM (1992) Anaerobic psalteriomonad amoeboflagellates. Thesis, University of Nijmegen, Nijmegen

Brul S, Veltman RH, Lombardo MC, Vogels GD (1994) Molecular cloning of hydrogenosomal ferredoxin cDNA from the anaerobic amoeboflagellate *Psalteriomonas lanterna*. Biochim Biophys Acta 1183:544–546

Bui ET, Bradley PJ, Johnson PJ (1996) A common evolutionary origin for mitochondria and hydrogenosomes. Proc Natl Acad Sci USA 93:9651–9656

Campuzano V, Montermini L, Molto MD, Pianese L, Cossee M, Cavalcanti F, Monros E, Rodius F, Duclos F, Monticelli A, Zara F, Canizares J, Koutnikova H, Bidichandani SI, Gellera C, Brice A, Trouillas P, DeMichele G, Filla A, DeFrutos R, Palau F, Patel PI, DiDonato S, Mandel JL, Cocozza S, Koenig M, Pandolfo M (1996) Friedreich's ataxia: autosomal recessive disease caused by an intronic GAA triplet repeat expansion. Science 271:1423–1427

Cavalier-Smith T (1987a) Eukaryotes with no mitochondria [news]. Nature 326:332–333
Cavalier-Smith T (1987b) The origin of eukaryotic and archaebacterial cells. Ann N Y Acad Sci 503:17–54
Čerkasovová A, Lukášová G, Čerkasov J, Kulda J (1973) Biochemical characterization of large granule fraction of *Tritrichomonas foetus* (KV1 strain). J Protozool 20:537
Chan KW, Slotboom DJ, Cox S, Embley TM, Fabre O, van der Giezen M, Harding M, Horner DS, Kunji ERS, Leon-Avila G, Tovar J (2005) A novel ADP/ATP transporter in the mitosome of the microaerophilic human parasite *Entamoeba histolytica*. Curr Biol 15:737–742
Crossnoe CR, Germanas JP, LeMagueres P, Mustata G, Krause KL (2002) The crystal structure of *Trichomonas vaginalis* ferredoxin provides insight into metronidazole activation. J Mol Biol 318:503–518
Dacks JB, Davis, LAM, Sjögren ÅM, Andersoon JO, Roger AJ, Doolittle WF (2003) Evidence for cryptic Golgi in putatively 'Golgi-lacking' lineages. Proc Biol Sci 270:S168–171
Dailey HA, Dailey TA, Wu CK, Medlock AE, Wang KF, Rose JP, Wang BC (2000) Ferrochelatase at the millennium: structures, mechanisms and [2Fe-2S] clusters. Cell Mol Life Sci 57:1909–1926
Davidson EA, van der Giezen M, Horner DS, Embley TM, Howe CJ (2002) An [Fe] hydrogenase from the anaerobic hydrogenosome-containing fungus *Neocallimastix frontalis* L2. Gene 296:45–52
Decottignies A, Goffeau A (1997) Complete inventory of the yeast ABC proteins. Nat Genet 15:137–145
Doležal P, Šmíd O, Rada P, Zubčová Z, Bursac D, Šut'ák R, Nebesářová J, Lithgow T, Tachezy J (2005) Giardia mitosomes and trichomonad hydrogenosomes share a common mode of protein targeting. Proc Natl Acad Sci USA 102:10924–10929
Dong J-S, Lai R, Nielsen K, Fekete CA, Qiu H-F, Hinnebusch AG (2004) The essential ATP-binding cassette protein Rli1 functions in translation by promoting preinitiation complex assembly. J Biol Chem 274:42157–42168
Dutkiewicz R, Schilke B, Knieszner H, Walter W, Craig EA, Marszalek J (2003) Ssq1, a mitochondrial Hsp70 involved in iron-sulfur (Fe/S) center biogenesis – similarities to and differences from its bacterial counterpart. J Biol Chem 278:29719–29727
Dyall SD, Johnson PJ (2000) Origins of hydrogenosomes and mitochondria: evolution and organelle biogenesis. Curr Opin Microbiol 3:404–411
Dyall SD, Brown MT, Johnson PJ (2004a) Ancient invasions: from endosymbionts to organelles. Science 304:253–257
Dyall SD, Yan W, Gadillo-Correa MG, Lunceford A, Loo JA, Clarke CF, Johnson PJ (2004b) Non-mitochondrial complex I proteins in a hydrogenosomal oxidoreductase complex. Nature 431:1103–1107
Embley TM, Finlay BJ, Dyal PL, Hirt RP, Wilkinson M, Williams AG (1995) Multiple origins of anaerobic ciliates with hydrogenosomes within the radiation of aerobic ciliates. Proc R Soc Lond B Biol Sci 262:87–93
Embley TM, van der Giezen M, Horner DS, Dyal PL, Bell S, Foster PG (2003a) Hydrogenosomes, mitochondria and early eukaryotic evolution. IUBMB Life 55:387–395
Embley TM, van der Giezen M, Horner DS, Dyal PL, Foster P (2003b) Mitochondria and hydrogenosomes are two forms of the same fundamental organelle. Philos Trans R Soc Lond B Biol Sci 358:191–202
Emelyanov VV (2003) Phylogenetic affinity of a *Giardia lamblia* cysteine desulfurase conforms to canonical pattern of mitochondrial ancestry. FEMS Microbiol Lett 226:257–266
Esser C, Ahmadinejad N, Wiegand C, Rotte C, Sebastiani F, Gelius-Dietrich G, Henze K, Kretschmann E, Richly E, Leister D, Bryant D, Steel MA, Lockhart PJ, Penny D, Martin W (2004) A genome phylogeny for mitochondria among alpha-proteobacteria and a predominantly eubacterial ancestry of yeast nuclear genes. Mol Biol Evol 21:1643–1660
Fenchel T, Finlay BJ (1994) The evolution of life without oxygen. Am Sci 82:22–29
Finel M (1998) Organization and evolution of structural elements within complex I. Biochim Biophys Acta 1364:112–121

Gakh O, Adamec J, Gacy AM, Twesten RD, Owen WG, Isaya G (2002) Physical evidence that yeast frataxin is an iron storage protein. Biochemistry 41:6798–6804

Germot A, Philippe H, LeGuyader H (1996) Presence of a mitochondrial-type 70-kDa heat shock protein in *Trichomonas vaginalis* suggests a very early mitochondrial endosymbiosis in eukaryotes. Proc Natl Acad Sci USA 93:14614–14617

Gray MW, Burger G, Lang BF (1999) Mitochondrial evolution. Science 283:1476–1481

Grieshaber MK, Volkel S (1998) Animal adaptations for tolerance and exploitation of poisonous sulfide. Annu Rev Physiol 60:33–53

Hackstein JHP, Akhmanova A, Boxma B, Harhangi HR, Voncken FGJ (1999) Hydrogenosomes: eukaryotic adaptations to anaerobic environments. Trends Microbiol 7:441–447

Han KL, Yong TS, Ryu JS, Hwang UW, Park SJ (2004) Identification of the hybrid cluster protein, HCP, from amitochondriate eukaryotes and its phylogenetic implications. J Microbiol Biotechnol 14:134–139

Hashimoto T (1998) Secondary absence of mitochondria in *Giardia lamblia* and *Trichomonas vaginalis* revealed by valyl-tRNA synthetase phylogeny. Proc Natl Acad Sci USA 95:6860–6865

Hausmann A, Guilar Netz DJ, Balk J, Pierik AJ, Mühlenhoff U, Lill R (2005) The eukaryotic P loop NTPase Nbp35: an essential component of the cytosolic and nuclear iron-sulfur protein assembly machinery. Proc Natl Acad Sci USA 102:3266–3271

Horner DS, Foster PG, Embley TM (2000) Iron hydrogenases and the evolution of anaerobic eukaryotes. Mol Biol Evol 17:1695–1709

Horner DS, Heil B, Happe T, Embley TM (2002) Iron hydrogenases - ancient enzymes in modern eukaryotes. Trends Biochem Sci 27:148–153

Horner DS, Hirt RP, Kilvington S, Lloyd D, Embley TM (1996) Molecular data suggest an early acquisition of the mitochondrion endosymbiont. Proc R Soc Lond B Biol Sci 263:1053–1059

Hrdý I, Hirt RP, Doležal P, Bardoňová L, Foster PG, Tachezy J, Embley TM (2004) Trichomonas hydrogenosomes contain the NADH dehydrogenase module of mitochondrial complex I. Nature 432:618–622

Huber C, Wächtershäuser G (1997) Activated acetic acid by carbon fixation on (Fe,Ni)S under primordial conditions. Science 276:245–247

Jacobson MR, Cash VL, Weiss MC, Laird NF, Newton WE, Dean DR (1989) Biochemical and genetic analysis of the *nif* USVWZM cluster from *Azotobacter vinelandii*. Mol Gen Genet 219:49–57

Johnson DC, Dean DR, Smith AD, Johnson MK (2005) Structure, function, and formation of biological iron-sulfur clusters. Annu Rev Biochem 74:247–281

Johnson PJ, d'Oliveira CE, Gorrell TE, Müller M (1990) Molecular analysis of the hydrogenosomal ferredoxin of the anaerobic protist Trichomonas vaginalis. Proc Natl Acad Sci USA 87:6097–6101

Kapazoglou A, Mould RM, Gray JC (2000) Assembly of the Rieske iron-sulphur protein into the cytochrome bf complex in thylakoid membranes of isolated pea chloroplasts. Eur J Biochem 267:352–360

Katinka MD, Duprat S, Cornillot E, Metenier G, Thomarat F, Prensier G, Barbe V, Peyretaillade E, Brottier P, Wincker P, Delbac F, El Alaoui H, Peyret P, Saurin W, Gouy M, Weissenbach J, Vivares CP (2001) Genome sequence and gene compaction of the eukaryote parasite *Encephalitozoon cuniculi*. Nature 414:450–453

Keithly JS, Langreth SG, Buttle KF, Mannella CA (2005) Electron tomographic and ultrastructural analysis of the *Cryptosporidium parvum* relict mitochondrion, its associated membranes, and organelles. J Eukaryot Microbiol 52:132–140

Kennedy C, Dean D (1992) The nifU, nifS and nifV gene products are required for activity of all three nitrogenases of *Azotobacter vinelandii*. Mol Gen Genet 231:494–498

Kispal G, Csere P, Prohl C, Lill R (1999) The mitochondrial proteins Atm1p and Nfs1p are essential for biogenesis of cytosolic Fe/S proteins. EMBO J 18:3981–3989

Kispal G, Sipos K, Lange H, Fekete Z, Bedekovics T, Janaky T, Bassler J, Guilar Netz DJ, Balk J, Rotte C, Lill R (2005) Biogenesis of cytosolic ribosomes requires the essential iron-sulphur protein Rli1p and mitochondria. EMBO J 24:589–598

Kita K, Hirawake H, Takamiya S (1997) Cytochromes in the respiratory chain of helminth mitochondria. Int J Parasitol 27:617–630

Kulda J (1999) Trichomonads, hydrogenosomes and drug resistance. Int J Parasitol 29:199–212

Kurland CG, Andersson SGE (2000) Origin and evolution of the mitochondrial proteome. Microbiol Mol Biol Rev 64:786–820

LaGier MJ, Tachezy J, Stejskal F, Kutišová K, Keithly JS (2003) Mitochondrial-type iron-sulfur cluster biosynthesis genes (IscS and IscU) in the apicomplexan *Cryptosporidium parvum*. Microbiol-SGM 149:3519–3530

Land KM, Gadillo-Correa MG, Tachezy J, Vaňáčová S, Hsieh CL, Šut'ák R and Johnson PJ (2004) Targeted gene replacement of a ferredoxin gene in *Trichomonas vaginalis* does not lead to metronidazole resistance. Mol Microbiol 51:115–122

Lange H, Lisowsky T, Gerber J, Mühlenhoff U, Kispal G, Lill R (2001) An essential function of the mitochondrial sulfhydryl oxidase Ervlp/ALR in the maturation of cytosolic Fe/S proteins. EMBO Rep 2:715–720

Leon-Avila G, Tovar J (2004) Mitosomes of *Entamoeba histolytica* are abundant mitochondrion-related remnant organelles that lack a detectable organellar genome. Microbiology 150:1245–1250

Lill R, Kispal G (2000) Maturation of cellular Fe-S proteins: an essential function of mitochondria. Trends Biochem Sci 25:352–356

Lill R, Mühlenhoff U (2005) Iron-sulfur-protein biogenesis in eukaryotes. Trends Biochem Sci 30:133–141

Lill R, Diekert K, Kaut A, Lange H, Pelzer W, Prohl C, Kispal G (1999) The essential role of mitochondria in the biogenesis of cellular iron-sulfur proteins. Biol Chem 380:1157–1166

Lindmark DG, Müller M (1973) Hydrogenosome, a cytoplasmic organelle of the anaerobic flagellate *Tritrichomonas foetus,* and its role in pyruvate metabolism. J Biol Chem 248:7724–7728

Lindmark DG, Müller M, Shio H (1975) Hydrogenosomes in *Trichomonas vaginalis.* J Parasitol 61:552–554

Lloyd D, Harris JC (2002) Giardia: highly evolved parasite or early branching eukaryote? Trends Microbiol 10:122–127

Lloyd D, Ralphs JR Harris JC (2002) *Giardia intestinalis*, a eukaryote without hydrogenosomes, produces hydrogen. Microbiology 148:727–733

Lopez-Garcia P, Moreira D (1999) Metabolic symbiosis at the origin of eukaryotes. Trends Biochem Sci 24:88–93

Mai Z, Ghosh S, Frisardi M, Rosenthal B, Rogers R, Samuelson J (1999) Hsp60 is targeted to a cryptic mitochondrion-derived organelle ("crypton") in the microaerophilic protozoan parasite *Entamoeba histolytica.* Mol Cell Biol 19:2198–2205

Marczak R, Gorrell TE, Müller M (1983) Hydrogenosomal ferredoxin of the anaerobic protozoon, *Tritrichomonas foetus.* J Biol Chem 258:12427–12433

Martin W, Müller M (1998) The hydrogen hypothesis for the first eukaryote. Nature 392:37–41

Marvin Sikkema FD, Pedro Gomes TM, Grivet JP, Gottschal JC, Prins RA (1993) Characterization of hydrogenosomes and their role in glucose metabolism of *Neocallimastix* sp. L2. Arch Microbiol 160:388–396

McGoldrick JP, Yeh YC, Solomon M, Essigmann JM, Lu AL (1995) Characterization of a mammalian homolog of the Escherichia-coli Muty mismatch repair protein. Mol Cell Biol 15:989–996

Moreira D, Lopez-Garcia P (1998) Symbiosis between methanogenic archaea and delta-proteobacteria as the origin of eukaryotes: the syntrophic hypothesis. J Mol Evol 47:517–530

Mühlenhoff U, Lill R (2000) Biogenesis of iron-sulfur proteins in eukaryotes: a novel task of mitochondria that is inherited from bacteria. Biochim Biophys Acta 1459:370–382

Mühlenhoff U, Gerber J, Richhardt N, Lill R (2003) Components involved in assembly and dislocation of iron-sulfur clusters on the scaffold protein Isu1p. EMBO J 22:4815–4825

Müller M (1988) Energy metabolism of protozoa without mitochondria. Annu Rev Microbiol 42:465–488

Müller M (2003). Energy metabolism. Part I: anaerobic protozoa. In: Marr JJ, Nilsen TW, Komuniecki RW (eds) Molecular medical parasitology. Academic, New York, pp 125–139

Nixon JEJ, Filed J, McArthur AG, Sogin ML, Yarlett N, Loftus BJ, Samuelson J (2003) Iron-dependent hydrogenases of *Entamoeba histolytica* and *Giardia lamblia*: activity of the recombinant entamoebic enzyme and evidence for lateral gene transfer. Biol Bull 204:1–9

Nixon JEJ, Wang A, Field J, Morrison HG, McArthur AG, Sogin ML, Loftus BJ, Samuelson J (2002a) Evidence for lateral transfer of genes encoding ferredoxins, nitroreductases, NADH oxidase, and alcohol dehydrogenase 3 from anaerobic prokaryotes to *Giardia lamblia* and *Entamoeba histolytica*. Eukaryot Cell 1:181–190

Paul RG, Williams AG, Butler RD (1990) Hydrogenosomes in the rumen entodiniomorphid ciliate *Polyplastron multivesiculatum*. J Gen Microbiol 136:1981–1989

Payne MJ, Chapman A, Cammack R (1993) Evidence for an [Fe]-type hydrogenase in the parasitic protozoan *Trichomonas vaginalis*. FEBS Lett 317:101–104

Pilkington SJ, Skehel JM, Gennis RB, Walker JE (1991) Relationship between mitochondrial NADH-ubiquinone reductase and a bacterial NAD-reducing hydrogenase. Biochemistry 30:2166–2175

Posewitz MC, King PW, Smolinski SL, Zhang LP, Seibert M, Ghirardi ML (2004) Discovery of two novel radical S-adenosylmethionine proteins required for the assembly of an active [Fe] hydrogenase. J Biol Chem 279:25711–25720

Putignani L, Tait A, Smith HV, Horner D, Tovar J, Tetley L, Wastling JM (2004) Characterization of a mitochondrion-like organelle in *Cryptosporidium parvum*. Parasitol 129:1–18

Pütz S, Doležal P, Gelius-Dietrich G, Tachezy J, Henze K (2006) [Fe]-hydrogenase maturases in the hydrogenosomes of *Trichomonas vaginalis*. Eukaryot Cell 5:579–586

Regoes A, Zourmpanou D, Leon-Avila G, van der Giezen M, Tovar J, Hehl AB (2005) Protein import, replication and inheritance of a vestigial mitochondrion. J Biol Chem 280:30557–30563

Riordan CE, Langreth SG, Sánchez LB, Kayser O, Keithly JS (1999) Preliminary evidence for a mitochondrion in *Cryptosporidium parvum*: Phylogenetic and therapeutic implications. J Eukaryot Microbiol 46:52S–55S

Riordan CE, Ault JG, Langreth SG, Keithly JS (2003) *Cryptosporidium parvum* Cpn60 targets a relict organelle. Curr Genet 44:138–147

Roger AJ, Svard SG, Tovar J, Clark CG, Smith MW, Gillin FD, Sogin ML (1998) A mitochondrial-like chaperonin 60 gene in *Giardia lamblia*: Evidence that diplomonads once harbored an endosymbiont related to the progenitor of mitochondria. Proc Natl Acad Sci USA 95:229–234

Roger AJ, Sandblom O, Doolittle WF, Philippe H (1999) An evaluation of elongation factor 1 alpha as a phylogenetic marker for eukaryotes. Mol Biol Evol 16:218–233

Rotte C, Stejskal F, Zhu G, Keithly JS, Martin W (2001) Pyruvate : NADP(+) oxidoreductase from the mitochondrion of *Euglena gracilis* and from the apicomplexan *Cryptosporidium parvum*: A biochemical relic linking pyruvate metabolism in mitochondriate and amitochondriate protists. Mol Biol Evol 18:710–720

Rouault TA, Tong WH (2005) Iron-sulphur cluster biogenesis and mitochondrial iron homeostasis. Nat Rev Mol Cell Biol 6:345–351

Roy A, Solodovnikova N, Nicholson T, Antholine W, Walden WE (2003) A novel eukaryotic factor for cytosolic Fe-S cluster assembly. EMBO J 22:4826–4835

Santos R, Buisson N, Knight SAB, Dancis A, Camadro JM, Lesuisse E (2004) *Candida albicans* lacking the frataxin homologue: a relevant yeast model for studying the role of frataxin. Mol Microbiol 54:507–519

Schilke B, Voisine C, Beinert H, Craig E (1999) Evidence for a conserved system for iron metabolism in the mitochondria of *Saccharomyces cerevisiae*. Proc Natl Acad Sci USA 96:10206–10211

Sickmann A, Reinders J, Wagner Y, Joppich C, Zahedi R, Meyer HE, Schonfisch B, Perschil I, Chacinska A, Guiard B, Rehling P, Pfanner N, Meisinger C (2003) The proteome of *Saccharomyces cerevisiae* mitochondria. Proc Natl Acad Sci USA 100:13207–13212

Simpson AGB, Patterson DJ (1999) The ultrastructure of Carpediemonas membranifera (Eukaryota) with reference to the "excavate hypothesis". Eur J Protistol 35:353–370

Simpson AGB, VandenHoff J, Bernard C, Burton HR, Patterson DJ (1997) The ultrastructure and systematic position of the euglenozoon *Postgaardi mariagerensis*, Fenchel et al. Arch Protist 147:213–225

Sipos K, Lange H, Fekete Z, Ullmann P, Lill R, Kispal G (2002) Maturation of cytosolic iron-sulfur proteins requires glutathione. J Biol Chem 277:26944–26949

Šlapeta J, Keithly JS (2004) *Cryptosporidium parvum* mitochondrial-type HSP70 targets homologous and heterologous mitochondria. Eukaryot Cell 3:483–494

Smith AD, Agar JN, Johnson KA, Frazzon J, Amster IJ, Dean DR, Johnson MK (2001) Sulfur transfer from IscS to IscU: the first step in iron-sulfur cluster biosynthesis. J Am Chem Soc 123:11103–11104

Snyers L, Hellings P, Bovy-Kesler C, Thines-Sempoux D (1982) Occurence of hydrogenosomes in the rumen ciliates *Ophryoscolecidae*. FEBS Lett 137:35–39

Steinbüchel A, Müller M (1986) Anaerobic pyruvate metabolism of *Tritrichomonas foetus* and *Trichomonas vaginalis* hydrogenosomes. Mol Biochem Parasitol 20:57–65

Stejskal F, Šlapeta J, Čtrnáctá V, Keithly JS (2003) A Narf-like gene from *Cryptosporidium parvum* resembles homologues observed in aerobic protists and higher eukaryotes. FEMS Microbiol Lett 229:91–96

Suchan P, Vyoral D, Petrák J, Šuťák R, Rasoloson D, Nohýnkova E, Doležal P, Tachezy J (2003) Incorporation of iron into *Tritrichomonas foetus* cell compartments reveals ferredoxin as a major iron-binding protein in hydrogenosomes. Microbiology 149:1911–1921

Šuťák R, Doležal P, Fiumera HL, Hrdý I, Dancis A, Gadillo-Correa M, Johnson PJ, Müller M, Tachezy J (2004) Mitochondrial-type assembly of FeS centers in the hydrogenosomes of the amitochondriate eukaryote *Trichomonas vaginalis*. Proc Natl Acad Sci USA 101:10368–10373

Tachezy J, Sánchez LB, Müller M (2001) Mitochondrial type iron-sulfur cluster assembly in the amitochondriate eukaryotes *Trichomonas vaginalis* and *Giardia intestinalis*, as indicated by the phylogeny of IscS. Mol Biol Evol 18:1919–1928

Takahashi Y, Tokumoto U (2002) A third bacterial system for the assembly of iron-sulfur clusters with homologs in archaea and plastids. J Biol Chem 277:28380–28383

Thong KW, Coombs GH (1987) Comparative study of ferredoxin-linked and oxygen-metabolizing enzymes of trichomonads. Comp Biochem Physiol B 87:637–641

Tielens AGM, Rotte C, van Hellemond JJ, Martin W (2002) Mitochondria as we don't know them. Trends Biochem Sci 27:564–572

Tokumoto U, Kitamura S, Fukuyama K, Takahashi Y (2004) Interchangeability and distinct properties of bacterial Fe-S cluster assembly systems: functional replacement of the isc and suf operons in *Escherichia coli* with the nifSU-like operon from *Helicobacter pylori*. J Biochem 136:199–209

Tovar J, Fischer A, Clark CG (1999) The mitosome, a novel organelle related to mitochondria in the amitochondrial parasite *Entamoeba histolytica*. Mol Microbiol 32:1013–1021

Tovar J, Leon-Avila G, Sánchez LB, Šuťák R, Tachezy J, van der GM, Hernandez M, Müller M, Lucocq JM (2003) Mitochondrial remnant organelles of Giardia function in iron-sulphur protein maturation. Nature 426:172–176

Upcroft P, Upcroft JA (2001) Drug targets and mechanisms of resistance in the anaerobic protozoa. Clin Microbiol Rev 14:150–164

van den Berg WAM, Hagen WR, van Dongen WMAM (2000) The hybrid-cluster protein ('prismane protein') from Escherichia coli – characterization of the hybrid-cluster protein, redox properties of the [2Fe-2S] and [4Fe-2S-20] clusters and identification of an associated NADH oxidoreductase containing FAD and [2Fe-2S]. Eur J Biochem 267:666–676

van der Giezen M, Cox S, Tovar J (2004) The iron-sulfur cluster assembly genes *iscS* and *iscU* of *Entamoeba histolytica* were acquired by horizontal gene transfer. BMC Evol Biol 4:7

van der Giezen M, Leon-Avila G, Tovar J (2005) Characterization of chaperonin 10 (Cpn10) from the intestinal human pathogen *Entamoeba histolytica*. Microbiology 151:3107–3115 van Weelden SW, van Hellemond JJ, Opperdoes FR, Tielens AG (2005) New functions for parts of the Krebs cycle in procyclic *Trypanosoma brucei*, a cycle not operating as a cycle. J Biol Chem 280:12451–12460

Vignais PM, Billoud B, Meyer J (2001) Classification and phylogeny of hydrogenases. FEMS Microbiol Rev 25:455–501

Vollmer M, Thomsen N, Wiek S, Seeber F (2001) Apicomplexan parasites possess distinct nuclear-encoded, but apicoplast-localized, plant-type ferredoxin-NADP(+) reductase and ferredoxin. J Biol Chem 276:5483–5490

Voncken FGJ, Boxma B, van Hoek AHAM, Akhmanova AS, Vogels GD, Huynen M, Veenhuis M, Hackstein JHP (2002) A hydrogenosomal [Fe]-hydrogenase from the anaerobic chytrid *Neocallimastix* sp L2. Gene 284:103–112

Williams BAP, Hirt RP, Lucocq JM, Embley TM (2002) A mitochondrial remnant in the microsporidian *Trachipleistophora hominis*. Nature 418:865–869

Xu P, Widmer G, Wang YP, Ozaki LS, Alves JM, Serrano MG, Puiu D, Manque P, Akiyoshi D, Mackey AJ, Pearson WR, Dear PH, Bankier AT, Peterson DL, Abrahamsen MS, Kapur V, Tzipori S, Buck GA (2004) The genome of *Cryptosporidium hominis*. Nature 431:1107–1112

Yang D, Oyaizu Y, Oyaizu H, Olsen GJ, Woese CR (1985) Mitochondrial origins. Proc Natl Acad Sci USA 82:4443–4447

Yano T (2002) The energy-transducing NADH:quinone oxidoreductase, complex I. Mol Asp Med 23:345–368

Yarlett N, Hackstein JHP (2005) Hydrogenosomes: one organelle, multiple origins. BioScience 55:657–667

Yarlett N, Hann AC, Lloyd D, Williams A (1981) Hydrogenosomes in the rumen protozoon *Dasytricha ruminantium* Schuberg. Biochem J 200:365–372

Yarlett N, Hann AC, Lloyd D, Williams AG (1983) Hydrogenosomes in a mixed isolate of *Isotricha prostoma* and *Isotricha intestinalis* from ovine rumen contents. Comp Biochem Physiol B 74:357–364

Yarlett N, Orpin CG, Munn EA, Yarlett NC, Greenwood CA (1986a) Hydrogenosomes in the rumen fungus *Neocallimastix patriciarum*. Biochem J 236:729–739

Yarlett N, Yarlett NC, Lloyd D (1986b) Ferredoxin-dependent reduction of nitroimidazole derivatives in drug-resistant and susceptible strains of *Trichomonas vaginalis*. Biochem Pharmacol 35:1703–1708

Yarunin A, Panse VG, Petfalski E, Dez C, Tollervey D, Hurt EC (2005) Functional link between ribosome formation and biogenesis of iron-sulfur proteins. EMBO J 24:580–588

Zheng L, Cash VL, Flint DH, Dean DR (1998) Assembly of iron-sulfur clusters. Identification of an iscSUA-hscBA-fdx gene cluster from *Azotobacter vinelandii*. J Biol Chem 273: 13264–13272

# 7 Hydrogenosomes (and Related Organelles, Either) Are Not the Same

Johannes H.P. Hackstein, Joachim Tjaden, Werner J.H. Koopman, Martijn Huynen

## 7.1 Introduction

Hydrogenosomes are membrane-bounded organelles, approximately 1–2 μm in size, that compartmentalize the terminal reactions of anaerobic cellular energy metabolism in eukaryotes. They were first described in the parabasilid flagellate, *Tritrichomonas foetus*, in an influential publication by Lindmark and Müller (1973) as subcellular particles that produce hydrogen and ATP. Since that time hydrogenosomes have been described in a number of rather different unicellular eukaryotes adapted to microaerobic or anoxic environments (Roger 1999; Yarlett 2004; Yarlett and Hackstein 2005). Hydrogenosomes seem to be related to a very diverse family of organelles such as mitosomes or mitochondrial remnants that are believed to share a common ancestor with present-day mitochondria. Such "textbook" mitochondria are regarded as "powerhouses" of the eukaryotic cell, supplying it with ATP. This ATP is generated by an ATP synthase, which exploits a proton gradient that is generated by a membrane-bound electron-transport chain that drives three different proton pumps, i.e. the mitochondrial complexes I, III, and IV. This electron-transport chain depends on molecular oxygen as a terminal electron acceptor (Saraste 1999).

However, mitochondria are not the "static" textbook organelles; they are very diverse and dynamic, and quite a number of "genuine" mitochondria function in the absence of oxygen using alternative electron acceptors, such nitrite, nitrate, or fumarate (Yaffe 1999; Tielens et al. 2002; Tielens and van Hellemond, Chap. 5 in this volume). Although mitochondria are metabolically much more diverse than depicted in textbooks, all mitochondria studied so far have retained a genome, which documents unequivocally their descent from an α-proteobacterium (Lang et al. 1997; Gray et al. 1999; Gabaldon and Huynen 2003, 2004; Esser et al. 2004). The mitochondrial genome has been reduced dramatically in size during organelle evolution by up to 2 orders of magnitude. Only a few genes, e.g. between five in *Plasmodium* sp. and 97 in *Reclinomonas americana*, have been retained in the mitochondrial genome (Lang et al. 1997; Feagin 2000; Burger et al. 2003; Nosek and Tomaska 2003; Timmis et al. 2004). Textbook hydrogenosomes, but also mitochondrial remnants such as mitosomes or cryptons, which are found in various unicellular parasites, lost their genome completely,

Origin of Mitochondria and Hydrogenosomes (ed. by William F. Martin and Miklós Müller)

prohibiting any straightforward assessment of their ancestry and evolutionary relationships (Leon-Avila and Tovar 2004; van der Giezen and Tovar 2005).

Moreover, neither textbook hydrogenosomes nor mitochondrial remnants possess an electron-transport chain, and while hydrogenosomes produce ATP by substrate-level phosphorylation, mitochondrial remnants such as mitosomes or cryptons do not produce any ATP at all (Müller 1993, 1998; van der Giezen et al. 1997; Embley and Martin 1998; Martin and Müller 1998; Tovar et al. 1999, 2003; Clemens and Johnson 2000; Martin et al. 2001; Williams et al. 2002; Dyall et al. 2004a).

Thus, both mitosomes and hydrogenosomes lack a genome and an electron-transport chain, the most important characteristics of textbook mitochondria. Moreover, both mitosomes and hydrogenosomes evolved in rather distinct lineages of unicellular organisms, suggesting that neither all mitosomes nor all hydrogenosomes are the same (Fig. 7.1). This reinforces the question as to whether all these organelles share a common ancestry. It is likely that all these organelles arose repeatedly by evolutionary tinkering as an adaptation to the particular requirements of their hosts, which thrive in rather different environments. Since the only characteristic known so far that

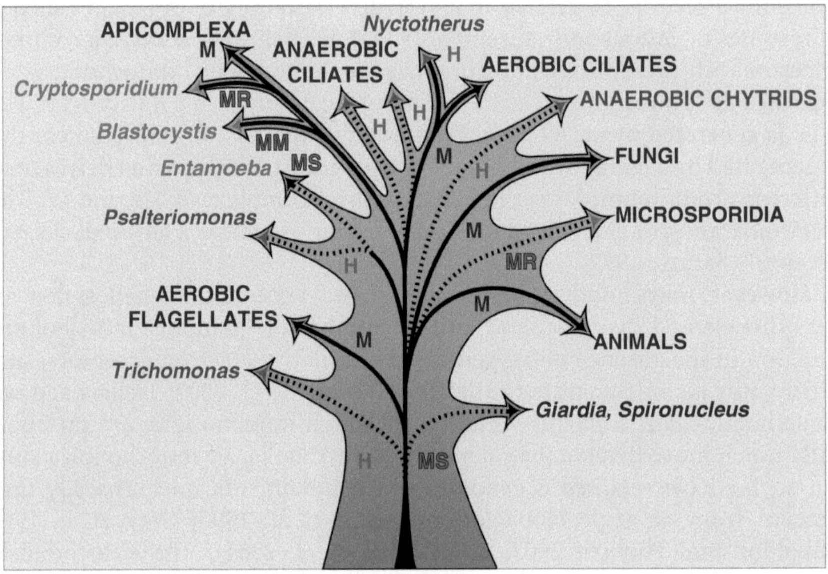


**Fig. 7.1.** Phylogenetic relationships between aerobic and anaerobic protists (based on a variety of molecular data). Superimposed is a tentative evolutionary tree of mitochondria (*M*), modified mitochondria (*MM*), mitochondrial remnants (*MR*), mitosomes (*MS*), and hydrogenosomes (*H*). The *solid black lines* indicate phylogenetic relationships that are based on the analysis of "mitochondrial" genomes. *Dashed lines* indicate the loss of organellar genomes. Mitosomes and hydrogenosomes evolved in rather distinct lineages of unicellular organisms, suggesting that neither all mitosomes nor all hydrogenosomes are the same

is shared by all mitosomes, hydrogenosomes, and mitochondria is the synthesis of iron–sulphur clusters, uncovering the ancestry of these different organelles is far from trivial (Williams et al. 2002; Vivares et al. 2002; Vanacova et al. 2003; Abrahamsen et al. 2004; Balk and Lill 2004; Tovar et al. 2003; Gabaldon and Huynen 2004, van der Giezen et al. 2004; Regoes et al. 2005; Lill and Mühlenhoff 2005).

A proteomics approach to analyse the protein composition of hydrogenosomes and mitosomes, which would allow a straightforward comparison of the organelles and a conclusive analysis of the evolution of these organelles, has not been published so far. Proteomics of isolated mitochondria and large-scale bioinformatic analysis of complete eukaryotic and α-proteobacterial genomes revealed the presence of 630 eukaryotic proteins that were likely to be derived from the α-proteobacterial ancestor of the mitochondria (Gabaldon and Huynen 2003). However, most of these proteins are not located in mitochondria but elsewhere in the cell: significantly less than 20% of the mitochondrial proteome is of α-proteobacterial origin (Gabaldon and Huynen 2003, 2004), a figure very similar to that for the peroxisomal proteome, which contains a comparable fraction of proteins of α-proteobacterial origin (Gabaldon 2005). In addition, mitochondrial proteomics has revealed that even textbook mitochondria are very different – not only in the size of their proteomes, but also by virtue of their protein composition and function (Gabaldon and Huynen 2004). Thus, the vast majority of mitochondrial proteins did not originate from the ancestral endosymbiont but from a variety of eukaryotic, eubacterial, and archaeal sources (Gabaldon and Huynen 2004; Esser et al. 2004; Timmis et al. 2004; Fig. 7.2). This means that besides a dramatic differential gene loss from the organelle genome, a substantial gain of proteins shapes the evolution of mitochondria. Although comparable proteomics data for mitosomes and hydrogenosomes are lacking until now, it is reasonable to anticipate similar processes in the evolution of mitochondria, mitosomes, and hydrogenosomes. Clearly, the lack of organelle genomes and the scarcity of genomic data from anaerobic, unicellular eukaryotes have hampered any straightforward analysis of the evolution of these organelles until now. Moreover, the anticipated enormous diversity of both mitosomes and hydrogenosomes has been hardly addressed so far, and the discovery of organelles, which have retained a genome, cannot be excluded; one example, the hydrogenosomes of the ciliate *Nyctotherus ovalis*, will be described in detail later (Bullerwell and Lang 2005; Hackstein 2005; Yarlett and Hackstein 2005; van der Giezen et al. 2005; van der Giezen and Tovar 2005).

Mitosomes, hydrogenosomes, and mitochondria all share enzymes involved in the synthesis and assembly of iron–sulphur clusters, but neither an electron-transport chain nor substrate-level ATP synthesis (Gabaldon and Huynen 2003, 2004) is common to them all. Moreover, since the original endosymbiont for sure lacked ATP-exporting proteins, it is likely that the original endosymbiosis was not driven by providing ATP to the host (Martin

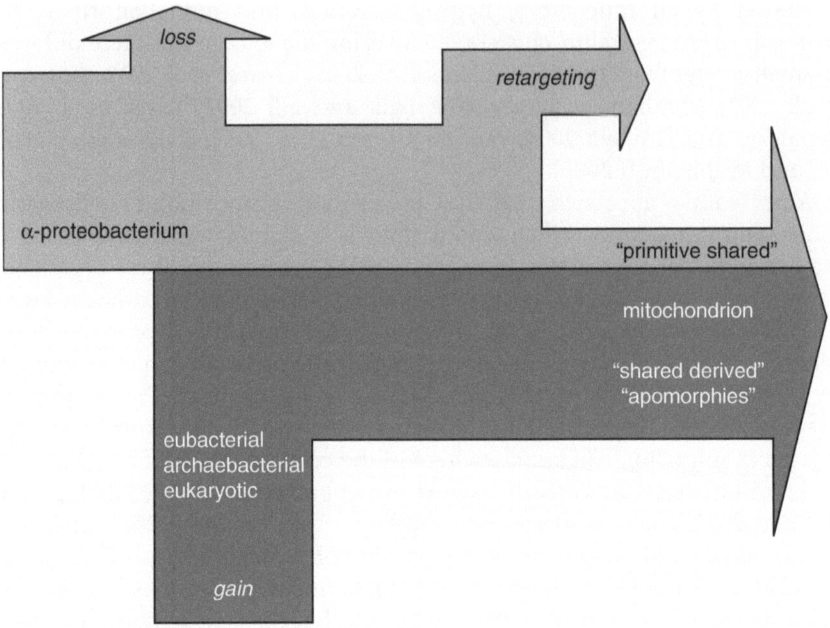


**Fig. 7.2.** The vast majority of mitochondrial proteins did not originate from the ancestral α-proteobacterial endosymbiont but from a variety of eukaryotic, eubacterial, and archaeal sources (Gabaldon and Huynen 2004; Esser et al. 2004; Timmis et al. 2004). The evolution of mitochondria was not only accompanied by a substantial loss of superfluous genes, but also by the transfer of many α-proteobacterial genes into the nucleus. As a consequence, many α-proteobacterial proteins were retargeted to other cellular compartments. Consequently, mitochondria and peroxisomes possess a similar fraction of proteins of α-proteobacterial proteins (Gabaldon et al. 2006), whereas up to 80% of the mitochondrial proteome can be made up from proteins that have various non-α-proteobacterial origins. (Modified from Gabaldon and Huynen 2004)

and Müller 1998). The very limited set of (presumably) primitive-shared properties of mitochondria, hydrogenosomes, and mitochondrial remnants (mitosomes) suggests that an evaluation of the "shared-derived" characteristics might be much more informative than an analysis of the "primitive-shared" characteristics, i.e. traits that are common to all descendants of the original endosymbiont. Candidates for such shared-derived properties are the "mitochondrial" ADP/ATP carriers (AACs), which are clearly of eukaryotic and not of α-proteobacterial origin (Palmieri 1994; Winkler and Neuhaus 1999; Amiri et al. 2003). Mitochondrial-type AACs are characteristic for all "real" mitochondria studied so far, suggesting that they evolved before the divergence of all present-day mitochondria organisms. This suggests that certain eukaryotic, in particular the anaerobic "amitochondrial" lineages, might have evolved "alternative" or "premitochondrial" AACs that could highlight the evolution of the original endosymbiont into an organelle. Thus, in the absence of organelle genomes, AACs are the second-best proteins to analyse the evolution of mitosomes and hydrogenosomes. Therefore, in this review we will focus on hydrogenosomes that evolved in different lineages, comparing genomes, AACs, and metabolic traits that demonstrate the distinctness, and also the common evolutionary roots of the various hydrogenosomes and mitosomes.

## 7.2 Hydrogenosomes and Mitochondrial-Remnant Organelles Evolved Repeatedly: Evidence from ADP/ATP Carriers

Initially, the arguments that both mitosomes, the mitochondrial-remnant organelles, and hydrogenosomes evolved several times were based on the observation that hydrogenosomes and mitosomes, respectively, were found in a broad spectrum of rather unrelated taxa of unicellular organisms, such as trichomonads, diplomonads, sarcodina (entamoebids), flagellates, apicomplexa, ciliates, and chytrids (Biagini et al. 1997a; Embley et al. 1997; Roger 1999; Abrahamsen et al. 2004; Yarlett and Hackstein 2005; van der Giezen and Tovar 2005; van der Giezen et al. 2005; Fig. 7.1). The presence of an anaerobic mitochondrion-like organelle in *Blastocystis hominis* (Stramenopiles) has been proposed – mainly based upon redox-sensitive dyes (Nasirudeen and Tan 2004). However, since most of these organelles did not retain a genome, the only common diagnostic characters seemed to be (1) the presence of "mitochondrial-type" chaperonins (Clark and Roger 1995; Bui et al. 1996, Germot et al. 1996, 1997; Roger et al. 1998; Riordan et al. 2003), (2) the fact that these organelles were membrane-bounded, and (3), in the case of hydrogenosomes, that these organelles produced ATP (Müller 1993, 1998). The phylogenetic analysis of HSP (cpn)60 supported different "mitochondrial" origins of the various organelles (Voncken 2001; Horner and Embley 2001; Voncken et al. 2002a; van der Giezen et al. 2003). Interestingly, the HSP60s of *Trichomonas*, *Giardia*, and *Entamoeba* formed a moderately supported cluster distinct from "mitochondriate" organisms, whereas the HSPs of the apicomplexa, anaerobic chytrids, and ciliates clustered with high statistical support with the homologues from their aerobic, mitochondriate relatives (Fig. 7.3).

Notably, *Trichomonas*, the anaerobic chytrids, and the anaerobic ciliates possess hydrogenosomes, whereas *Giardia* and *Entamoeba* possess mitosomes. Also, a phylogenetic analysis of the AACs has so far revealed that only the hydrogenosomes of chytrids and ciliates possess genuine mitochondrial AACs, which cluster with the mitochondrial homologues of their aerobic, mitochondria-bearing relatives (Voncken 2001; Voncken et al. 2002a; van der Giezen et al. 2002; Tjaden et al. 2004). *Trichomonas* and *Entamoeba* use alternative members of the mitochondrial carrier family for the transport of ATP across its hydrogenosomal/mitosomal membranes (Dyall et al. 2000; Tjaden et al. 2004; Chan et al. 2005; Tjaden and Leroch, unpublished results; Fig. 7.4),

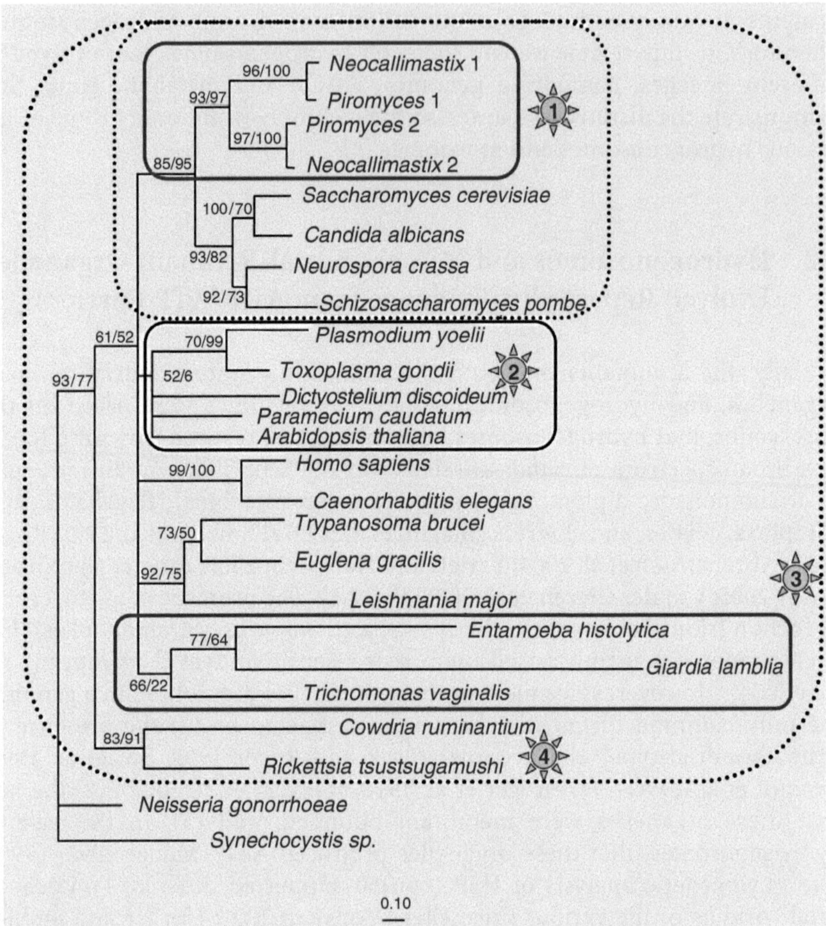


**Fig. 7.3.** Phylogenetic tree of HSP (cpn)60 (neighbour-joining). The HSP60s of the anaerobic chytrids *Piromyces* sp. E2 (and *Neocallimastix* sp.) cluster with high bootstrap values with the mitochondrial HSP60s of aerobic yeasts and fungi (**1**). The HSP60 of the ciliate *Paramecium caudatum* clusters with low resolution with other animals (**2**), and the parabasalian flagellate *Trichomonas vaginalis*, the diplomonad *Giardia lamblia*, and the parasite *Entamoeba histolytica* form another, moderately supported cluster (**3**). Notably, *Trichomonas*, the anaerobic chytrids, and the anaerobic ciliates possess hydrogenosomes, whereas *Giardia* and *Entamoeba* possess mitosomes. *Rickettsia* and other α-proteobacteria form a sister group (**4**). (Modified from Voncken et al. 2002a)

whereas any evidence for the presence of a potential (mitochondrial-type) AAC in *Giardia* is lacking so far. The biochemical analysis of these carriers reveals that they are also functionally distinct (see later). Moreover, *Trichomonas* hydrogenosomes generate ATP, whereas the mitosomes of both *Entamoeba* and *Giardia* do not. The mitochondrial remnants (mitosomes?) of microsporidians, e.g. *Encephalitozoon*, do not possess potential AACs that belong to the genuine mitochondrial AAC cluster; they possess completely

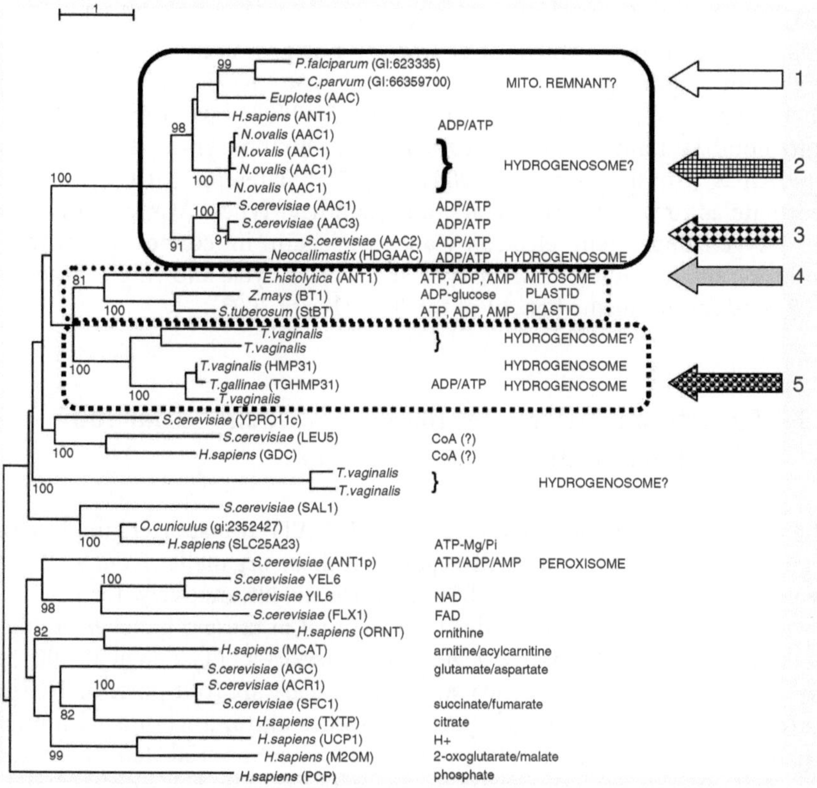


**Fig. 7.4.** Phylogenetic analysis of the ADP/ATP carriers (AACs) has so far revealed that only the hydrogenosomes of chytrids and ciliates possess genuine mitochondrial AACs, which cluster with the mitochondrial homologues of their aerobic, mitochondria-bearing relatives (**2, 3**) (Voncken 2001; Voncken et al. 2002a; van der Giezen et al. 2002; Tjaden et al. 2004). The AAC of the mitochondrial remnant of *Cryptosporidium* clusters with the mitochondrial transporters of the distantly related *Plasmodium* (the closest relative from which a complete genome DNA sequence is available) (**1**). *Trichomonas* (**5**) and *Entamoeba* (**4**) use alternative members of the mitochondrial carrier family for the transport of ATP across the hydrogenosomal/mitosomal membranes (Dyall et al. 2000; Tjaden et al. 2004; Chan et al. 2005; Tjaden and Leroch, unpublished results)

different adenosine nucleotide transporters of bacterial (or plastid) origin that are likely to import ATP (Katinka et al. 2001; Vivares et al. 2002). Therefore, the presence of different, alternative AACs in *Trichomonas* and *Entamoeba* and their absence in *Giardia* and *Encephalitozoon* cannot easily be explained by the absence of organellar ATP synthesis in *Encephalitozoon*, *Entamoeba*, and *Giardia*, which are likely to require import of ATP for iron–sulphur cluster synthesis (Lill and Mühlenhoff 2005).

The genome projects of *Encephalitozoon*, *Giardia*, and *Entamoeba* could not reveal the presence of true mitochondrial AACs (see http://www.nibi.gov). Thus, there is no evidence for the presence of true mitochondrial

AACs in any of the mitosomes/cryptons or mitochondrial-remnant organelles until now, which, of course, do not produce ATP, and therefore might not require genuine "mitochondrial-type" AACs (Tovar et al. 1999, 2003; Katinka et al. 2001; Williams et al. 2002). An exception is the mitochondrial remnant of *Cryptosporidium*, which possesses mitochondrial-type AACs (Abrahamsen et al. 2004; Fig. 7.3). Therefore, it is tempting to speculate as to whether the "real" mitochondrial-type AACs coevolved with the oxygen-dependent electron-transport chain in connection with the evolution of a complete tricarboxylic acid (TCA) cycle allowing a large-scale ATP production for the benefit of the host (Berry 2003).

## 7.3 Functional Differences Between Mitochondrial and Alternative ADP/ATP Transporters

The phylogenetic analysis of the various ADP/ATP transporters might either argue for a deep evolutionary divergence of these organelles from a (facultatively) anaerobic, premitochondrial ancestor or, alternatively, for a secondary loss of all "true" mitochondrial AACs in all of these organelles, followed by the evolution of alternative ADP/ATP transporters (Tjaden et al. 2004). In any case, phylogenetic evidence supports at least six (but potentially much more) independent origins of hydrogenosomes and mitochondrial-remnant organelles from organelles belonging to the "mitochondrial" family – regardless of whether these ancestral organelles were strictly aerobic or facultatively anaerobic (Martin et al. 2001; Embley et al. 2003; Gabaldon and Huynen 2003, 2004; Figs. 7.1, 7.3).

Although remarkable progress has been made in matching the secondary structure and the function in the various members of the mitochondrial carrier family (del Arco and Satrustegui 2005), satisfying information about the function of potential ADP/ATP transporters can only come from a detailed functional analysis (see later). Several methods are currently available: reconstitution in liposomes, in vivo expression, rescue experiments, and functional tests in bacteria (*Escherichia coli*, *Lactococcus lactis*). Such tests involve competition experiments, effector and inhibitor studies (Voncken 2001; Voncken et al. 2002a; Haferkamp et al. 2002; van der Giezen et al. 2002; Tjaden et al. 2004; Chan et al. 2005; Leroch 2006).

Table 7.1 shows that one mitochondrial, two hydrogenosomal, and one mitosomal ADP/ATP transporter all possess different $K_m$ values and have different requirements for a membrane potential. Also, the spectrum of nucleotides which can be transported by the various proteins is quite peculiar: at present, the transporter of *Entamoeba* is the only ADP/ATP transporter that also transports AMP, just like its relative (*brittle-1*) in the plastids of *Solanum tuberosum* (Leroch et al. 2005; Table 7.2). In addition, the different ADP/ATP transporters exhibit a peculiar spectrum of sensitivity, or resistance,

**Table 7.1.** $K_m$ values for ADP and ATP of several heterologously expressed adenine nucleotide carriers determined on intact *Escherichia coli* cells under various energy conditions (coupled and uncoupled). $K_m$ is given in nanomoles per milligram of protein per hour), *E. coli* cells were preincubated with 100 μM carbonyl cyanide 3-chlorophenylhydrazone (CCCP) for 2 min for uncoupling. Uptake studies were performed as described in Haferkamp et al. (2002), Tjaden et al. (2004) Leroch (2006), and Leroch et al. (2005)

| | AAC2(*A.t.*) | AAC1(*N.*sp.) | HMP31(*T.g.*) | ANT1(*E.h.*) |
|---|---|---|---|---|
| $K_m$ ADP (−CCCP) | 14 | 165 | 64 | 502 |
| $K_m$ ADP (+CCCP) | 8 | 155 | 42 | Unchanged |
| $K_m$ ATP (−CCCP) | 22 | 2,325 | 134 | 1,170 |
| $K_m$ ATP (+CCCP) | 6 | 226 | 47 | Unchanged |

*A.t. Arabidopsis thaliana* (mitochondrial), *N.sp. Neocallimastix* sp. L2 (hydrogenosomal), *T.g. Trichomonas gallinae* (hydrogenosomal), *E.h. Entamoeba histolytica* (mitosomal)

**Table 7.2.** Effects of various metabolites on [$\alpha^{32}$P]ADP uptake into *E. coli* cells expressing several adenine nucleotide carriers. For uptake experiments, effectors were always present in a fivefold higher concentration than the given uptake substrates. [$\alpha^{32}$P]ADP uptake by AAC2(*A.t.*), AAC1(*N.*sp.), HMP31(*T.g.*), and ANT1(*E.h.*) was measured at a substrate concentration of 10, 100, 50, and 200 μM, respectively (Voncken et al. 2002; Tjaden et al. 2004; Leroch et al. 2005; Leroch 2006)

| Effector | Percentage of control | | | |
|---|---|---|---|---|
| | AAC2(*A.t.*) | AAC1(*N.*sp.) | HMP31(*T.g.*) | ANT1(*E.h.*) |
| None | 100.0 | 100.0 | 100.0 | 100.0 |
| ADP | 21.6 | 6.8 | 18.3 | 32.6 |
| ATP | 36.6 | 48.8 | 49.5 | 42.6 |
| AMP | 104.7 | 96.4 | 83.7 | 58.1 |
| UTP | 99.2 | 104.5 | 91.6 | 90.3 |
| CTP | 113.7 | 100.0 | 87.3 | 95.0 |
| GTP | 115.5 | 100.8 | 90.2 | 92.1 |
| UMP | 110.3 | 106.4 | 103.9 | 101.2 |
| NADH | 100.2 | 103.4 | 87.1 | 91.8 |
| dATP | 90.0 | 88.3 | 94.6 | 49.8 |
| dTTP | 100.8 | 98.1 | 100.4 | 111.3 |
| dGTP | 87.5 | 91.3 | 82.4 | 97.1 |
| dCTP | 89.0 | 99.7 | 95.8 | 106.5 |
| UDP-Glc | 111.3 | 90.0 | 112.4 | 91.4 |
| ADP-Glc | 109.2 | 94.2 | 96.2 | 92.4 |
| Coenzyme A | 95.8 | 97.4 | 96.3 | 92.8 |

respectively, against bongkrekic acid (BKA), carboxyatractyloside (CAT), *N*-ethylmaleimide (NEM), pyridoxal 5′-phosphate, and mersalyl (Table 7.3). One conclusion might be possible: all true mitochondrial transporters are sensitive to BKA, albeit to a different degree (Table 7.3). Notably, this does not hold true for CAT, which, in general, is believed to exhibit a sensitivity spectrum similar to that of BKA, but clearly binds to different epitopes. In general, all ADP/ATP transporters tested so far have rather characteristic substrate specificities, and specific $K_m$ values. The same is true for their inhibitor spectrum. Thus, the various ADP/ATP transporters are not only different at the DNA level, they are also functionally different. This, and the observation that *Trichomonas* possesses several closely related copies of HMP31 plus two transporters of unknown function, supports the hypothesis that in different evolutionary lines of organisms, different, alternative ADP/ATP transporters evolved, each with specific biochemical properties (Fig. 7.4).

## 7.4 Evolutionary Tinkering in the Evolution of Hydrogenosomes

### 7.4.1 Hydrogenosomes of *Trichomonas vaginalis*

The hydrogenosomes of the trichomonads (Parabasalia) have been studied intensively for more than 30 years (Lindmark and Müller 1973; Müller 1993). Upon initial inspection, these organelles were considered, both morpho-

**Table 7.3.** Effects of various inhibitors on [$\alpha^{32}$P]ADP uptake into *E. coli* cells expressing several adenine nucleotide carriers. *E. coli* cells were preincubated for 10 min with lysozyme (2.5 mg/ml) to allow penetration of the reagents across the outer membrane. The inhibitors used were bongkrekic acid (BKA, 50 μM), carboxyatractyloside (CAT, 1 mM), *N*-ethylmaleimide (NEM, 1 mM), pyridoxal 5′-phosphate (PLP, 2 mM), and mersalyl (100 μM). [$\alpha^{32}$P]ADP uptake by AAC2(*A.t.*), AAC1(*N.*sp.), HMP31(*T.g.*), and ANT1(*E.h.*) was measured at a substrate concentration of 10, 100, 50, and 200 μM, respectively (Voncken et al. 2002; Tjaden et al. 2004; Leroch et al. 2005; Leroch 2006)

| Reagents | Percentage of control | | | |
|---|---|---|---|---|
| | AAC2(*A.t.*) | AAC1(*N.*sp.) | HMP31(*T.g.*) | ANT1(*E.h.*) |
| None | 100.0 | 100.0 | 100.0 | 100.0 |
| BKA | 17.4 | 64.2 | 98.2 | 101.2 |
| CAT | 38.6 | 100.9 | 99.1 | 97.8 |
| NEM | 65.6 | 105.9 | 92.7 | 33.3 |
| PLP | 104.6 | 54.5 | 98.8 | 40.9 |
| Mersalyl | 100.4 | 95.7 | 100.7 | 98.9 |

logically and biochemically, distinct from mitochondria. Subsequent biochemical and molecular studies have changed this view, since it appeared that trichomonad hydrogenosomes possess mitochondrial-like chaperonins, HSP 10, HSP 60, and HSP 70 (Clark and Roger 1995; Bui et al 1996, Germot et al 1996), proteins of the mitochondrial carrier family (HMP31; Dyall et al. 2000, 2004a; Tjaden et al. 2004), and proteins with a certain homology to the 24- and 51-kD subunits of a "mitochondrial-type" complex I (Hrdy et al. 2004; see however Dyall et al. 2004b for an alternative view).

The N-terminal extensions of many hydrogenosomal proteins favoured the assumption that a mitochondrial-type import machinery would facilitate the import of nuclear-encoded hydrogenosomal proteins into the organelle (Dyall and Johnson 2000; Dyall et al. 2000, 2004a; Embley et al. 2003; van der Giezen et al. 2005). Also, the presence of acetate:succinate coenzyme A (CoA)transferase (ASCT) activity (Müller 1993, 1998; van Hellemond et al. 1998), an enzyme activity that is shared by these organelles and certain mitochondria, seemed to suggest a "mitochondrial" ancestry for the hydrogenosomes of *Trichomonas* (Müller 1993, 1998; Dyall and Johnson 2000, Rotte et al. 2000). However, trichomonad hydrogenosomes are clearly different from mitochondria since they lack a genome, ribosomes, cytochromes, an electron-transport chain, cardiolipin, and cristae (Müller 1993, 1998; Benchimol et al. 1996; Clemens and Johnson 2000; Voncken et al. 2002a; Benchimol and Engelke 2003). Also, the proteins with homology to the mitochondrial 24 kD (nuoE) and 51 kD (nuoF) subunits are unlikely to function in a rudimentary mitochondrial complex I, since there are no indications for the presence of other components of a mitochondrial-type electron-transfer chain. Rather the 24 kD (nuoE) and 51 kD (nuoF) subunits (also named Ndh24/Ndh51 or Tvh-22/Tvh-47, respectively, Dyall et al. 2004b; Hrdy et al. 2004) serve as a diaphorase or, alternatively, as an NADH dehydrogenase that allows the reoxidation of NADH and flavin mononucleotide (FMN) and the transfer of the electrons to the organellar hydrogenase as postulated for ciliate and chytrid hydrogenosomes (Akhmanova et al. 1998a; Hackstein et al. 1999; Hrdy et al. 2004; Dyall et al. 2004b; see later).

Whereas the rudiments of a "mitochondrial"-type electron-transport chain might reflect an α-proteobacterial heritage, followed by "loss and modification", the organellar import machinery and the ADP/ATP translocators evolved "later", as essential acquisitions allowing the establishment of the original symbiont as an organelle. Therefore, it is not surprising that the hydrogenosomal import machinery of *Trichomonas* hydrogenosomes appears to exhibit certain peculiar characteristics that are not shared with textbook mitochondria (Dyall et al. 2003, 2004a). Further, *Trichomonas* hydrogenosomes possess ADP/ATP translocators that do not belong to the "mitochondrial" type of AACs – they are distinct both phylogenetically and biochemically (see before; Fig. 7.4, Tables 7.1–7.3). This is exactly what would be predicted for the evolution of shared-derived characters.

On the other hand, the original symbiont possessed a broad spectrum of metabolite transporters (Gabaldon and Huynen 2003, 2004). Consequently, the import of pyruvate and malate by the trichomonad hydrogenosomes is not surprising, albeit these organelles neither use pyruvate dehydrogenase (PDH) nor a TCA cycle for the catabolism of pyruvate. These hydrogenosomes metabolize pyruvate through pyruvate:ferredoxin oxidoreductase (PFO) and hydrogenase to acetate, carbon dioxide, and hydrogen (Müller 1993, 1998). Acetate formation from acetyl-CoA is believed to be coupled to the substrate-level phosphorylation of succinate via the enzyme ASCT (Müller 1993); this route should yield one ATP per mole of pyruvate consumed. Although this (ASCT) is one of the few enzymatic reactions that are known to be shared between hydrogenosomes and certain mitochondria, confirmation of whether or not the genes encoding proteins with ASCT activity are the same in trichomonads and kinetoplastids could not been provided until now (cf. Rivière et al. 2004). The generation of a proton motive force (PMF) has not yet been studied in detail, but the generation of a proton gradient/PMF by trichomonad hydrogenosomes cannot be excluded (Turner and Lushbaugh 1991; Humphreys et al. 1998). Notably, potential "mitochondrial" $F_1F_0$ ATP synthases of *Trichomonas* have not been identified so far, and the observation that trichomonad hydrogenosomes can serve as cellular $Ca^{2+}$ stores (Biagini et al. 1997a) is rather circumstantial with respect to a "mitochondrial" ancestry. Thus, the relationship between the hydrogenosomes of *Trichomonas* and *aerobically* functioning mitochondria is much less evident than suggested in many publications: rather, the hydrogenosomes of *Trichomonas* represent a peculiar type of organelle that is characteristic for an adaptation to an anaerobic environment in one out of the many eukaryotic evolutionary lines – starting from a facultatively aerobic, primitive organelle that did not yet possess the whole set of "textbook-mitochondrial" functions (Gabaldon and Huynen 2003). Clearly, the hydrogenosomes of *Trichomonas* are *not* a blueprint for all the various hydrogenosomes found in other taxa of unicellular eukaryotes (see later).

### 7.4.2 Hydrogenosomes of Anaerobic Chytrids: an Alternative Way to Adapt to Anaerobic Environments

Anaerobic chytrids are important symbionts in the gastrointestinal tract of many herbivorous mammals. Their life cycle consists of an alternating flagellated zoospore stage and a vegetative phase when a multinucleated mycelium is formed. The hyphae of the rhizomycelial system attach to the digesta and secrete a broad spectrum of fibrolytic enzymes that are very efficient in digesting plant polymers (Teunissen et al. 1991; Orpin 1994; Yarlett 1994). These organisms are highly adapted to intestinal environments; their optimal growth temperature coincides with the body temperature of their mammalian hosts, and during almost their whole life cycle, they live and multiply under anoxic conditions (Orpin 1994). The anaerobic chytrids evolved from mitochondria-bearing ancestors, since DNA sequence analysis reveals a clustering of both aerobic yeasts and fungi with the anaerobic chytrids (Bowman et al. 1992; cf. Paquin et al. 1995; Paquin and Lang 1996; Hackstein et al. 1999). Also, an analysis of biochemical and morphological traits consistently establishes a close relationship between chytrids and other fungi (Ragan and Chapman 1978). Akhmanova et al. (1998b) demonstrated that several enzymes of mitochondrial origin, which lack putative targeting signals, were retargeted to the cytoplasm (in active form) and were no longer present in the hydrogenosomes. Consequently, there is little doubt that the chytrids living in the gastrointestinal tract of herbivorous mammals have secondarily adopted an anaerobic life style (Hackstein et al. 1999).

Consequently, anaerobic chytrids such as, for example, *Neocallimastix* and *Piromyces*, possess hydrogenosomes that are structurally and functionally clearly different from the hydrogenosomes of the ciliate *N. ovalis*, the amoeboflagellate *Psalteriomonas lanterna*, and the parabasilid *Trichomonas vaginalis* (Coombs and Hackstein 1995; Hackstein et al. 1999, 2001; Voncken 2001). Like the hydrogenosomes of the amoeboflagellate *P. lanterna* (Hackstein, unpublished results), and of the parabasilid *T. vaginalis* (Clemens and Johnson 2000), the hydrogenosomes of *Neocallimastix* and *Piromyces* lack a genome (van der Giezen et al. 1997; Hackstein, unpublished results). But unlike *T. vaginalis* hydrogenosomes, the chytrid hydrogenosomes rely on malate and not pyruvate for hydrogen formation. The imported malate is oxidatively decarboxylated by a hydrogenosomal malic enzyme, and it had been postulated that the resulting pyruvate is oxidized further by PFO to acetyl-CoA. The reducing equivalents should be transferred via ferredoxin to hydrogenase, thus maintaining the redox balance (Marvin-Sikkema et al. 1992, 1993, 1994). However, Akhmanova et al. (1999) and Boxma et al. (2004) showed that the hydrogenosomes of anaerobic chytrids perform a bacterial-type mixed-acid fermentation during which pyruvate is split into acetyl-CoA and formate by a pyruvate:formate lyase (PFL), and is *not* oxidatively decarboxylated by a PFO. Accordingly, genes encoding PFL-activating (Gelius-Dietrich and Henze 2004) and PFL-inactivating enzymes (Boxma et al. 2004) have been identified. Using PFL instead of PFO allows the formation of reduced equivalents to be avoided, and consequently the chytrid hydrogenosome excretes formate and acetate as end products of its energy metabolism (Fig. 7.5). Moreover, the vast majority of the carbon flow through the hydrogenosome is mediated by pyruvate, which is imported from the cytosol and metabolized in the hydrogenosome *without* hydrogen formation (Boxma et al. 2004). Obviously, the hydrogenosomes of anaerobic chytrids followed a different strategy when adapting to anaerobic environments. Avoiding the formation of reduced equivalents renders hydrogen production a rudimentary metabolic activity in these organelles, which might rely on the hydrogenase activity solely to cope with varying concentrations of NADH that might result from metabolic imbalances (see later).

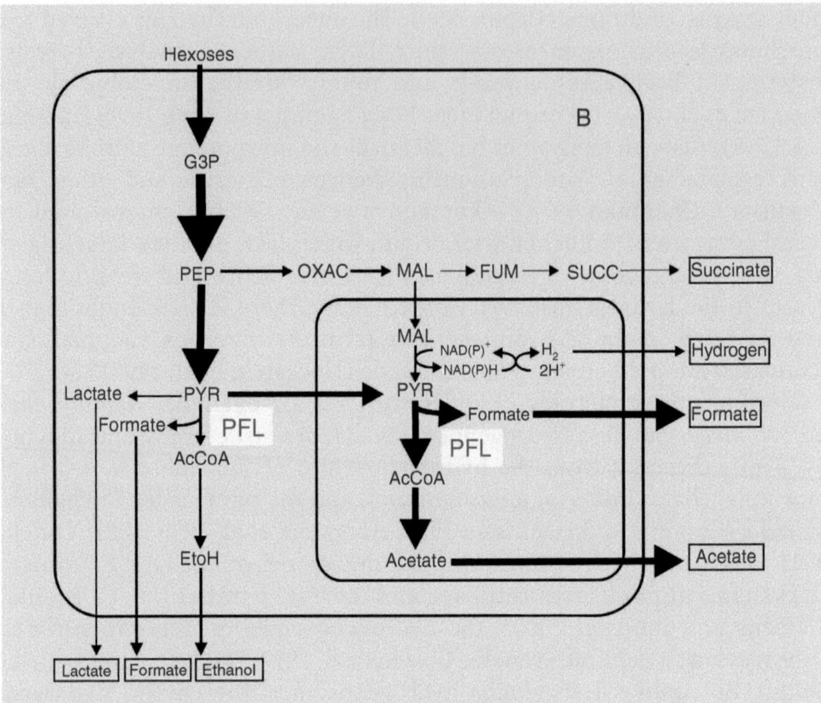


**Fig. 7.5.** Pyruvate:formate lyase (PFL) and not pyruvate:ferredoxin oxidoreductase (PFO) is the key enzyme in the hydrogenosomal energy metabolism of the anaerobic chytrid of *Piromyces* sp. E2 (Akhmanova et al. 1999; Boxma et al. 2004). Hydrogen formation depends solely on malate; it can become marginal under certain metabolic conditions. The relative fluxes through the pathways are indicated by the thickness of the arrows, which are proportional to the calculated fluxes in the presence of 0.3% fructose. Use of PFL instead of PFO allows the formation of reduced equivalents to be avoided. Consequently, chytrid hydrogenosomes excrete formate and acetate as end products of their energy metabolism. (From Boxma et al. 2004)

As mentioned earlier, the absence of a hydrogenosomal genome hampers a straightforward analysis of the origin of the organelle, but the functional *and* phylogenetic analyses of the AACs from anaerobic chytrid hydrogenosomes clearly support a fungal mitochondrial origin for these organelles (see before; Figs. 7.3, 7.4, Tables 7.1–7.3) (van der Giezen et al. 2002; Voncken 2001; Voncken et al. 2002a; Haferkamp et al. 2002; Tjaden et al. 2004). Given that chytrid hydrogenosomes lack a genome, the AACs (being of eukaryotic origin) and HSP 60 (being of α-proteobacterial origin) are the "second-best" markers for tracing the evolutionary history of these organelles (Andersson and Kurland 1999). Phylogenetic analysis of both genes unequivocally reveals a descent from fungal mitochondria (Figs. 7.3, 7.4) (Voncken 2001; van der Giezen et al. 2002, 2003; Voncken et al. 2002a), in agreement with the earlier finding that typical mitochondrial enzymes had been retargeted to the cytoplasm in the course of the evolution of the chytrid hydrogenosomes

(Akhmanova et al. 1998b). Chytrid hydrogenosomes are therefore clearly distinct from the hydrogenosomes of *Trichomonas,* and of those from anaerobic ciliates (see later).

### 7.4.3 Hydrogenosomes of Anaerobic Ciliates: At Least One Appears to Be a Missing Link

Ciliates represent a very diverse and complex group of unicellular eukaryotes. Anaerobic species were identified in at least eight of the 22 orders of ciliates (Fenchel and Finlay 1995). There is a certain agreement that anaerobic ciliates evolved secondarily from aerobic ancestors since some higher ciliate taxa comprise both aerobic and anaerobic species (Fenchel and Finlay 1995; Embley et al. 1995; Embley et al. 1997, 2003; Hackstein et al. 2001, 2002). Bona fide hydrogenosomes are present in seven of the 22 orders, but their identification as "hydrogenosomes" is nearly exclusively based on ultrastructural inference or the presence of methanogenic symbionts (Vogels et al. 1980; van Bruggen et al. 1983, 1984, 1986; Goosen 1988, 1990a, b; Zwart et al. 1988; Gijzen et al. 1991; Biagini et al. 1997b). Also, the evidence that these hydrogenosomes evolved independently and repeatedly from mitochondria is rather circumstantial (Embley et al. 1995). Notably, Akhmanova et al. (1998a) and van Hoek et al. (2000a) have presented straightforward evidence for the presence of a mitochondrial genome in the hydrogenosomes of *N. ovalis*, an anaerobic, heterotrichous ciliate that inhabits the intestinal tract of cockroaches (Akhmanova et al. 1998a; van Hoek et al. 1998, 1999, 2000b).

This genome, initially identified by immunocytochemistry and cellular fractionation, hosts ribosomal RNA (rRNA) genes that are abundantly expressed, and phylogenetic analysis revealed a clustering among the mitochondrial rRNA genes of aerobic ciliates (van Hoek et al. 2000a; Hackstein et al. 2001; Boxma et al. 2005). The phylogenies of the nuclear 18S rRNA genes of the ciliates are congruent with the small subunit rRNA genes of their mitochondria and hydrogenosomes (Akhmanova et al. 1998a; van Hoek et al. 1998, 2000a, b; Hackstein et al. 2001). Moreover, the hydrogenosomes of *N. ovalis* possess cristae and electron-dense structures similar in size to 70S ribosomes (Akhmanova et al. 1998a; Boxma et al. 2005). Bioinformatical analysis of a 14-kb fragment of the hydrogenosomal genome revealed the presence of two genes encoding ribosomal proteins (*rpl2* and *rpl14*), one transfer RNA gene, and four genes encoding proteins of a mitochondrial complex I, i.e. *nad2, nad4L, nad5*, and *nad7*. Three additional components of a mitochondrial complex I appeared to be encoded by nuclear genes (24, 51, and 75 kD). Also, two of the four components of mitochondrial complex II (SDH-α and SDH-β, also known as the Fp and Ip subunits), were identified, but no components of the mitochondrial complexes III and IV have been so far (Boxma et al. 2005; Hackstein, unpublished results).

A reciprocal Smith–Waterman (Smith and Waterman 1981) sequence comparison between about 2,000 six-frame-translated clones from our genomic DNA library of *N. ovalis* and the yeast (Cotter et al. 2004) and human (Sickmann et al. 2003) mitochondrial proteomes revealed the presence of more than 50 additional nuclear genes encoding potential mitochondrial proteins (Boxma et al. 2005). These proteins encompass PDH, representatives of the mitochondrial carrier family, including several AACs (Voncken et al. 2002a; Fig. 7.4), components of a mitochondrial import machinery, the TCA cycle, and proteins involved in a mitochondrial protein, fatty acid, and amino acid metabolism. Phylogenetic analysis consistently revealed a ciliate origin – in cases where DNA sequences of ciliate homologues were available; thus, a ciliate mitochondrial origin of the hydrogenosome of *N. ovalis* is obvious, with the exception of the [Fe]-hydrogenase (see later). The presence of a hydrogenase, which identifies the organelles as hydrogenosomes, together with components of a mitochondrial electron-transport chain raises questions about the function of such a chimera.

As mentioned before, seven out of the 14 "minimal" components of a mitochondrial complex I have been identified, together with two out of the four components of a mitochondrial complex II. The lack of evidence for the presence of the mitochondrial complexes III and IV matches the lack (or very low concentration) of oxygen in the gut environment, where *N. ovalis* thrives. Oxygen is unlikely to serve as a terminal electron acceptor, and prolonged exposure to oxygen kills *N. ovalis*. The insensitivity against cyanide and salicylhydroxamic acid (SHAM), an inhibitor of alternative oxidases, reinforces the conclusion that a potential activity of mitochondrial complexes III and IV is unlikely (Fig. 7.6). In contrast, a function of mitochondrial complexes I and II is obvious, since the organelles stain with rhodamine 123, indicating the presence of a proton gradient generated by the activity of the inferred "mitochondrial" complex I. Complex I proton-pumping activity could be proven by the sensitivity of the rhodamine 123 staining against various approved inhibitors of mitochondrial complex I, which interfere with three different epitopes characteristic for a functional mitochondrial complex I (Boxma et al. 2005; Fig. 7.6). Moreover, the excretion of succinate, (besides acetate, lactate, and ethanol) strongly suggests a function of mitochondrial complex II as fumarate reductase, which is likely to accept electrons from complex I through a rhodoquinone 9 (Fig. 7.7).

Such a metabolism ("fumarate respiration") is well known from anaerobic mitochondria (Tielens et al. 2002; Tielens and van Hellemond, Chap. 6 in this volume), but is unique in combination with a hydrogenase that might compete with the fumarate reductase for the same substrates. This hydrogenase of *N. ovalis* represents a novel type of [Fe]-only or [FeFe]-hydrogenase that allows $H_2$ formation to be coupled directly to the reoxidation of NADH. The [Fe]-hydrogenase is linked covalently with a protein, which possesses NAD and FMN binding sites, and a ferredoxin-like FeS module that allows transfer

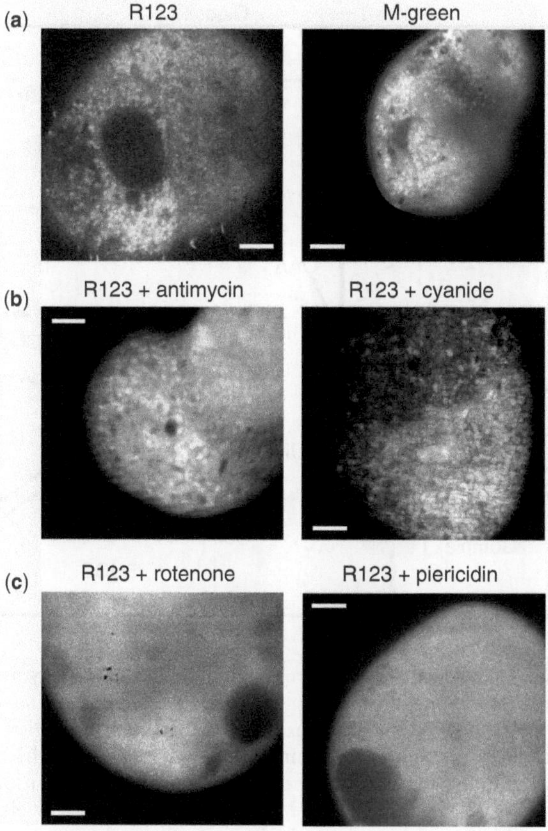


**Fig. 7.6.** Hydrogenosomes of *Nyctotherus ovalis* exhibit complex I activity. The organelles stain with rhodamine 123 (*R123*) and Mitotracker green FM (*M-green*) (**row A**). The R123 staining is not affected by treatment with cyanide or antimycin A (**row B**), but disappears after treatment with 5 mM carbonyl cyanide *p*-trifluoromethoxyphenylhydrazone (FCCP) (not shown). This indicates that the membrane potential of the organelles does not depend on the activity of complexes III and IV. On the other hand, the R123 staining disappears when the cells are treated with well-characterized inhibitors of mitochondrial complex I such as rotenone or piericidin (**row C**). Since also the complex I inhibitors 1-methyl-4-phenylpyridinium and fenazaquin interfere with R123 staining (not shown), it has to be concluded that the membrane potential of the hydrogenosomes of *N. ovalis* is caused by the activity of complex I (Boxma et al. 2005). *Scale bars* 10 μm

of electrons to the catalytic site of the hydrogenase (Akhmanova et al. 1998a; Vignais et al. 2001; Voncken et al. 2002b). It appears to be a mosaic of proteins of δ-proteobacterial and β-proteobacterial origins (Akhmanova et al. 1998a; Horner et al. 2000, 2002; Voncken et al. 2002b; Boxma et al. 2005). The β-proteobacterial origin of the 24 kD (*hoxF*) and 51 kD (*hoxU*) genes is statistically well supported and, besides this, made likely by the presence of the mitochondrial paralogues 24 kD (nuoE) and 51 kD (nuoF) which are subunits of the functional "mitochondrial" complex I. The 24 kD (*hoxF*) and 51 kD

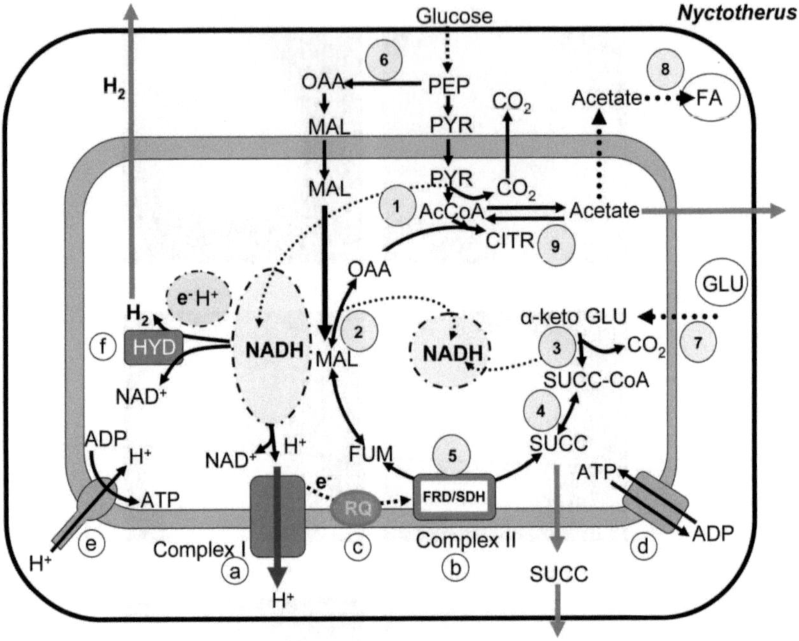


**Fig. 7.7.** The major end products of the hydrogenosome of *N. ovalis* are hydrogen, $CO_2$, acetate, and succinate. The excretion of succinate (besides acetate, lactate, and ethanol) strongly suggests a function of mitochondrial complex II as fumarate reductase, which is likely to accept electrons from complex I through a rhodoquinone 9 (Boxma et al. 2005, Tielens et al. 2002). The hydrogenosomes of *N. ovalis* perform a chimerical metabolism that exhibits mitochondrial and hydrogenosomal traits. The basis is provided by a ciliate mitochondrial metabolism. However, the tricarboxylic acid (*TCA*) cycle is incomplete, and it is likely that the TCA cycle is used in a reductive way in the hydrogenosomes of *N. ovalis*. *a*, *b* Active components of the mitochondrial electron transport chain, *c* rhodoquinone 9, *d* AAC, ADP/ATP translocator (carrier), mitochondrial type, *e* ATP synthase (hypothetical), **1** pyruvate dehydrogenase, **2** malate dehydrogenase, **3** α-ketoglutarate dehydrogenase, **4** succinyl coenzyme A (*CoA*) synthase, **5** succinate dehydrogenase/fumarate reductase, **6** phosphoenolpyruvate carboxykinase, **7** 2-oxoglutaraat/malate translocator, **8** acetyl/propionyl CoA carboxylase (a-subunit), **9** acetyl coenzyme A Synthase

(*hoxU*) subunits are NADH dehydrogenases, which are modules of a heterotetrameric [NiFe]-hydrogenase of *Ralstonia eutropha*; this [NiFe]-hydrogenase is especially resistant against oxygen (Bleijlevens et al. 2004).

## 7.5 Why an [Fe]-Only Hydrogenase?

The use of fumarate as an endogenous electron acceptor requires a well-controlled balance between the various catabolic and anabolic reactions in the cell. Depending on the metabolic state of the cell, the NADH pool might be subject to large fluctuations. The presence of alternative oxidases in anaerobic mitochondria demonstrates how the cell can cope with such fluctuations in the NADH pool (Tielens and van Hellemond, Chap. 6 in this volume). Such an alternative oxidase appears to be absent in N. ovalis, and the hydrogenase could fulfil the task of regulating the NADH pool. More detailed knowledge of the hydrogenosomal metabolism of *N. ovalis*, however, is required to judge the various metabolic roles of the hydrogenosome. The current data (Boxma et al. 2005) support a role in amino acid and lipid metabolism – besides its role in energy metabolism. Therefore, changes in the NADH pool in the response to varying metabolic activities are likely that would require a homeostatic adjustment.

In the case of the hydrogenosomes of *N. ovalis* all our data identify this organelle as a ciliate-type mitochondrion that produces hydrogen. The presence of mitochondrial complex I and II respiratory-chain activity, in combination with hydrogen formation, characterizes the *N. ovalis* hydrogenosome as a true missing link in the evolution of mitochondria and hydrogenosomes. It has to be noted, however, that the *N. ovalis* type of hydrogenosomes is restricted to ciliates belonging to a particular taxonomic group, i.e. the Armophoridae and Clevelandellids (Hackstein et al. 2001; Boxma et al. 2005). Anaerobic ciliates belonging to other taxa seem to have quite different hydrogenosomes.

## 7.6 Conclusions

We have reviewed here the evidence that all the three hydrogenosomes that have been studied so far in more detail are functionally and structurally distinct. They evolved independently in the various evolutionary lines of unicellular eukaryotes in the course of adaptation to anaerobic environments. Nevertheless, all available evidence supports the interpretation that all these three types of hydrogenosomes are the consequence of evolutionary tinkering starting with a mitochondrial-type organelle, which, however, was clearly different from the present-day textbook mitochondria (cf. Gabaldon and Huynen 2003, 2004). The discovery of a missing link, which combines a mitochondrial with a hydrogenosomal metabolism (Boxma et al. 2005), terminates the discussions about an inherent weakness of the endosymbiont hypothesis because it reveals the existence of anaerobic eukaryotes which combine an anaerobic and an aerobic metabolism as postulated in the hydrogen hypothesis (Martin and Müller 1998; Martin 2005). These findings allow an extrapolation to the more exotic, more degenerated members of the "mitochondrial" family: mitosomes and mitochondrial remnants (Fig. 7.1). Also these organelles are much more likely to be retrograde, highly adapted members of the mitochondrial family than the results of different, independent endosymbioses.

## References


Abrahamsen MS, Templeton TJ, Enomoto S, Abrahante JE, Zhu G, Lancto CA, Deng M, Liu C, Widmer G, Tzipori S, Buck GA, Xu P, Bankier AT, Dear PH, Konfortov BA, Spriggs HF, Iyer L, Anantharaman V, Aravind L, Kapur V (2004) Complete genome sequence of the Apicomplexan, *Cryptosporidium parvum*. Science 304:441–445

Akhmanova A, Voncken FG, van Alen A, van Hoek A, Boxma B, Vogels GD, Veenhuis M, Hackstein JHP (1998a) A hydrogenosome with a genome. Nature 396:527–528

Akhmanova A, Voncken FGJ, Harhangi H, Hosea KM, Vogels GD, Hackstein JHP (1998b) Cytosolic enzymes with a mitochondrial ancestry from the anaerobic chytrid *Piromyces* sp E2. Mol Microbiol 30:1017–1027

Akhmanova A, Voncken FGJ, Hosea KM, Harhangi H, Keltjens JT, op den Camp HJ, Vogels GD, Hackstein JHP (1999) A hydrogenosome with pyruvate formate-lyase: anaerobic chytrid fungi use an alternative route for pyruvate catabolism. Mol Microbiol 32:1103–1114

Amiri H, Karlberg O, Andersson SG (2003) Deep origin of plastid/parasite ATP/ADP translocases. J Mol Evol 56:137–150

Andersson SGE, Kurland CG (1999) Origin of mitochondria and hydrogenosomes. Curr Opin Microbiol 2:535–541

Balk J, Lill R (2004) The cell's cookbook for iron-sulfur clusters: recipes for fool's gold? Chem Bio Chem 5:1044–1049

Benchimol M, Engelke F (2003) Hydrogenosome behavior during the cell cycle in *Tritrichomonas foetus*. Biol Cell 95:283–293

Benchimol M, Johnson PJ, de Souza W (1996) Morphogenesis of the hydrogenosome: an ultrastructural study. Biol Cell 87:197–205

Berry S (2003) Endosymbiosis and the design of eukaryotic electron transport. Biochim Biophys Acta 1606:57–72

Biagini GA, Finlay BJ, Lloyd D (1997a) Evolution of the hydrogenosome. FEMS Microbiol Lett 155:133–140

Biagini GA, Hayes AJ, Suller MTE, Winters C, Finlay BJ, Lloyd D (1997b) Hydrogenosomes of *Metopus contortus* physiologically resemble mitochondria. Microbiology 143:1623–1629

Bleijlevens B, Buhrke T, van der Linden E, Friedrich B, Albracht SPJ (2004) The auxiliary protein HypX provides oxygen tolerance to the soluble [NiFe]-hydrogenase of *Ralstonia eutropha* H16 by way of a cyanide ligand to nickel. J Biol Chem 279:46686–46691

Bowman BH, Taylor JW, Brownlee AG, Lee J, Lu SD, White TJ (1992) Molecular evolution of the fungi: relationship of the Basidiomycetes, Ascomycetes, and Chytridiomycetes. Mol Biol Evol 9:285–296

Boxma B, Voncken F, Jannink S, van Alen T, Akhmanova A, van Weelden SWH, van Hellemond JJ, Ricard G, Huynen M, Tielens AGM, Hackstein JHP (2004) The anaerobic chytridiomycete fungus *Piromyces sp* E2 produces ethanol via pyruvate:formate lyase and an alcohol dehydrogenase E. Mol Microbiol 51:1389–1399

Boxma B, de Graaf RM, van der Staay GWM, van Alen TA, Ricard G, Gabaldon T, van Hoek AHAM, Moon-van der Staay SY, Koopman WJH, van Hellemond JJ, Tielens AGM, Friedrich T, Veenhuis M, Huynen MA, Hackstein JHP (2005) An anaerobic mitochondrion that produces hydrogen. Nature 434:74–79

Bui ETN, Bradley PJ, Johnson PJ (1996) A common evolutionary origin for the mitochondria and hydrogenosomes. Proc Natl Acad Sci USA 93:9651–9656

Bullerwell CE, Lang BF (2005) Fungal evolution: the case of the vanishing mitochondrion. Curr Opin Microbiol 8:362–369

Burger G, Gray MW, Lang BF (2003) Mitochondrial genomes: anything goes. Trends Genet 19:709–716

Chan KW, Slotboom DJ, Cox S, Embley TM, Fabre O, van der Giezen M, Harding M, Horner DS, Kunji ERS, Leon-Avila G, Tovar J (2005) A novel ADP/ATP transporter in the mitosome of the microaerophilic human parasite *Entamoeba histolytica*. Curr Biol 15:737–742

Clark CG, Roger AJ (1995) Direct evidence for secondary loss of mitochondria in *Entamoeba histolytica*. Proc Natl Acad Sci USA 92:6518–6521

Clemens DL, Johnson PJ (2000) Failure to detect DNA in hydrogenosomes of *Trichomonas vaginalis* by nick translation and immunomicroscopy. Mol Biochem Parasitol 106:307–313

Coombs GH, Hackstein JHP (1995) Anaerobic protists and anaerobic ecosystems. In: Brugerolle G, Mignot J-P (eds) Protistological actualities. Proceedings of the second European congress of protistology, Clermont-Ferrand, France. Universite Blaise Pascal, Clermont-Ferrand, pp 90–101

Cotter D, Guda P, Fahy E, Subramaniam S (2004) MitoProteome: mitochondrial protein sequence database and annotation system. Nucleic Acids Res 32:D463–D467

del Arco A, Satrustegui J (2005) New mitochondrial carriers: an overview. Cell Mol Life Sci 62:2204–2227

Dyall SD, Johnson PJ (2000) Origins of hydrogenosomes and mitochondria: evolution and organelle biogenesis. Curr Opin Microbiol 3:404–411

Dyall SD, Koehler CM, Delgadillo-Correa MG, Bradley PJ, Plümper E, Leuenberger D, Turck CW, Johnson PJ (2000) Presence of a member of the mitochondrial carrier family in hydrogenosomes: conservation of membrane-targeting pathways between hydrogenosomes and mitochondria. Mol Cell Biol 20:2488–2497

Dyall SD, Lester DC, Schneider RE, Delgadillo-Correa MG, Plümper E, Martinez A, Koehler CM, Johnson PJ (2003) *Trichomonas vaginalis* HMP35, a putative pore-forming hydrogenosomal membrane protein, can form a complex in yeast mitochondria. J Biol Chem 278:30548–30561

Dyall SD, Brown, MT, Johnson PJ (2004a) Ancient invasions: from endosymbionts to organelles. Science 304:253–257

Dyall SD, Yan WH, Delgadillo-Correa MG, Lunceford A, Loo JA, Clarke CF, Johnson PJ (2004b) Non-mitochondrial complex I proteins in a hydrogenosomal oxidoreductase complex. Nature 431:1103–1107

Embley TM, Martin W (1998) A hydrogen-producing mitochondrion. Nature 396:517–519

Embley TM, Finlay BJ, Dyal PL, Hirt RP, Wilkinson M, Williams AG (1995) Multiple origins of anaerobic ciliates with hydrogenosomes within the radiation of aerobic ciliates. Proc R Soc Lond B Biol Sci 262:87–93

Embley TM, Horner DA, Hirt RP (1997) Anaerobic eukaryote evolution: hydrogenosomes as biochemically modified mitochondria? Trends Ecol Evol 12:437–441

Embley TM, van der Giezen M, Horner DS, Dyal PL, Bell S, Foster PG (2003) Hydrogenosomes, mitochondria and early eukaryotic evolution. IUBMB Life 55:387–395

Esser C, Ahmadinejad N, Wiegand C, Rotte C, Sebastiani F, Gelius-Dietrich G, Henze K, Kretschmann E, Richly E, Leister D, Bryant D, Steel MA, Lockhart PJ, Penny D, Martin W (2004) A genome phylogeny for mitochondria among alpha-proteobacteria and a predominantly eubacterial ancestry of yeast nuclear genes. Mol Biol Evol 21:1643–1660

Feagin JE (2000) Mitochondrial genome diversity in parasites. Int J Parasit 30:371–390

Fenchel T, Finlay BJ (1995) Ecology and evolution in anoxic worlds. Oxford University Press, Oxford

Gabaldon T (2005) Origin and evolution of the mitochondrial proteome. Applications for protein function prediction in the eukaryotes. Thesis, Nijmegen

Gabaldon T, Huynen MA (2003) Reconstruction of the proto-mitochondrial metabolism. Science 301:609–609

Gabaldon T, Huynen MA (2004) Shaping the mitochondrial proteome. Biochim Biophys Acta 1659:212–220

Gabaldon T, Snel B, van Zimmeren F, Memrika W, Tabak H, Huynen MA (2006) Origin and evolution of the peroxisomal proteome. Biol Direct 1:8

Gelius-Dietrich G, Henze K (2004) Pyruvate formate lyase (PFL) and PFL activating enzyme in the chytrid fungus Neocallimastix frontalis: a free-radical enzyme system conserved across divergent eukaryotic lineages. J Eukaryot Microbiol 51:456–463

Germot A, Philippe H, LeGuyader H (1996) Presence of a mitochondrial-type 70-kDa heat shock protein in *Trichomonas vaginalis* suggests a very early mitochondrial endosymbiosis in eukaryotes. Proc Natl Acad Sci USA 93:14614–1467

Germot A, Philippe H, LeGuyader H (1997) Evidence for loss of mitochondria in Microsporidia from a mitochondrial-type HSP70 in *Nosema locustae*. Mol Biochem Parasitol 87:159–168

Gijzen HJ, Broers CAM, Barughare M, Stumm CK (1991) Methanogenic bacteria as endosymbionts of the ciliate *Nyctotherus ovalis* in the cockroach hindgut. Appl Environ Microbiol 57:1630–1634

Goosen NK, Horemans AMC, Hillebrand SJW, Stumm CK, Vogels GD (1988) Arch Microbiol 150:165–170

Goosen NK, van der Drift C, Stumm CK, Vogels GD (1990a) End products of metabolism in the anaerobic ciliate *Trimyema compressum*. FEMS Microbiol Lett 69:171–175

Goosen NK, Wagener S, Stumm CK (1990b) A comparison of 2 strains of the anaerobic ciliate *Trimyema compressum* Arch Microbiol 153:187–192

Gray MW, Burger G, Lang BF (1999) Mitochondrial evolution. Science: 283:1476–1481

Hackstein JHP (2005) Eukaryotic Fe-hydrogenases – old eukaryotic heritage or adaptive acquisitions? Biochem Soc Trans 33:47–50

Hackstein JHP, Akhmanova A, Boxma B, Harhangi HR, Voncken FG (1999) Hydrogenosomes: eukaryotic adaptations to anaerobic environments. Trends Microbiol 7:441–447

Hackstein JHP, Akhmanova A, Voncken F, van Hoek A, van Alen T, Boxma B, Moon-van der Staay SY, van der Staay G, Leunissen J, Huynen M, Rosenberg J, Veenhuis M (2001) Hydrogenosomes: convergent adaptations of mitochondria to anaerobic environments. Zoology 104:290–302

Hackstein JHP, van Hoek AHAM, Leunissen JAM, Huynen M (2002) Anaerobic ciliates and their methanogenic endosymbionts. In: Seckbach J (ed) Symbiosis: mechanisms and model systems. Kluwer, Dordrecht, pp 451–464

Haferkamp I, Hackstein JHP, Voncken FGJ, Schmit G, Tjaden J (2002) Functional integration of mitochondrial and hydrogenosomal ADP/ATP carriers in the *Escherichia coli* membrane reveals different biochemical characteristics for plants, mammals and anaerobic chytrids. Eur J Biochem 269:3172–3181

Horner DS, Foster PG, Embley TM (2000) Iron hydrogenases and the evolution of anaerobic eukaryotes. Mol Biol Evol 17:1695–1705

Horner DS, Embley TM (2001) Chaperonin 60 phylogeny provides further evidence for secondary loss of mitochondria among putative early-branching eukaryotes. Mol Biol Evol 18:1970–1975

Horner DS, Heil B, Happe T, Embley TM (2002) Iron hydrogenases – ancient enzymes in modern eukaryotes. Trends Biochem Sci 27:148–153

Hrdy I, Hirt RP, Dolezal P, Bardonova L, Foster PG, Tachezy J, Embley TM (2004) *Trichomonas* hydrogenosomes contain the NADH dehydrogenase module of mitochondrial complex I. Nature 432:618–622

Humphreys M, Ralphs J, Durrant L, Lloyd D (1998) Confocal laser scanning microscopy of trichomonads: Hydrogenosomes store calcium and show a membrane potential. Eur J Protistol 34:356–362

Katinka MD, Duprat S, Cornillot E, Metenier G, Thomarat F, Prensier G, Barbe V, Peyretaillade E, Brottier P, Wincker P, Delbac F, El Alaoui H, Peyret P, Saurin W, Gouy M, Weissenbach J, Vivares CP (2001) Genome sequence and gene compaction of the eukaryote parasite *Encephalitozoon cuniculi*. Nature 414:450–453

Lang BF, Burger G, OKelly CJ, Cedergren R, Golding GB, Lemieux C, Sankoff D, Turmel M, Gray MW (1997) An ancestral mitochondrial DNA resembling a eubacterial genome in miniature. Nature: 387:493–497

Leon-Avila G, Tovar J (2004) Mitosomes of *Entamoeba histolytica* are abundant mitochondrion-related remnant organelles that lack a detectable organellar genome. Microbiology 150:1245–1250

Leroch M (2006) Molekulare, biochemische und physiologische Eigenschaften von Transportproteinen aus der MCF (mitochondrial carrier family) in Pflanzen und Protisten. PhD thesis, Technische Universität Kaiserslautern, Germany

Leroch M, Kirchberger S, Haferkamp I, Wahl M, Neuhaus HE, Tjaden J (2005) Identification and characterization of a novel plastidic adenine nucleotide uniporter from *Solanum tuberosum*. J Biol Chem 280:17992–18000

Lill R, Mühlenhoff U (2005) Iron-sulfur-protein biogenesis in eukaryotes. Trends Biochem Sci 30:133–141

Lindmark DG, Muller M (1973) Hydrogenosome, a cytoplasmic organelle of the anaerobic flagellate *Tritrichomonas foetus*, and its role in pyruvate metabolism. J Biol Chem 248:7724–7728

Martin W (2005) The missing link between hydrogenosomes and mitochondria. Trends Mirobiol 13:457–459

Martin W, Müller M (1998) The hydrogen hypothesis for the first eukaryote. Nature 392:37–41

Martin W, Hoffmeister M, Rotte C, Henze K (2001) An overview of endosymbiotic models for the origins of eukaryotes, their ATP producing organelles (mitochondria and hydrogenosomes), and their heterotrophic lifestyle. Biol Chem 382:1521–1539

Marvin-Sikkema FD, Lahpor GA, Kraak MN, Gottschal JC, Prins RA (1992) Characterization of an anaerobic fungus from llama faeces. J Gen Microbiol 138:2235–2241

Marvin-Sikkema FD, Kraak MN, Veenhuis M, Gottschal JC, Prins RA (1993) The hydrogenosomal enzyme hydrogenase from the anaerobic fungus *Neocallimastix* sp. L2 is recognized by antibodies directed against the C-terminal microbody targeting signal SKL. Eur J Cell Biol 61:86–91

Marvin-Sikkema FD, Driessen AJM, Gottschal JC, Prins RA (1994) Metabolic energy generation in hydrogenosomes of the anaerobic fungus *Neocallimastix*: Evidence for a functional relatioship with mitochondria. Mycol Res 98:205–212

Müller M (1993) The hydrogenosome. J Gen Microbiol 139:2879–2889

Müller M (1998) Enzymes and compartmentalization of core energy metabolism of anaerobic protists – a special case in eukaryotic evolution? In: Coombs GH, Vickerman K, Sleigh MA, Warren A (eds) Evolutionary relationships among protozoa. The Systematics Association, special volume series 56. Kluwer, Dordrecht, pp 109–132

Nasirudeen AMA, Tan KSW (2004) Isolation and characterization of the mitochondrion-like organelle from *Blastocystis hominis*. J Microbiol Methods 58:101–109

Nosek J, Tomaska L (2003) Mitochondrial genome diversity: evolution of the molecular architecture and replication strategy. Curr Genet 44:73–84

Orpin CG (1994) Anaerobic fungi: taxonomy, biology and distribution in nature. In: Mountfort DO, Orpin CG (eds) Anaerobic fungi. Dekker, New York, pp 1–45

Palmieri F (1994) Mitochondrial carrier proteins. FEBS Lett 346:48–54

Paquin B, Forget L, Roewer I, Lang BF (1995) Molecular phylogeny of *Allomyces macrogynus*: congruency between nuclear ribosomal RNA- and mitochondrial protein-based trees. J Mol Evol 41:657–665

Paquin B, Lang BF (1996) The mitochondrial DNA of *Allomyces macrogynus*: the complete genomic sequence from an ancestral fungus. J Mol Biol 255:688–701

Ragan MA, Chapman DJ (1978) A biochemical phylogeny of the protists. Academic, New York

Regoes A, Zourmpanou D, Leon-Avila G, van der Giezen M, Tovar J, Hehl AB (2005) Protein import, replication, and inheritance of a vestigial mitochondrion. J Biol Chem 280:30557–30563

Riordan CE, Ault JG, Langreth SG, Keithley JG (2003) *Cryptosporidium parvum* Cpn60 targets a relict organelle. Curr Genet 44:138–147

Rivière L, van Weelden SWH, Glass P, Vegh P, Coustou V, Biran M, van Hellemond JJ, Bringaud F, Tielens AGM, Boshart M (2004) Acetyl:succinate CoA-transferase in procyclic *Trypanosoma brucei* – gene identification and role in carbohydrate metabolism. J Biol Chem 279:45337–45346

Roger AJ (1999) Reconstructing early events in eukaryotic evolution. Am Nat 154:S146–S163

Roger AJ, Svard SG, Tovar J, Clark CG, Smith MW, Gillin FD, Sogin ML (1998) A mitochondrial-like chaperonin 60 gene in *Giardia lamblia*: Evidence that diplomonads once harbored an endosymbiont related to the progenitor of mitochondria. Proc Natl Acad Sci USA 95:229–234

Rotte C, Henze K, Müller M, Martin W (2000) Origins of hydrogenosomes and mitochondria – commentary. Curr Opin Microbiol 3:481–486

Saraste M (1999) Oxidative phosphorylation at the *fin de siecle*. Science 283:1488–1493

Sickmann A, Reinders J, Wagner Y, Joppich C, Zahedi R, Meyer HE, Schonfisch B, Perschil I, Chacinska A, Guiard B, Rehling P, Pfanner N, Meisinger C (2003) The proteome of *Saccharomyces cerevisiae* mitochondria. Proc Natl Acad Sci USA 100:13207–13212

Smith TF, Waterman MS (1981) Identification of common molecular subsequences. J Mol Biol 147:195–197

Teunissen MJ, Smits AA, Op den Camp HJ, Huis in't Veld JH, Vogels GD (1991) Fermentation of cellulose and production of cellulolytic and xylanolytic enzymes by anaerobic fungi from ruminant and non-ruminant herbivores. Arch Microbiol 156:290–296

Tielens AG, Rotte C, van Hellemond JJ, Martin W (2002) Mitochondria as we don't know them. Trends Biochem Sci 27:564–572

Tjaden J, Haferkamp I, Boxma B, Tielens AGM, Huynen M, Hackstein JHP (2004) A divergent ADP/ATP carrier in the hydrogenosomes of *Trichomonas gallinae* argues for an independent origin of these organelles. Mol Microbiol 51:1439–1446

Timmis JN, Ayliffe MA, Huang CY, Martin W (2004) Endosymbiotic gene transfer: Organelle genomes forge eukaryotic chromosomes. Nat Rev Genet 5:123–135

Tovar J, Fischer A, Clark CG (1999) The mitosome, a novel organelle related to mitochondria in the amitochondrial parasite *Entamoeba histolytica*. Mol Microbiol 32:1013–1021

Tovar J, Leon-Avila G, Sanchez LB, Sutak R, Tachezy J, van der Giezen M, Hernandez Müller M, Lucocq JM (2003) Mitochondrial remnant organelles of *Giardia* fuction in iron sulfur protein maturation. Nature 426:172–176

Turner AC, Lushbaugh WB (1991) Three aspecific ATPases in *Trichomonas vaginalis*. Comp Biochem Physiol B 100:691–696

Vanacova S, Liston DR, Tachezy J, Johnson PJ (2003) Molecular biology of the amitochondriate parasites, *Giardia intestinalis, Entamoeba histolytica and Trichomonas vaginalis*. Int J Parasitol 33:235–255

van Bruggen JJA, Stumm CK, Vogels GD (1983) Symbiosis of methanogenic bacteria and sapropelic protozoa. Arch Microbiol 136:89–95

van Bruggen JJA, Zwart KB, van Assema RM, Stumm CK, Vogels GD (1984) *Methanobacterium formicicum*, an endosymbiont of the anaerobic ciliate *Metopus striatus* McMurrich. Arch Microbiol 139:1–7

van Bruggen JJA, Zwart KB, Herman JGF, van Hove EM, Assema RM, Stumm CK, Vogels GD (1986) Isolation and characterization of *Methanoplanus endosymbiosus* sp.nov., an endosymbiont of the marine sapropelic ciliate *Metopus contortus* Quennerstedt. Arch Microbiol 144:367–374

van der Giezen M, Sjollema KA, Artz RRE, Alkema W, Prins RA (1997) Hydrogenosomes in the anaerobic fungus *Neocallimastix frontalis* have a double membrane but lack an associated organelle genome FEBS Lett 408:147–150

van der Giezen M, Slotboom DJ, Horner DS, Dyal PL, Harding M, Xue GP, Embley TM, Kunji ERS (2002) Conserved properties of hydrogenosomal and mitochondrial ADP/ATP carriers: a common origin for both organelles. EMBO J 21:572–579

van der Giezen M, Birdsey GM, Horner DS, Lucocq J, Dyal PL, Benchimol M, Danpure CJ, Embley TM (2003) Fungal hydrogenosomes contain mitochondrial heat-shock proteins. Mol Biol Evol 20:1051–1061

van der Giezen M, Cox S, Tovar J (2004) The iron-sulfur cluster assembly genes iscS and iscU of Entamoeba histolytica were acquired by horizontal gene transfer. BMC Evol Biol 4:7

van der Giezen M, Tovar J, Clark CG (2005) Mitochondrion-derived organelles in protists and fungi. Int Rev Cytol 244:175–225

van der Giezen M, Tovar J (2005) Degenerate mitochondria. EMBO Rep 6:525–530
van Hellemond JJ, Opperdoes FR, Tielens AGM (1998) Trypanosomatidae produce acetate via a mitochondrial acetate:succinate CoA transferase. Proc Natl Acad Sci USA 95:3036–3041
van Hoek AHAM, van Alen TA, Sprakel VSI, Hackstein JHP, Vogels GD (1998) Evolution of anaerobic ciliates from the gastrointestinal tract: phylogenetic analysis of the ribosomal repeat from *Nyctotherus ovalis* and its relatives. Mol Biol Evol 15:1195–1206
van Hoek AHAM, Sprakel VSI, van Alen TA, Theuvenet APR, Vogels GD, Hackstein JHP (1999) Voltage dependent reversal of anodic galvanotaxis in *Nyctotherus ovalis*. J Euk Microbiol 46:427–433
van Hoek AHAM, Akhmanova AS, Huynen M, Hackstein JHP (2000a) A mitochondrial ancestry of the hydrogenosomes of *Nyctotherus ovalis*. Mol Biol Evol 17:202–206
van Hoek AHAM, van Alen TA, Sprakel VSI, Leunissen JAM, Brigge T, Vogels GD, Hackstein JHP (2000b) Multiple acquisition of methanogenic archaeal symbionts by anaerobic ciliates. Mol Biol Evol 17:251–258
Vignais PM, Billoud B, Meyer J (2001) Classification and phylogeny of hydrogenases. FEMS Microbiol Rev 25:455–501
Vivares CP, Gouy M, Thomaratb F, Météniera G (2002) Functional and evolutionary analysis of a eukaryotic parasitic genome. Curr Opin Microbiol 5:499–505
Vogels GD, Hoppe WF, Stumm CK (1980) Association of methanogenic bacteria with rumen ciliates. Appl Environ Microbiol 40:608–612
Voncken FGJ (2001) Hydrogenosomes: eukaryotic adaptations to anaerobic environments. Thesis, Nijmegen. Ponsen and Looien, Wageningen
Voncken FGJ, Boxma B, Tjaden J, Akhmanova AS, Huynen M, Verbeek F, Tielens AGM, Haferkamp I, Neuhaus HE, Vogels G, Veenhuis M, Hackstein JHP (2002a) Multiple origins of hydrogenosomes: functional and phylogenetic evidence from the ADP/ATP carrier of the anaerobic chytrid *Neocallimastix* sp. Mol Microbiol 44:1441–1454
Voncken FGJ, Boxma B, van Hoek AHAM, Akhmanova AS, Vogels GD, Huynen M, Veenhuis M, Hackstein JHP (2002b) A hydrogenosomal [Fe]-hydrogenase from the anaerobic chytrid *Neocallimastix* sp L2. Gene 284:103–112
Williams BAP, Hirt RP, Lucocq JM, Embley TM (2002) A mitochondrial remnant in the microsporidian *Trachipleistophora hominis*. Nature 418:865–869
Winkler HH, Neuhaus HE (1999) Non-mitochondrial ATP transport. Trends Biochem Sci 24:64–68
Yaffe MP (1999) The machinery of mitochondrial inheritance and behavior. Science 283:1493–1497
Yarlett N (1994) Fermentation product generation in rumen chytridiomycetes. In: Mountfort DO, Orpin CG (eds) Anaerobic fungi. Dekker, New York, pp 129–146
Yarlett N (2004) Anaerobic protists and hidden mitochondria. Microbiology 150:127–129
Yarlett N, Hackstein JHP (2005) Hydrogenosomes: one organelle, multiple origins. Bioscience 55:657–668
Zwart KB, Goosen NK, van Schijndel MW, Broers CAM, Stumm CK, Vogels GD (1988) Cytochemical-localization of hydrogenase activity in the anaerobic protozoa *Trichomonas vaginalis*, *Plagiopyla nasuta* and *Trimyema compressum*. J Gen Microbiol 134:2165–2170

# 8 The Chimaeric Origin of Mitochondria: Photosynthetic Cell Enslavement, Gene-Transfer Pressure, and Compartmentation Efficiency

THOMAS CAVALIER-SMITH

## Summary

Less than a billion years ago a protoeukaryote host enslaved a probably photosynthetic α-proteobacterium to form the first mitochondrion. After a phagocytosed α-proteobacterium escaped from the phagosome into the host cytoplasm, host carrier proteins of probable peroxisomal origin entered its periplasm via pre-existing β-barrel outer-membrane (OM) proteins related to the usher proteins of proteobacteria and ancestral to Tom40 – the OM protein translocase. Carrier insertion into the bacterial inner membrane (IM) then extracted photosynthesate for the host, giving it an immediate strong selective advantage for permanently enslaving the bacterium by inserting additional proteins into its envelope to improve carrier insertion, notably the Tim22 complex. Pre-existing bacterial periplasmic chaperones evolved into the periplasmic small Tims, whilst the bacterial Omp85 complex evolved into the Sam50 mitochondrial complex for the insertion of β-barrel proteins into the mitochondrial OM from the periplasm. Concomitant massive transfer of thousands of bacterial protein genes into the nucleus was initially very harmful to the host, especially by inserting many bacterial proteins with signal sequences into its endoplasmic reticulum (ER). This disaster was circumvented by the evolution of positive charges on one face of the signal sequences, preventing their entry into the ER and allowing recognition by Tom22 receptors added to the TOM complex, moving them into the mitochondrial periplasm, and by the origin, by gene duplication of Tim22, of the Tim23/Tim17 complex that translocated them into the mitochondrial matrix, where an additional adaptor Tim44 was added to pass them smoothly to the pre-existing bacterial chaperone Hsp70, ATP-driven, motor to pull them efficiently into the matrix; IM proteins were then inserted via the Oxa1 machinery, derived from proteobacterial YidC or a posibacterial relative. Only then did matrix proteins acquire presequences, allowing massive mitochondrial genome reduction through selection for efficiency. I discuss the molecular mechanisms and selective advantages of these major changes that endowed early eukaryote cells with two novel genetic membranes and a radically improved respiratory machinery. I also discuss the relative contribution of symbiont and host genes to the chimaeric mitochondrion and to its now

Origin of Mitochondria and Hydrogenosomes (ed. by William F. Martin and Miklós Müller)

similarly chimaeric slave-owner 'host', and attempt to dispel some misconceptions about the power of sequence-based bioinformatic methods to elucidate the most important changes during the origin of eukaryotes.

## 8.1 Key Early Ideas

"Mitochondria and chloroplasts did not have a purely symbiotic origin but are chimaeric structures incorporating proteins from the host as well as the symbiont". (Cavalier-Smith 1987c)

The origin of mitochondria by the essentially permanent internal enslavement of a purple non-sulphur bacterium ($\alpha$-proteobacterium) (John and Whatley 1975, 1977) by a host protoeukaryotic cell (Whatley et al. 1979) is now well established. There are, however, still many differences of opinion about the details of this major event in cell evolution, as readers of other chapters in this book will readily appreciate. These concern the nature of the host, the precise nature of the symbiont, the evolutionary driving force for enslavement, the molecular mechanisms of the process of enslavement and conversion of symbiont into organelle, details of the diversification of mitochondria and their polyphyletic conversion into hydrogenosomes (anaerobic energy-generating organelles) and mitosomes (even simpler two-membrane organelles), and the timing of their origin. Here I focus on the aerobic origin of mitochondria rather than their secondary anaerobic diversification into hydrogenosomes and mitosomes. I start with a bit of history to introduce my perspective.

Initially it was assumed that the host was a phagotrophic eukaryote, which phagocytosed a bacterium with a single bounding membrane into a food vacuole (Schnepf 1964); Schnepf (1964) suggested that the double mitochondrial envelope was a chimaera of symbiont cytoplasmic membrane and host phagosomal membrane. Sagan (1967) (later Margulis 1970) promoted the chimaeric origin of the envelope for many years but suggested instead that the host was itself a wall-less bacterium and that phagotrophy evolved after its uptake using genes provided by the symbiont (Margulis 1981). This mutualistic symbiotic theory of the origin of the eukaryotic cell and the supposed ER source of the mitochondrial OM became widely popular (Whatley et al. 1979), and still is in some textbooks, but failed to explain the origins of non-mitochondrial features of the eukaryote cell. More seriously, it overlooked the important fact that purple bacteria, like all other *Negibacteria* (all *Eubacteria* except *Posibacteria*; Cavalier-Smith 2002a), have an envelope of two membranes and that it is much more parsimonious to suppose that the bacterial OM directly became the mitochondrial OM and retained its bacterial biogenesis and division mechanisms, rather than being replaced by a radically different host endomembrane that would have to reacquire these properties (Cavalier-Smith 1983a). I argued that the known greater permeability of the OM of mitochondria and chloroplasts was mediated by porins and other OM proteins (Omps) inherited directly from those of the negibacterial OM (Cavalier-Smith 1983a). This view of the tremendous conservatism of the negibacterial envelope was applied earlier to the symbiogenetic origin of the chloroplast (Cavalier-Smith 1982), in opposition to prevailing chimaeric assumptions. It is now vindicated for both organelles by the discovery that the OM Omp85 homologues play a key role in OM biogenesis in mitochondria, chloroplasts, and all negibacteria (Gentle et al. 2005). Negibacterial Omp85, mitochondrial Sam50, and chloroplast Toc75 are related β-barrel proteins essential for inserting all OM β-barrel membrane proteins; β-barrel proteins do not occur in any host membranes or in those of unibacteria, the ancestors of the host part of the eukaryote chimaera (Fig. 8.1).

Considering that phagotrophy and the economic – division-of-labour – advantages of cell compartmentation were the major forces driving the origin of the eukaryote cell (Cavalier-Smith 1975), I attempted to dissociate the symbiogenetic origin of mitochondria from the most fundamental changes in eukaryogenesis (origin of the endomembrane system, nuclear pore complexes, and cytoskeleton) by proposing the Archezoa theory, which postulated that these features evolved substantially prior to the enslavement of mitochondria and that some premitochondrial eukaryotes may still survive (Cavalier-Smith 1983b). The latter part of this hypothesis has been decisively disproved by molecular phylogeny and the discovery of mitosomes (see Chaps. 6, 10, 11 by Tachezy and Doležal, Barberà et al., and Tovar, respectively), though Margulis now adheres to it instead of the prokaryotic host theory, having adopted it after I abandoned it (Margulis et al. 2000). Her original prokaryote host theory had several drawbacks (Cavalier-Smith 1983a), many shared by later prokaryotic host ideas (Martin and Müller 1998). I now consider that the host was neither a fully developed eukaryote cell nor a prokaryote, but a transitional intermediate that had already evolved phagocytosis, and at least a rudimentary endomembrane system and cytoskeleton, and probably also nucleus, mitosis, sex and cilium, but they might not have completed their evolving to their modern form (Cavalier-Smith 2002b). Following Whatley et al. (1979), this may be called the protoeukaryote theory of the mitochondrial host. In this theory compartmentation of respiratory activity played a key role in eukaryogenesis just as supposed by autogenous theories of mitochondrial origin (Cavalier-Smith 1975). Either the three respiratory organelles of eukaryotes – mitochondria, peroxisomes, and ER (all of them consume oxygen, the classical definition of respiration) overlapped temporally in their origins or (perhaps most likely) mitochondria arose almost immediately after the ER and peroxisomes. The two latter probably evolved by autogenous diversification of the protoendomembrane system (Cavalier-Smith 1975, 2006a), involving only vertical inheritance in contrast to the origin of mitochondria by foreign cell enslavement.

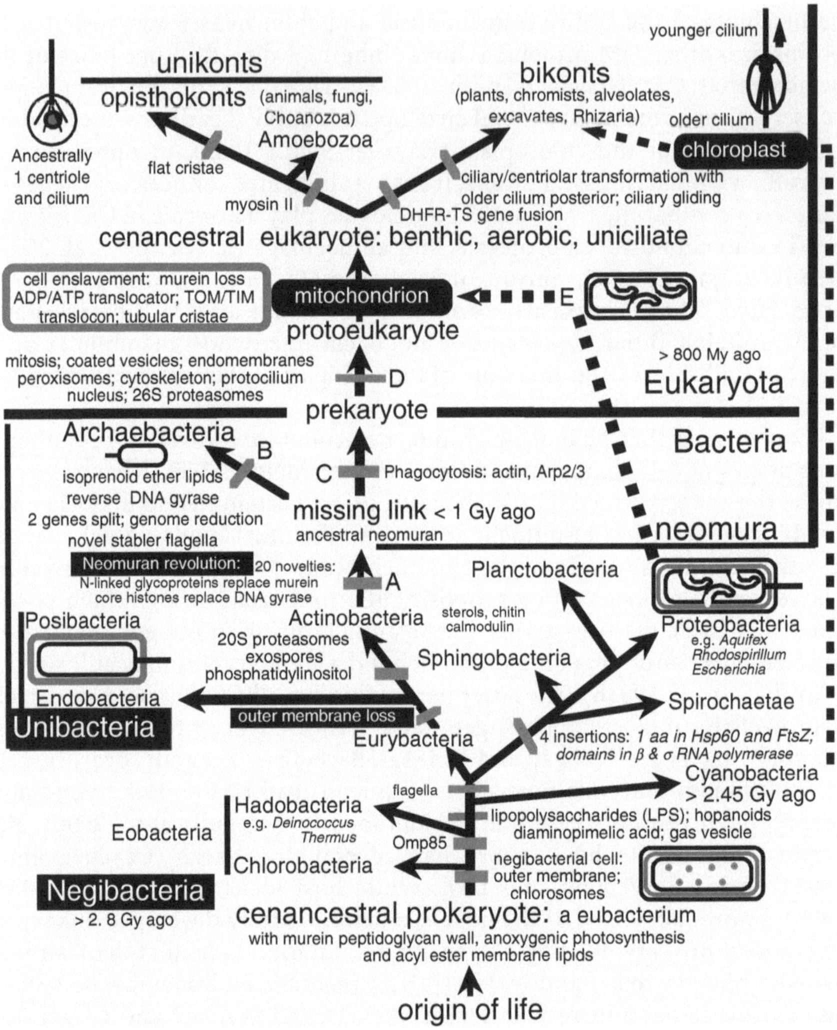


**Fig. 8.1.** The probable phylogenetic context of the internal enslavement of a proteobacterium to make the first mitochondrion. The root of the tree of life is considered to be among negibacteria, between *Chlorobacteria* and *Hadobacteria* (see Cavalier-Smith 2002b, 2006a, c for detailed reviews of the evidence). The Omp85 targeting mechanism for outer-membrane (*OM*) β-barrel proteins evolved after *Chlorobacteria* diverged and was arguably never lost, except incidentally when the negibacterial OM itself was lost to form the ancestral unibacterium, probably by murein hypertrophy; Omp85 was retained and modified by mitochondria for targeting their OM proteins, though in both organelles outer-leaflet lipopolysaccharide was replaced by host phosphatidylcholine making a chimaeric OM. The protoeukaryote was formed from the ancestral neomuran by massive gene duplication and dramatic divergence of thousands of genes (**D**) to form the endomembrane system, cytoskeleton, and nucleus during and following the origin of phagocytosis (Cavalier-Smith 2006a) (**C**). Their sister stem archaebacterial lineage originated reverse DNA gyrase and underwent lipid and flagellar replacement and approximately threefold genome reduction and massive loss of metabolic enzymes as adaptations to hyperthermophily (**B**).

The primary mechanistic step in the origin of mitochondria was originally thought to be the insertion of the host-coded ATP carrier into its IM (Cavalier-Smith 1983a; John and Whatley 1975); the host could then freely extract its ATP for its own use. The key step in integrating the enslaved bacterium as a genuine organelle was the origin of a generalised protein-import mechanism able to import any proteins with the appropriate N-terminal presequence recognised by this import machinery (the TOM/TIM system) provided that the rest of the molecule had no structures impeding its import and that could not be overridden by chaperones (Cavalier-Smith 1983a). Thinking about the origins of protein import for both mitochondria and chloroplasts was as important as the improving molecular phylogenetic evidence in causing me to accept the symbiogenetic theory of their origin (Cavalier-Smith 1982, 1983a), which I earlier thought unparsimonious and insufficiently well supported (Cavalier-Smith 1975, 1977): 'post translational import ... could have evolved more simply by the reversal of the transenvelope protein export mechanism of a prokaryotic symbiont than by autogenous development' (Cavalier-Smith 1987c). For chloroplasts the discovery of both peptidoglycan and the small organelle genome of *Cyanophora* chloroplasts was the clinching factor, but no such intermediate has been found for mitochondria, so I became convinced of their symbiogenetic origin only after it had been established that the eukaryote host was more closely related to archaebacteria than to purple bacteria (Fox et al. 1980; Van Valen and Maiorana 1980).

However, the common view that archaebacteria were the hosts or directly ancestral to eukaryotes (Martin and Müller 1998; Van Valen and Maiorana 1980) is almost certainly mistaken. Instead I consider that eukaryotes are sisters to archaebacteria and that they diverged from a prokaryotic common ancestor after it acquired about 20 major novel properties not found in eubacteria (Cavalier-Smith 2002a). As these common properties include

Most novel eukaryote-specific genes became so markedly different from their neomuran ancestors that prokaryote homologues are undetectable by sequence comparison. By contrast, when the protoeukaryote enslaved the α-proteobacterium (E) it kept over 1,000 of its genes, mostly for less novel functions that did not require sufficient divergence to conceal their eubacterial ancestry – though only the least divergent of such genes can be recognised specifically as proteobacterial. Only genes that underwent immense quantum evolution at (A), but became exceptionally conservative thereafter (e.g. ribosomal RNA, RNA polymerase) show robust evidence of the sister relationship of eukaryotes and archaebacteria and thus neomuran holophyly. Neomuran genes for metabolic enzymes probably often underwent markedly less change in eukaryotes than in archaebacteria, not having to adapt to hyperthermophily; the least changed may retain actinobacterial signatures, but more changed ones should be hard to distinguish from similarly diverged ones from the enslaved proteobacterium. The Tom/Tim protein-import machinery is probably a chimaera of proteobacterial and host proteins; its origin allowed many host proteins and many proteobacterial enzymes encoded by genes transferred to the nucleus to be translocated into the mitochondrion after evolving topogenic presequences. (Modified from Cavalier-Smith 2006a, c) replacement of the wall peptidoglycan murein that is widespread in eubacteria by novel cotranslationally synthesised N-linked glycoproteins, I refer to archaebacteria and eukaryotes collectively as neomura, and their immediate common ancestor as the cenancestral neomuran (Cavalier-Smith 1987b, 2002a). There is substantial evidence that this neomuran ancestor itself evolved from an actinobacterium during the dramatic neomuran revolution caused by the simultaneous origin of N-linked glycoproteins and two core histones and the coevolutionary changes these caused in ribosomes and DNA-handling enzymes (Cavalier-Smith 2002a). Thus, as Fig. 8.1 indicates, eukaryotes are not chimaeras of a eubacterium and an archaebacterium, as often incorrectly stated, but chimaeras of a eubacterium (the purple bacterium) and a missing link: the protoeukaryote. This protoeukaryote missing link, in turn evolved via a prekaryote (Cavalier-Smith 2002b, 2006a) with primitive phagocytosis but no nucleus from a somewhat earlier missing link, the cenancestral neomuran, a now extinct intermediate between eubacteria and archaebacteria that possessed all 20 properties shared by eukaryotes and archaebacteria (e.g. N-linked glycoproteins, helix 6 in signal recognition particle (SRP) RNA, core histones, and greatly modified DNA-handling enzymes), but none of those many unique to eukaryotes (e.g. endomembrane system and cytoskeleton) or those few unique to archaebacteria (notably isoprenoid ether membrane lipids, special flagella, reverse DNA gyrase, splits in RNA polymerase A and glutamate synthetase) (Cavalier-Smith 2002a). The specific archaebacterial characters evolved only after the stem pre-eukaryote and prearchaebacterial lineages diverged from the neomuran cenancestor (Fig. 8.1). These distinctions are subtle but vital for understanding the origins of eukaryotes, mitochondria, and archaebacteria.

The central issue in the origin of mitochondria is the origin of the TOM complex (protein translocase of the mitochondrial OM, MOM), the multiprotein assembly that endows mitochondria with their individuality and is the fundamental determinant of mitochondrial membrane heredity (Cavalier-Smith 2004b). This problem of its origin has now been largely solved by the identification of a probable bacterial ancestor of its core protein, Tom40, and a unified scenario placing carrier insertion into the IM as the primary focus of mitochondrial enslavement (Cavalier-Smith 2006b), as I elaborate later.

## 8.2 The Host Was a Protoeukaryote Not an Archaebacterium

Ever since Woese and Fox (1977) suggested that the last common ancestor of all life was a precellular incompetent 'progenote' and Van Valen and Maiorana (1980) suggested that eukaryotes evolved from archaebacteria there has been confusion over this issue. Woese and Fox's never remotely tenable idea of the cenancestor as a simple precellular entity has been adequately refuted numerous times by many authors (e.g. Cavalier-Smith 1981, 1987a,b; Peretó et al. 2004) but to this day continues to mislead some who ignore cell biology (Martin and Russell 2003). Since Van Valen and Maiorana (1980) it has been clear that, although some eukaryote genes more closely resemble those of archaebacteria, others are more like those of eubacteria. This has been demonstrated with steadily increasing comprehensiveness and precision with successive advances in DNA sequencing and genomics (Brown and Doolittle 1997; Esser et al. 2004; Golding and Gupta 1995). Too often, genes more similar to those of archaebacteria are called 'archaebacterial'. This is misleading as it begs the question whether archaebacteria are actually ancestors of eukaryotes (i.e. paraphyletic) as Van Valen and Maiorana (1980) suggested, or really their sisters, as ribosomal RNA (rRNA) trees long suggested and I later argued from a cell biological viewpoint (Cavalier-Smith 1987b); if archaebacteria are holophyletic, those genes are neomuran, not archaebacterial. I have strongly opposed the idea of archaebacterial paraphyly (Van Valen and Maiorana 1980) as it entails two changeovers, not just one, in the nature of surface membrane lipids during their history. I proposed the neomuran theory, with the root of the universal tree being in eubacteria, and neomura being holophyletic and derived from posibacterial ancestors, as it involved only one change in membrane lipids and one change in the number of bounding membranes in the history of life and is compatible with the firm evidence from the fossil record that eukaryotes are much younger than eubacteria (probably over 3 times younger; Cavalier-Smith 2006a). In addition, archaebacterial holophyly and eubacterial paraphyly is the only relationship between the three domains compatible not only with the fossil record and a simple interpretation of cell evolution, but also with the mixture of eubacteria-like and archaebacteria-like genes in eukaryotes, without multiplying assumptions devoid of evidence. Despite these major advantages, the theory was usually ignored, until my later discussion (Cavalier-Smith 2002a) that gave greatly increased evidence for polarising the eubacteria/neomuran transition thus, not in reverse. I argued that three gene splits demonstrate archaebacterial holophyly and that the absence of many genes from archaebacteria that are widespread in the other domains was caused by loss in their common ancestor.

As the legend to Fig. 8.1 indicates, the neomuran theory simply explains the observed pattern of gene similarity between eukaryotes and prokaryotes. The theory was recently seriously misrepresented by a claim that it predicts that 'eukaryote nuclear genes should bear greatest overall similarity to their homologues from actinobacteria' (Esser et al. 2004). I have never made that prediction and fail to understand how anyone could deduce it from my theory. I refute it later in this chapter after first discussing the more important issue of how the $\alpha$-proteobacterium was actually enslaved to make the mitochondrion.

There are no known surviving primitively amitochondrial eukaryotes, making it likely that mitochondrial enslavement took place immediately after the evolution of the protoeukaryote, significantly after the origin of phagocytosis, which provided the mechanism of its ingestion, but possibly during the later stages of perfection of such cenancestral eukaryotic characters as efficient nucleocytoplasmic transport and ciliary motion. It is now highly probable that the cenancestral eukaryote had a centriole and cilium and that there are no primitively non-ciliate eukaryotes (Cavalier-Smith et al. 2004; Kudryavtsev et al. 2005; Richards and Cavalier-Smith 2005). Given also the arguments for a simultaneous origin of nuclei and cilia to generate a uniciliate eukaryote cenancestor (similar to *Phalansterium* or *Mastigamoeba* without mitochondria or mitosomes; Cavalier-Smith 1987b), it is likely that the protoeukaryote also was uniciliate, not a simple amoeba – contrary to early ideas (Whatley et al. 1979). Because the root of the eukaryote tree is between bikonts and unikonts (Richards and Cavalier-Smith 2005; Stechmann and Cavalier Smith 2003), in each of which hydrogenosomes and mitosomes evolved polyphyletically from aerobic cristate mitochondria, the notion that hydrogenosomes are relics of early anaerobic eukaryote evolution is mistaken. All of them evolved from fully aerobic mitochondria, as I first argued (Cavalier-Smith 1987c) when opposing their less parsimonious origin by a separate bacterial enslavement (Whatley et al. 1979). Representatives of every protozoan phylum can be isolated from anaerobic as well as aerobic habitats. The capacity to grow aerobically, microaerophilically, or anaerobically is very widespread and probably was ancestral. Thus, mitochondria probably primitively had considerable potential to become hydrogenosomes in some lineages, or more strictly aerobic in others. As other chapters deal with the secondary polyphyletic origins of hydrogenosomes in detail, here I simply point out that they probably all retain the ancestral mitochondrial protein-import mechanism, despite in most cases having entirely lost the proteobacterial intraorganellar genomes (but not all their transferred genes) hundreds of millions of years ago, testifying to reliability of membrane heredity in general and the evolutionary longevity of their two genetic membranes that, though acquired from α-proteobacteria, became new kinds of genetic membrane by the origins of novel protein-import machinery (Cavalier-Smith 2004b).

## 8.3 Was the Slave Initially Photosynthetic?

The more secure new rooting of the eukaryotic tree (Fig. 8.1) is between bikonts, which almost certainly ancestrally had tubular mitochondrial cristae, and unikonts, which comprise the Amoebozoa with tubular cristae and opisthokonts with flat cristae (Richards and Cavalier-Smith 2005; Stechmann and Cavalier Smith 2003). The cenancestral eukaryote probably therefore had tubular mitochondrial cristae. The majority of photosynthetic α-proteobacteria have tubular chromatophores, which carry the dual photosynthetic/respiratory chain and closely resemble the tubular cristae of many ciliate protozoa. Molecular phylogeny does not yet clearly establish whether the closest relative to mitochondria was a phototroph or a heterotroph, though a recent careful analysis suggests that *Rhodospirillum rubrum*, which has tubular chromatophores, is at least as close as any for which there is genomic data (Esser et al. 2004). It does not support the idea that mitochondria are closer to the genomically reduced endosymbionts, e.g. rickettsias (Andersson et al. 2003); as the latter probably did not evolve until after fully developed eukaryotes had arisen, providing them with hosts, they are probably irrelevant to mitochondrial origin. As heterotrophs like *Paracoccus* typically lack invaginations of the cytoplasmic membrane (homologue of the mitochondrial IM), the simplest explanation of the origin of cristae is that they evolved directly from tubular chromatophores of a photosynthetic proteobacterium, not de novo in an enslaved heterotroph (Cavalier-Smith 2002b).

*R. rubrum* alternates between anaerobic photosynthesis with tubular chromatophores and aerobic respiration, but other phototrophs, e.g. *Roseobacter*, have been discovered that photosynthesise under aerobic conditions. It would be worth screening strains of these to see how many can maintain the chromatophores during both aerobic respiration and aerobic anoxygenic photosynthesis and to search among those that can for the strain closest to mitochondria. The major reason for suggesting that the symbiont was initially photosynthetic, however, is not to simplify the origin of cristae, but is because it would have provided a much stronger selective advantage to the host for enslaving it than if both partners were heterotrophic aerobes with oxidative phosphorylation (Cavalier-Smith 2006b). The idea that the symbiont was originally photosynthetic is not new (Woese 1977; Searcy 1992), but the reasons for now favouring it are. Instead of the host benefiting from organic carbon as argued here, Searcy (1992) postulated an implausible intermediate stage where it supposedly benefited instead from sulphur produced by oxidation of hydrogen sulphide by proteobacterial photosynthesis. Not only is it unclear that such a benefit ever existed for a protoeukaryote, but how it might be translated into a selective force for insertion of IM carriers of organic molecules is obscure.

## 8.4 Three Phases of $\alpha$-proteobacterial Enslavement

It is most likely that the protoeukaryote host and enslaved purple symbiont were both facultative aerobes, able to live under anaerobic or aerobic conditions (Cavalier-Smith 2002b); this makes their coming together easy to understand and fits all we know of the diversity of mitochondrial/hydrogenosomal properties as well as of the rest of the eukaryotic cell, especially in the basal kingdom Protozoa. Given that most actinobacteria are aerobes, it is likely that prior to the origin of mitochondria protoeukaryotes had a surface membrane-associated respiratory chain (Cavalier-Smith 2002b), of which the ER respiratory chain may be a relic, and which may originally have been associated with oxidative phosphorylation by the ancestor of the vacuolar ATPase. But after the enslavement of purple bacteria and endomembrane diversification into ER, Golgi, and lysosomes, the vacuolar ATPase specialised in proton pumping to acidify the new intracellular digestive compartments. On that view the ultimate driving force for mitochondrial enslavement was not the dramatic acquisition of an entire novel function – oxidative phosphorylation – by a previously incapable host (John and Whatley 1975; Margulis 1981), but a more prosaic gain in efficiency by a newly phagotrophic cell by compartmentation of different physiological functions between the newly evolved endomembrane system (protein digestion and secretion), peroxisomes (β-oxidation of lipids mainly from the prey) and mitochondria (Cavalier-Smith 2002b).

Recently, however, I argued that, although the ultimate driving force for mitochondrial enslavement was efficiency through cell compartmentation, the proximate driving force probably was the acquisition of a novel function, photosynthesis, but that this was only temporary. According to this novel phagotrophic-host/photosynthetic-symbiont theory, enslavement took place in three distinct phases, each with very distinct primary selective forces and each evolving a mechanistically distinct part of the mitochondrial protein-import machinery (Cavalier-Smith 2006b; Fig. 8.2). First was the insertion of numerous carriers of host origin into the proteobacterial IM; this modified and supplemented bacterial protein-export machinery to yield an efficient import machinery for IM carriers, involving the Tom40 import channel, Tom70 receptor, and IM Tim22 insertion machinery, thus creating a mutualistic endosymbiosis between the facultatively aerobic phagotrophic host and the facultatively aerobic photosynthetic endosymbionts – like that now seen between ciliates and endogenous photosynthetic purple bacteria (Fenchel and Bernard 1993). According to this theory, the ATP/ADP translocator was probably not the first carrier to evolve; more likely another carrier first provided net organic photosynthesate to the host. This would have been especially beneficial when prey was scarce, as in the coral/dinoflagellate endosymbioses that power coral reefs in oligotrophic waters. Carrier import does not require N-terminal presequences or the receptors that recognise them: Tom22 and Tom20. Contrary to all previous theories, I argued that import of proteins with removable presequences evolved only in the second phase of mitochondrial evolution, after evolution of the simpler carrier-import machinery that required only minor modifications to pre-existing bacterial envelope biogenesis mechanisms (Cavalier-Smith 2006b).

The driving force of the second phase was probably completely different – correction of harmful phenotypic costs inevitably stemming from the success of the first phase: once the α-proteobacterium had become a permanent

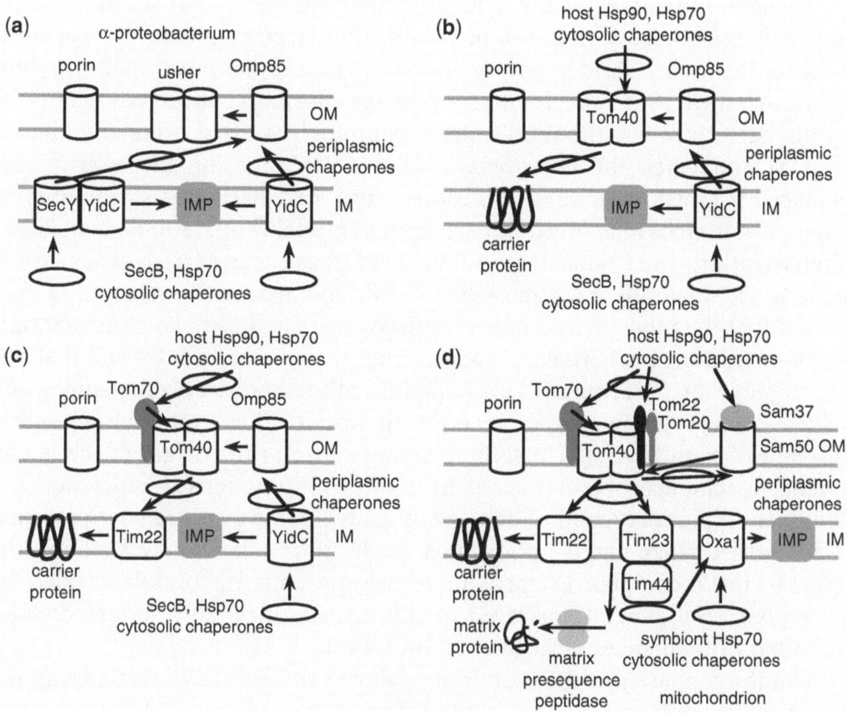


**Fig. 8.2.** Origin of the mitochondrial protein-import systems (**d**) from the ancestral α-proteobacterial protein-export systems (**a**). The OM β-barrel proteins (Sam50, Tom40, porins) all evolved directly from symbiont OM proteins, and Oxa1/Oxa2 evolved from the inner-membrane (IM) protein YidC (or an actinobacterial relative of it). **b** Enslavement was initiated by the insertion of novel carrier proteins able to enter through Tom40 with the help of pre-existing host and periplasmic chaperone proteins and they inserted themselves into the IM. The first carrier, possibly descended from the peroxisomal ATP importer, exported ATP or photosynthesate to the host and gave rise to over 30 other carriers by gene duplication. Their import was made efficient by evolution of a receptor Tom70 inserting itself by its N-terminal membrane-spanning α-helix, helped by Tom40, and Tim22 entering via the Tom40 pore (**c**). Following massive transfer of genes from symbiont to nucleus, the presequence mechanism evolved by gene duplication of Tim22 to Tim23 and Tim17 to generate the IM translocase and the addition of Tom22 and Tom20 to recognise the hydrophobic and positively charged parts, respectively, of the presequences, of Tim44 to transfer the symbiont/matrix chaperone Hsp70 more efficiently onto the emerging nascent proteins, and a matrix peptidase to remove the presequence. Pre-existing periplasmic chaperones diversified and adapted to prevent periplasmic aggregation and improve transfer between membrane–protein complexes, and various less key proteins were added to each to increase stability and improve transfer rates and efficiency. Sec machinery was retained by the excavate jakobid protozoa but lost by other mitochondria (Gray et al. 2004). Acquisition of presequences by about 1,000 proteobacterial genes transferred to the nucleus allowed loss of their symbiont versions and huge genome reduction, raising efficiency by increasing space for matrix enzymes, reducing nutrient and energy use for wasteful multiple copies of their DNA. IMP: inner membrane proteins other than carriers and translocons. (After Cavalier-Smith 2006b)

mutualistic inhabitant of the cytosol, it was inevitable that sooner or later accidental breakage of its envelope would allow large fragments of its genome to enter the nucleus and be incorporated by recombinational repair into host chromosomes. Such accidental breakage perhaps occurred when the proteobacterium lost the rigid murein peptidoglycan that originally divided its OM and when the new dynamin-based OM division mechanism that replaced it (Hashimoto 2003; McFadden and Ralf 2003) was still inefficient (dynamin may have evolved in part from bacterial MinD; Amos et al. 2004). Transcription and translation of many of these transferred genes would usually be wasteful, sometimes beneficial, and frequently harmful. I suggested that the most serious harm would come from insertion of former bacterial membrane proteins with signal sequences into the host ER and that this was reduced by the sequences evolving into mitochondrial presequences, and new receptors being added to TOM to import them into mitochondria (Cavalier-Smith 2006b). Thus, the primary cause of the origin of the presequence mechanism was not acquisition of novel functions or efficiency, but the phenotypic correction of damage by unavoidable gene-transfer pressure. After presequence-based import was made more efficient by evolving the Tim23/Tim17 complex IM protein translocase, matrix proteins coded by proteobacterial genes transferred to the nucleus inevitably sooner or later acquired presequences by accidental mutation.

The third phase of mitochondrial enslavement, partially overlapping the first two, but predominantly occurring after carrier import and presequence-based import were efficient, was the loss of most proteobacterial genes from mitochondrial DNA and of some host genes duplicated by newly acquired proteobacterial genes. The originally mutualistic symbiosis became enslavement as soon as the mitochondrion lost genes essential for free living and permanent as soon as the host lost its original genes for functions eventually restricted to the mitochondrion. The driving force for this third phase was selection for efficiency, by reduction of duplication and specialisation of labour through compartmentation.

After the origin of organelles by cell enslavement, evolution of organelle-specific protein import allows the host to insert some of its own proteins into it, so such organelles are essentially chimaeric – not purely symbiogenetic (Cavalier-Smith 1987c). Such incorporation of host proteins may be either the addition of novel functions or the replacement of existing ones by functionally similar host proteins: in principle such replacement could be total and largely conceal the symbiogenetic origin of an organelle (Cavalier-Smith 1990), but this has not occurred for mitochondria or chloroplasts. Both the mitochondrion and the rest of the eukaryote cell are chimaeras of protoeukaryote and proteobacterial genes, but establishment of these chimaeras was very asymmetric. Enslavement of a foreign cell by the evolution of generalised protein-import mechanisms creates a one-way gene transfer ratchet (Cavalier-Smith 2003a): not only can any slave genes accidentally transferred to the nucleus become useful within the host cytosol, without any retargeting having to evolve, but others will inevitably acquire import sequences and their proteins retain their original function in the mitochondrion; selection for efficiency will reduce duplication by miniaturising the slave genome. Thus, the α-proteobacterium donated some functions to the host that are now still located in the mitochondrion and others now located elsewhere in the cell because by chance the transferred genes did not acquire mitochondrial import sequences. The host donated many proteins but almost no genes to the mitochondrion. Thus mitochondrial symbiogenesis was not merely modification of the symbiont but involved a real shake-up of the host.

## 8.5 Did Syntrophy or Endosymbiosis Precede Enslavement?

Syntrophic consortia of prokaryotes (Overmann and Schubert 2002) were probably not important in the origin of eukaryotes, despite frequent suggestions of this (Margulis 1981; Martin and Müller 1998), because unlike phagotrophy (Cavalier-Smith 1987b, 2002b) they would not have provided sufficient impetus and means for the origin of the major eukaryote innovations. Symbiotic associations between protozoa and prokaryotes are much more common (Finlay and Esteban 2001) and once the protoeukaryote stage was reached might have been an intermediate stage for organelle origin by enslavement. Endosymbioses like that of a purple non-sulphur bacterium (α-proteobacterium) with the ciliate *Strombidium purpureum* (Fenchel and Bernard 1993) are better models than the rarer ectosymbioses; this consortium can photosynthesise under anaerobic or microaerophilic conditions and respire aerobically when oxygen is present, proving that even to this day such entities really have a niche. Such mutualistic symbioses are now initiated by phagocytosis plus escape from digestion by lysosomes. As soon as phagocytosis originated, such consortia would inevitably be established with high frequency and could undergo mutation in both the host and the symbiont to improve their overall reproduction rate and efficiency compared with unmutated consortia or others that arose independently.

Sooner or later one would evolve the capacity for protein insertion into the symbiont by the host, thereby moving it across the threshold from mutualistic consortium to helotism and thus converting the symbiont to an organelle by permanent internal enslavement. An early key step during the symbiotic phase was the liberation of the symbiont from the food vacuole into the cytoplasm. Although some parasitic α-proteobacteria have evolved specific mechanisms of rapid escape from the phagosome (e.g. *Caedibacter*, rickettsias), such reduced parasites are not likely candidates for mitochondrial ancestry as it is hard to envisage intermediate stages in conversion of a harmful heterotrophic parasite into a beneficial mitochondrion. Such a heterotroph would necessarily be a drain on gross resources. In the classical symbiogenesis theory, in which the host was assumed to have been anaerobic

(Whatley et al. 1979), the several-fold increase in efficiency of ATP generation under aerobic conditions would more than outweigh this; if the host was an aerobe capable both of oxidative phosphorylation and microaerophilic growth and surviving short-term complete anaerobic conditions, as now seems likely (Cavalier-Smith 2002b), the benefit of enslavement would arise from greater efficiency through compartmentation (and perhaps also from a more efficient respiratory chain). Even the latter could be more than sufficient benefit to favour enslavement. However, those benefits would only be reaped after the evolution of the ATP/ADP translocator, which in current cells is nuclear-encoded and inserted into the IM via the Tom22/Tim22 machinery. This raises a chicken and egg problem. If the selective advantage of inserting the translocator was provision of extra ATP to the host, that advantage would not be present before it was inserted, and thus could not have driven the origin of the Tom/Tim machinery: selection does not work for future benefits. The Tom22/Tim22 machinery could not evolve so that it could later insert the translocator. Either the translocator must originally have been inserted by a different mechanism or a different selective force was responsible for the origin of the Tom/Tim machinery.

Symbiosis between a heterotrophic host and photoheterotrophic symbiont provides a very strong initial selective advantage for improving the consortium to the point where TOM/TIM could evolve and later allow insertion of the ATP/ADP translocator. In the classical view of a heterotrophic symbiont like *Paracoccus* (John and Whatley 1975) there is no obvious selective advantage of the consortium prior to insertion of that translocator. The broad adaptive zone of phototrophic intracellular endosymbionts within eukaryote cells has been filled so often independently that its selective advantage and ease of evolution cannot be doubted (e.g. purple bacteria in *Strombidium*, dinoflagellates in corals, green and red algae in foraminifera, and green algae in *Paramecium*, hydra, the flatworm *Convoluta*, and radiozoans). In all such cases it is not ATP that is transferred to the host but small organic molecules. Many free-living phototrophs, including α-proteobacteria (Cavalier-Smith 2006b), are partially leaky to and extrude small organic molecules, so they are preadapted for becoming such endosymbionts, just as phagotrophy by eukaryotic hosts preadapted them for initiating them.

Host mutations that increased permeability of the symbiont to photosynthesate and those in the symbiont that increased its ability to donate them and import products of host metabolism derived from digestion of prey would both be selected at the level of the whole consortium. Until the association was mutually obligate, each partner would also be selected separately; thus, symbiont mutations increasing its permeability when in the host would only survive for free-living stages if they were facultative and switched on or off according to their environment. The symbiont probably kept photosynthesis until after evolution of the ATP/ADP translocator made its respiratory contribution to the host greater than that from host enzymes – putatively originally located in the plasma membrane, but perhaps in the ER or peroxisomes instead by the time of enslavement. Thereafter the host plasma membrane became increasingly specialised for phagotrophy, losing its respiratory functions, and the symbiont chromatophores became specialised for respiration by losing photosynthesis, and the rest of its IM more efficient for organic molecule exchange by evolving new channels and transporters. It may not be possible to determine which of the dozens of carriers now present in the mitochondrial IM (MIM) evolved first, but the fact that most are more divergent than the ATP/ADP translocase (Amiri et al. 2003) is consistent with the idea that it was not the first (Cavalier-Smith 2006b). The mitochondrial carriers are dimers of six-helix proteins that diverged from a common ancestor by repeated gene duplication during mitochondriogenesis; their common ancestor may have evolved from the peroxisomal ATP importer (van Roermund et al. 2001), which is much more similar to them then are any bacterial carriers, which typically have 12 helices (Cavalier-Smith 2006b).

Intracellular endosymbiosis, probably initiated by primitive phagotrophy (Cavalier-Smith 1987c, 2002b), was an essential prerequisite for carrier insertion; extracellular syntrophy, sometimes postulated as the starting point (Martin and Müller 1998), would not have helped; carrier insertion would probably not be mechanistically possible; if it did occur, it would be disadvantageous by extruding proteobacterial metabolites into the environment not the host. The hypothesis that the host was an anaerobic methanogenic archaebacterium (Martin and Müller 1998), rather than a facultatively aerobic, phagotrophic protoeukaryote, is untenable for several previously explained reasons (Cavalier-Smith 2002b). Such a host also lacks as plausible a candidate for the ancestor of the mitochondrial carriers as the peroxisomal carrier.

The ATP/ADP translocase family of plastids, endoparasitic bacteria (rickettsias and chlamydias), and microsporidia is apparently unrelated to the mitochondrial carriers (Amiri et al. 2003) and probably irrelevant to their origin. Lateral transfer has been invoked to explain the scattered phylogenetic distribution of this gene family (Cavalier-Smith 2000b; Wolf et al. 1999) but this has been questioned on the grounds that the divergence within it is as deep as for the mitochondrial ATP/ADP translocase family (Amiri et al. 2003). However Fig. 4 of Amiri et al. (2003) indicates that within plants the parasite/plastid family is evolving twice as fast as the mitochondrial family; if true of the whole tree, this would imply that the whole family may be younger than eukaryotes. If so, it is hard to deny that lateral transfer has been involved. Most likely this gene family evolved from a distantly related bacterial homologue when either rickettsias or chlamydias first became endoparasites of eukaryotes (expected to be after eukaryotes originated) and was then laterally transferred to the other within a co-infected cell (Wolf et al. 1999); later, plastids and microsporidia probably independently acquired the gene from a chlamydia infecting the same cell. There is no reason to suppose that this occurred in the protoeukaryote as postulated by Amiri et al. (2003). Although chlamydias and rickettsias belong to *Planctobacteria* and

*Proteobacteria*, respectively, which are probably sisters (Fig. 8.1), it is unlikely that these transporters were present in their common ancestor (as sometimes postulated; Cavalier-Smith 2000b; Amiri et al. 2003) as this ancestor was probably nearly 3 times as ancient as the first eukaryote and such carriers are unknown from more basal free-living members of either phylum. Nor is the idea that mitochondrial enslavement was initiated because of its protective reduction of cytosolic oxygen levels (Andersson et al. 2003) plausible. The host probably already had two oxygen-using organelles (ER and peroxisomes) under its own control; thus, any such selective force would be weak or non-existent and not require or explain carrier insertion.

## 8.6 The Chimaeric Origin of Mitochondrial Protein Import and Targeting

The idea that mitochondrial and chloroplast protein-import machineries both evolved by 'relatively slight modifications' of the negibacterial protein-export machinery, with some key novelties provided by the host (Cavalier-Smith 1982, 1983a), has proved essentially correct. Clear relics of four major bacterial export systems have persisted: the specifically negibacterial Omp85 system for inserting β-barrel proteins into the OM, and the prokaryotic YidC system for posttranslational insertion into the IM have persisted throughout; the TAT system for native protein transfer across membranes is used by all chloroplast thylakoids, and though probably lost by unikont mitochondria is still present in most bikont mitochondria (occasionally even still mitochondrially encoded; Gray et al. 2004). The universal Sec system, however, was more substantially modified in chloroplasts and is unrecognisable in most mitochondria, except for the presence of SecY genes in the *Reclinomonas* mitochondrial genome, showing that it was present in the mitochondrial ancestor (Gray et al. 2004).

It is sometimes asserted that the TOM/TIM machinery evolved from nuclear genes of the host cell (Andersson et al. 2003); however, this is probably untrue for at least some core components. Thus, Tom40, the core of the OM translocation channel, is a β-barrel protein, requiring the Omp85-derivative Sam50 for its insertion; both are almost certainly negibacterial in origin, as host channels would all be bundles of α-helices. Negibacterial OM channels are always β-barrels, unlike any other membranes in the living world except those of mitochondria and chloroplasts. This compellingly supports the negibacterial OM rather than endomembrane origin of the organelle OMs and the homology of the intermembrane space with the negibacterial periplasm (Cavalier-Smith 1983a). Even the receptor for all mitochondrial carrier protein import, Tom70, might be of negibacterial origin (Cavalier-Smith 2006b), although a host origin, e.g. from a peroxisomal protein such as Pex5, cannot be excluded. I previously pointed out that 'a receptor could be transferred from the IM to the OM simply by transferring its gene into the nucleus', but that, if the Tom receptor originated before such transfer a host origin is more likely (Cavalier-Smith 1987c). Lateral transfer from a third donor is also possible. My preliminary BLAST analysis of mouse Tom70 found stronger hits with cyanobacterial proteins than with putative bikont homologues. This protein might have originated from cyanobacteria, either simply prey, or from cyanobacteria harboured within the prokaryote cytoplasm (Cavalier-Smith 2006b). I previously suggested that early eukaryotes might simultaneously have harboured oxygen-consuming and $CO_2$-generating purple bacteria and oxygen-generating cyanobacteria as a mutualistic intracellular ecosystem at the time of mitochondrial enslavement (Cavalier-Smith 1987c). Although it now seems that chloroplasts were enslaved somewhat later than mitochondria, in an early bikont (Cavalier-Smith 2002b; Cavalier-Smith 2003b), rather than simultaneously as then suggested, such tripartite symbioses probably came into existence soon after phagocytosis made them relatively easy to evolve.

Although most components of the import machinery evolve too fast for their origin to be found by comparative sequence analysis, progress in understanding the roles of the five major macromolecular complexes involved led to a detailed three-stage theory for its origin (Fig. 8.2). Initially, import evolved to insert protein carriers in the IM to tap proteobacterial photosynthesate and ATP. Such IM proteins do not have cleaved presequences or have to be moved across the IM; thus, the Tim23 complex (presequence translocase) probably did not evolve until after the proteobacterium had been enslaved by carrier insertion. The most essential innovations were therefore the Tom70 receptor that recognises all the highly hydrophobic IM carriers (including the ATP/ADP carrier) and the Tom40 channel that conveys them into the periplasm. As the two-pore Tom40 β-barrel channel is most similar to the two-pore OM usher that exports pilus proteins, I suggested that it evolved from a proteobacterial OM usher (Cavalier-Smith 2006b). If pili were lost through endosymbiosis, the first carrier proteins could probably have spontaneously entered the bacterial periplasm through usher. Once there, pre-existing proteobacterial chaperones could reduce their aggregation and allow them to insert themselves spontaneously into the IM. Initially it would not matter if insertion were slow and inefficient or even if carriers inserted themselves randomly into the OM and IM. Any level and rate of insertion that extracted more energy for the host from the symbiont than the cost of their synthesis would be sufficient to start the process. Then selection for faster and more directional insertion would begin; probably both were first achieved by the origin of the Tim22 channel in the IM, which would ensure that carriers entered the IM not the OM, and by modification of pre-existing periplasmic chaperones by gene duplication and divergence to form the various small periplasmic soluble Tims, notably Tim9/Tim10 that transfers the carriers to Tim22.

The key innovations that made mitochondrial enslavement possible and turned the former bacterial OM and IM into two novel genetic membranes (Cavalier-Smith 2004b) were the origin respectively of the TOM complex and the Tim22 complex, which inserts carriers from the periplasm into the IM. The Tom40/Tom22 complex is, like all other determinants of genetic membrane identity (Cavalier-Smith 2000b), autocatalytically self-inserting, presumably through protein self-complementarity. Likewise the Tim22 complex is self-inserting and the key determinant of MIM genetic identity. Today Tim22 requires Tom40 and Tom70 for import (just like the carriers), but Tom40 requires another OM β-barrel protein, Sam50, for its own correct insertion. Tom20 and Tom70 require Tom40 but not themselves for insertion into the OM (Ahting et al. 2005). The scenario of Fig. 8.2 solves the chicken-and-egg problem of the origin of these interdependencies in the protein-import machinery.

The central logic involves two major assumptions. First, that TOM evolved in two separate stages: Tom40 (and associated Tom7) first during the evolution of carrier insertion, and the Tom20/Tom22 presequence receptors distinctly later, after IM carrier insertion had been perfected. Tom22 is notably different between bikonts and opisthokonts, with an extra N-terminal domain in the latter (Macasev et al. 2004); this is consistent with the primary unikont/bikont divergence (Richards and Cavalier-Smith 2005; Stechmann and Cavalier-Smith 2003; Fig. 8.1), but unfortunately Tom22 evolves too rapidly for me to detect amoebozoan homologues by BLAST to find which type they most resemble (interestingly the N-terminal region of Tom7 is somewhat longer in unikonts than bikonts). Second, the OM β-barrel proteins Tom40 and Sam50 evolved from β-barrel proteins already in the OM (encoded by the symbiont genome) and they remained thus encoded during evolution of the carrier-inserting Tim22 complex. The polarity of these β-barrel proteins in the OM and insertion into it from the periplasmic space remained unchanged throughout evolution; only after the presequence-import mechanism and Tom22/Tom20 receptors evolved could they enter from the cytosol rather than the matrix and mitochondrial copies of their genes be deleted. The proteobacterial autocatalytically self-inserting YidC machinery initially served for inserting bacterial IM, periplasmic, and OM proteins; its gene could not be lost from the bacterial genome until after Tom20/Tom22 evolved and Tim23 allowed it (or another eubacterial replacement; posibacteria have relatives) to enter the mitochondrial matrix from the cytosol. It is immaterial to this explanation whether the other import-machinery proteins came from the host (likely for many; no retargeting needed), the symbiont (retargeting needed), or from a foreign source by lateral transfer.

Neither a clear evolutionary homologue nor a functional equivalent of Tom70 has been detected in bikonts, though one was recently found in Amoebozoa (Wojtkowska et al. 2005), so Tom70 goes back at least to the ancestral unikont and is not opisthokont-specific. While my suggestion that its function originated in the eukaryote cenancestor prior to the divergence of unikonts and bikonts (Cavalier-Smith 2006b) might be mistaken, making Tom70 a unikont-specific invention, it would be premature to exclude the simpler possibility, assumed in Fig. 8.2, that Tom70 homologues exist in all eukaryotes but diverged beyond easy recognition during the unikont/bikont divergence and/or the origin of chloroplasts. It is an exaggeration to imply that no proteins related in sequence exist in bikonts (Lister and Whelan 2006; Perry et al. 2006). Toc64 of the plastid OM and mtOM64 of the mitochondrial OM (Chew et al. 2004) have significant sequence similarity to Tom70, but as they have only three, not seven tricopeptide repeats (TPRs) and their functions are unknown it is unclear whether they are orthologous or more distant paralogues. The real problem is that there are so many rapidly evolving divergent eukaryotic TPR paralogues that sorting out their possible relationships is hard without functional and structural data, especially near the base of the tree. Functional and proteomic data are needed for the Tom complex in all the bikont supergroups to clarify this. My suggestion of a cyanobacterial origin for Tom70 (Cavalier-Smith 2006b) need not be correct, as one cannot exclude the possibility that it could have evolved by combining a transmembrane anchor sequence with a pre-existing soluble eukaryotic TPR protein, such as the peroxisome receptor Pex5 (with seven TPRs like Tom70) or an actinobacterial amidase (with strong sequence similarity to mtOM64 and likewise fewer TPRs).

Functional and proteomic data are also needed for the Tom complex in all bikont supergroups to clarify Tom20 origins. Perry et al. (2006) suggest that the green plant Tom20 receptor evolved convergently to that of opisthokonts (animals, fungi) by an independent origin of presequence binding in a different TPR paralogue, on the grounds that its transmembrane helix is near the C-terminus rather than the N-terminus, i.e. its domains are in reverse order compared with those of opisthokonts. However, the assertion that 'no genetic mechanisms are known that could generate such a reversal in the order of structural domains' (Lister and Whelan 2006) is erroneous. Duplications followed by deletions can easily do so. In my view, although an independent origin is conceivable given that Tom20 seems to have less core function than Tom22, it is more likely that Tom20 evolved just once, prior to the eukaryotic cenancestor and was thus rearranged prior to the green plant cenancestor; a tandem-gene triplication followed by four deletions within the cluster could have effected both this rearrangement and the duplication of the single TPR seen in opisthokonts to the double repeat of green plants – the first repeat is very similar in secondary structure to that of opisthokonts (Perry et al. 2006; the second could have diverged more after the duplication). Was this change just a non-adaptive evolutionary accident? A more interesting possibility is that it was selectively favoured by the symbiogenetic origin of chloroplasts, which would have introduced the novel selective advantage of a need for discrimination between mitochondrial presequences and chloroplast transit sequences to prevent chloroplast proteins entering the mitochondrion.

Doubling the number of presequence binding sites per Tom20 receptor might have increased discrimination. Functional and proteomic studies of the Tom complex from red algae and glaucophytes (the other plant groups stemming from the primary origin of chloroplasts), as well as the non-plant bikonts and Amoebozoa, are needed to test this; Tom20 is too short and rapidly evolving for bioinformatic methods alone to elucidate its early evolution. Chromalveolate algae and other photosynthetic bikonts that arose by secondary symbiogenesis should have no such discrimination problem as their transit sequence is downstream from a signal sequence that will ensure that nuclear-coded plastid proteins will be cotranslationally inserted into the rough ER before the transit sequences are made. Thus, if mitochondrion/plastid discrimination was an important selective agent for the putative rearrangement, Tom20 homologues of chromalveolates, excavates, rhizaria, and amoebozoa should be of the ancestral type, unlike in plants.

Tim22 strikingly can act as receptor, pore, and energy transducer for insertion – all in one (Rehling et al. 2003), so adding it alone would give a specific insertion mechanism for multitopic IM proteins like the carriers. With a basically efficient mechanism for importing carriers (Fig. 8.2c), extra proteins could be added to Tom40 to increase its stability (Tom6, Tom7) and similarly to Tim22 (e.g. Tim54, though this evolves so fast that one cannot be sure that it was actually present in the eukaryote cenancestor and not just an opisthokont invention). Insertion into the OM is thermodynamically spontaneous, whereas that into MIM requires the electrical potential across it as power, but no ATP. Once Tim22 evolved, the more complex Tim23/Tim17 machinery that requires presequences and the presequence translocase-associated motor (PAM) motor fuelled by matrix ATP could arise. Tim23/Tim17 are homologues of each other and of Tim22, from which I suggested both evolved by gene duplication after the Tim22 complex became efficient (Cavalier-Smith 2006b); thus, Tim22 could immediately insert the Tim23/Tim17 channel with high efficiency in the IM. Efficient import of membrane proteins (and any periplasmic proteins like cytochrome *c* not needing presequences for recognition) probably evolved thus before any matrix proteins were imported. Initially the Tim23 complex might merely have supplemented Tim22, but could have evolved translocase function as a result of the mutation pressure of copies of thousands of proteobacterial genes being suddenly incorporated into the host nucleus.

## 8.7 Stage 2: Recovery from Massive Organelle–Host Gene Transfer

If 1,000 proteobacterial genes were newly transcribed within the nucleus a significant fraction of their encoded proteins would inevitably be recognised by TOM and enter the periplasm. Assuming that 5% were thus predisposed (as modern proteobacterial proteins are for the modern TOM machinery; Lucattini et al. 2004), 50 proteins would pass into the periplasm. There they could uselessly clog the system and lower efficiency or be taken up by both Tim22 and Tim23 complexes and driven into the membrane by the membrane potential. Any that were originally membrane proteins would stay trapped in the membrane, but with opposite polarity to those coming from the proteobacterial cytosol. For some the reversed polarity would not matter, but for others, e.g. components of the respiratory chain and directional pumps, it could be very harmful and selected against. Other transferred proteins would have signal sequences and be recognised by SRP and be inserted into the ER, where they would often be harmful or wasteful. Such mistargeting is an almost inevitable intermediate stage in the evolution of novel genetic membranes, and its correction would be a much stronger selective force than the import of a single matrix protein (I cannot think of any one such protein that would immediately have a large selective advantage if imported; without this the presequence translocase could not evolve).

Selection would occur at two levels: for deletions of the harmfully transferred gene or for phenotypic suppressors of the harm. One such phenotypic suppression would be to transfer the protein right across MIM into the matrix, where it could be recognised by the YidC protein-insertion machinery and reinsert itself into MIM the correct way round. Such phenotypic correction of the deleterious transfer of numerous membrane protein genes to the nucleus was probably the driving force for evolving presequences and the Tim23 and PAM complexes, not the import of matrix proteins (Cavalier-Smith 2006b). Tom70 might initially recognise other multitopic proteins than the carriers and retarget them into the periplasm. It would be predominantly membrane proteins with a signal sequence rather than matrix proteins that would cause a problem after massive gene transfer. Soluble proteobacterial enzymes that merely duplicated what was already going on in the cytosol (or did it more efficiently) would be merely wasteful: sooner or later deletion would remove them - or the redundant host version, causing enzyme replacement. But some bacterial IM proteins could be more detrimental in the ER.

Modifying their hydrophobic signal sequences by making one face positively charged would prevent binding to SRP and insertion into ER and turn them into proto-presequences for mitochondrial import. Addition of Tom22 to the Tom40 complex to recognise the positive charge enabled targeting to the mitochondrial periplasm instead and then self-insertion by their presequence into MIM. Multispanning IM proteins like Oxa1 (a mitochondrial relative of proteobacterial YidC, needed for protein insertion from the mitochondrial matrix, Fig. 8.2d; whether Oxa1/Oxa2 were derived from proteobacterial YidC, Preuss et al. 2005, or a host version of the shorter homologue found in *Posibacteria*, Tjalsma et al. 2003, is unclear) would be inserted by Tim22; the fact that they transfer directly from TOM to Tim22 without needing periplasmic soluble Tims (Frazier et al. 2003) is consistent with their initial insertion from the cytosol evolving after that of the carriers. If Tom22, perhaps already added to Tom40 to increase its stability or help with carrier import, by chance had a negatively charged groove on its cytosolic domain preadapted to recognise the positive face of a presequence (mitochondrial presequences have one hydrophobic and one hydrophilic positively charged face; Wiedemann et al. 2004), no other innovation was necessary in the carrier-import machinery to allow recognition of presequence-containing IM proteins or their inefficient self-insertion into MIM. Any such protein, made in the host cytosol after gene transfer, would be correctly retargeted into MIM if its signal sequence became more positive. The benefit to the cell would be greatest for those membrane proteins most harmful in the ER. When the mechanism was efficient, any protein could be retargeted, even with little or no selective advantage, by chance mutations of its signal sequence into a presequence.

This phenotypic rescue theory has the same logic as that given to explain the origin of RNA editing, spliceosomal introns, and the elimination of internal sequences in ciliate macronuclei (Cavalier-Smith 1993a, 2004a; Covello and Gray 1993): harmful mutations generated by extreme mutation pressure can often be corrected phenotypically more easily than removing the source of the mutation pressure. Having enslaved the proteobacterium and become totally dependent on it (possibly by then having lost its own respiratory machinery), the host could escape such damaging gene-transfer pressure of thousands of gene products only by evolving a generalised protein-import mechanism to return them to their alien donor: the protomitochondrion. This rain of alien genes could have influenced other features of host evolution. It may have forced modifications onto host ribosomes and SRPs to reduce the chances of translating and inserting proteobacterial proteins into the ER or, probably just as bad, of host ER/secretory proteins into the mitochondrion. This gives a reason additional to those identified previously (Cavalier-Smith 2002a) why protoeukaryote ribosomes and SRPs diverged so markedly from their archaebacterial sisters. This implies coevolutionary divergence of host signal and presequence mechanisms.

Presequence import was made more efficient by adding Tom20 to recognise their hydrophobic face. Being loosely bound to Tom40, Tom20 could colonise the mitochondrial surface, catching potential importees for transfer to the Tom22/Tom40 complex for import. Tom20 probably evolved from a (host or symbiont) protein with a signal sequence that became its membrane-anchor helix: this helix is recognised by SRP but prevented from insertion into the ER by a downstream sequence (Kanaji et al. 2000). Although efficient import of pre-sequence-bearing IM proteins into the periplasm would prevent their harming the ER, it would not directly help mitochondrial function, for if they inserted themselves into the IM using the presequence they would have the wrong polarity (also they might insert themselves instead into the inner side of the OM). However, by partial extrusion into the matrix by Tim22, they could associate with the proteobacterial matrix chaperone Hsp70, be pulled further in, and then reinsert themselves into the membrane with the same polarity as when bacterially encoded and thus function normally. Once this began, however crudely, it was improved by gene duplication of Tim22 to generate a protoTim23/Tim17 complex (both components are homologous to Tim22; Meier et al. 2005) able to do it more efficiently and by addition of Tim50 to foster its direct binding to TOM, bypassing the periplasmic chaperones. Adding Tim44 increased transfer efficiency to matrix mtHsp70. This would yield an efficient generalised mechanism for importing any proteobacterial IM protein whose nuclear gene copy mutated to generate a presequence and whose other structure was compatible with entry through the machinery without clogging it. Once this was achieved, its mitochondrial version was lost. Only two membrane proteins (cytochrome *b* and Cox1) never achieved this (Gray et al. 2004); most did so in the eukaryote cenancestor, but for some by chance this occurred only in some later lineages.

Inevitably many transferred nuclear genes for matrix proteins would thereafter accidentally acquire presequences and be targeted to the matrix; some proteins are predisposed to have presequences (Lucattini et al. 2004), but that is irrelevant to the initial origin of targeting if that involved only carrier targeting. Sooner or later a duplicate signal peptidase would be similarly accidentally retargeted into the mitochondrial matrix. Any presequence it could recognise would be removed, generating a soluble protein; random mutation and selection made other matrix presequences cleavable. Thus, import of soluble proteins probably evolved last of all. As each protein's presequence became efficient, deletion of the mitochondrial version of the gene was favoured by selection for efficiency (Cavalier-Smith 1987c). Gene by gene the mitochondrial genome would diminish, eventually entirely disappearing from hydrogenosomes and mitosomes (Embley et al. 2003). Thus, MIM and MOM became self-perpetuating independently of the genomes they once harboured, but they retain some key bacterial membrane proteins in the same polarity and with the same basic insertion mechanism as used for two billion years of free-living life. Contrary to widespread assumptions (Allen et al. 2005), we need postulate no special reasons other than the difficulty of protein reimport, plus historical accident (Cavalier-Smith 2003a), why most mitochondrial lineages failed to transfer all vital proteins to the nucleus and similarly lose local genomes (de Grey 2005). Postulating general functional/adaptive reasons for gene retention fails to explain why most genes retained in *Reclinomonas* are not in other organisms. It amounts to saying that even cytochrome *b* and Cox1 were in fact successfully retargeted in the past but that whenever this happened it was always the nuclear gene that was effectively deleted, and that whenever the mitochondrial gene was deleted instead that organism was at a selective disadvantage compared with those where both nuclear and mitochondrial versions were retained or where the nuclear version was deleted. This is very unlikely. The fact that most mitochondrial proteins that were not retargeted in some organisms were in others shows that there cannot be a generalised adaptive explanation of retention applicable to all, and makes the implicit assumption that retargeting was generally so easy that it must frequently have been successful even for cytochrome *b* and Cox1; if it was never successful, any functional 'explanation' for retaining these genes in the mitochondrial protein is scientifically vacuous – there never was an intermediate on which such hypothetical selective forces might act. Optimisation arguments that ignore retargeting constraints are irrelevant. My argument that presequence-based import of IM proteins preceded that of matrix proteins is consistent with the greater complexity of the latter, additionally requiring Pam16 (Frazier et al. 2004), I suggest added last of all.

The Tim44 adaptor could only evolve after proteins with presequences were already being imported, but with lower success rate and more misfolding and waste. In contrast to other MIM proteins, Tim22 complex proteins do not traverse the periplasm with soluble chaperones but via a fixed complex between TOM and Tim23 mediated by the large periplasmic domain of Tim50, which may also recognise the presequence. That Tim22 also helps proteins move from TOM to Tim23, but Tim23 is not needed for Tim22 fits my suggestion that Tim22 evolved first. Tim50 is in all eukaryotes, but evolves too fast to detect a bacterial relative; Tim54 of the Tim22 complex might be its ancestor, but Tim54 sequences evolve too fast to be recognisable outside fungi, so tracing its bacterial ancestry may only be possible when its 3D structure is known.

A significant feature of protein targeting in modern mitochondria, at least in opisthokonts, is that it is often cotranslational not posttranslational (Mukhopadhyay et al. 2004): cytosolic ribosomes making mitochondrial proteins associate with the OM (Sylvestre et al. 2003). Most messengers encoding them are associated with the OM, which depends on sequences at their 3′ untranslated region, which is similar among all of them. These associations are functionally important (Margeot et al. 2002), but how they work is unclear. It is unlikely that they played a role in the first phase of mitochondrial origin as the mechanism of transenvelope import must have evolved before they had a selective advantage. Thereafter they served to increase the rate and efficiency of import, progressively more important as the number of genes with presequences increased.

Also important in later stages, but not initially, was lipid import and export, which depends on contacts with host ER (Achleitner et al. 1999; de Kroon et al. 2003). At first all lipids were made inside the mitochondrion by proteins encoded by its genome, but increasingly as protein targeting was perfected their genes were transferred to the nucleus. Either the encoded proteins were retargeted to the mitochondrion or this was avoided by importing their lipid products instead. There is a detailed story to be reconstructed as to how this happened when more is known of lipid exchange mechanisms. Most lipids in mitochondria are the same as those used elsewhere in the cell. Cardiolipin is unique to mitochondria (largely the IM) and is made by nuclear-encoded enzymes probably ultimately of proteobacterial origin (however, BLAST hits for *Arabidopsis* cardiolipin synthase are only slightly better for α-proteobacteria than for actinobacteria; phylogeny is needed to decide). I assume that outer-leaflet OM lipopolysaccharide was replaced relatively early during symbiogenesis by phosphatidylcholine from the host, and that most new OM proteins and their insertion mechanisms evolved in this new lipid environment and did not have to adjust to novel lipids, as the original OM proteins such as Tom40, Sam50/Omp85, and porins would have had to. Cardiolipin synthase's dependence on cardiolipin for acquiring its catalytic conformation (Nowicki et al. 2005) shows the potential importance of protein–lipid coevolution and conservatism.

## 8.8 Mitochondrial Diversification

In addition to their repeated secondary diversification to form anaerobic hydrogenosomes and mitosomes, there would have been many adaptive specialisations of aerobic mitochondria unique to particular lineages. We know far too little about mitochondrial function in protists to be able to reconstruct them. For example, although some ciliates can use nitrate as a terminal electron acceptor (Finlay et al. 1983), we do not know if this is an ancestral or a derived trait. The fact that animals do β-oxidation of lipids in mitochondria as well as peroxisomes may be a derived trait, as plants and fungi seem not to. However, evolutionary losses can confound such conclusions when taxon sampling is sparse. It is likely that the origin of plastids had repercussions on mitochondria through the origin of photorespiration that involves both organelles plus peroxisomes. Protists vary greatly in their preferences for different oxygen concentrations and are likely to have contrasting adaptations to exploit natural redox gradients. Such adaptations probably mainly involve nuclear-coded genes, often perhaps of host origin. The great diversity across lineages in which protein genes were retained in the mitochondrial genome (Gray et al. 2004) is likely mainly to reflect historical accidents in successfully overcoming the graded difficulty of retargeting their encoded proteins rather than adaptive factors. Lineages that by chance evolved transfer RNA (tRNA) import mechanisms (Bhattacharyya et al. 2003) could lose mitochondrial tRNA genes. Other differences among lineages might not be adaptive. For example, most lineages retained the proteobacterial FtsZ for dividing their IM (Kiefel et al. 2004; Miyagishima et al. 2004; Osteryoung and Nunnari 2003), but opisthokonts lost it, presumably after evolving an extra dynamin ring.

## 8.9 Conceptual Aspects of Megaevolution

A major innovation such as the origin of mitochondrial protein targeting typically involves the origins of scores of new genes: probably about 100 in this case. Their complexity accounts for the uniqueness of major innovations.

Understanding each case is greatly helped by recognising that only a very small subset of the changes involves fundamental key innovations, i.e. those that initiate a really novel process or structure and new kind of selective force. For macromolecular assemblies like those involved in protein import, some molecules are central and others peripheral (often geometrically as well as conceptually, e.g. in TOM the key innovation was the central channel, as was the gated voltage-sensitive Tim22 for its complex in MIM). The fundamental conceptual problems in understanding megaevolutionary change are to identify the key enabling innovations, to order them correctly in time, and to explain the selective advantage of each. Understanding the peripheral add-on molecules is easier once this is done; their main role is to speed up the process or improve its specificity or the stability of the complex. Such increase in efficiency is standard in macro- and in microevolution, the domain of population genetics and gradualism. The distinction between key enabling inventions and minor, but cumulatively important, general improvements is seen also in human technology, e.g. for the automobile industry the internal combustion engine and hollow rubber tyres were key innovations from which much else stemmed.

Phylogeny is important to identify the cellular and molecular players, but it does not explain the changes. Indeed it is unavoidably biased away from explaining novelty, by focusing on homology, conserved features, and the least changed genes. This bias and limitation applies both to sequence-based phylogenetic reconstruction and classical cladistic analysis of rare characters, which also necessarily sidesteps the problems of major innovation by relying on conservative aspects of lineage history to establish homology and relationships. The greatest conceptual problems in understanding innovation arise when that innovation is so radical that entirely new genes are formed or pre-existing genes diverge beyond the point of recognition by present bioinformatics methods or structures arise without previously apparent homologues. Likewise traditional population genetics, by focusing on changes in allele frequency of pre-existing genes and recurrent statistical processes, largely sidesteps the fundamental qualitative problems of the origin of unique novel characters and genes. For these, imaginative but critical historical and conceptual analysis of multidimensional aspects of the problem are essential; such transition analysis of major innovation (Cavalier-Smith 1991) does not reject the more fashionable, and easier, approaches of phylogenetics and population genetics, but adds to them in an effort to achieve a more comprehensive synthesis (Cavalier-Smith 2006c). It needs to be more widely used, but thoroughly, with proper depth and breadth; if done superficially it can discredit the whole approach by generating proposals that are obviously inconsistent with much that the scientific community already knows, yet not apparent to their proposer.

Haeckel's classic dictum 'ontogeny recapitulates phylogeny' has many exceptions in the complex world of animal development because of interpolation of novel stages, secondary loss or modification of early stages, and positive selection for novel larval characters that were never present in the ancestor, so is discredited as a means of reconstructing ancestral stages (De Beer 1940). However, for macromolecular assemblies the potential for radically changing early stages of assembly may be much lower than for the more complex and flexible development of whole animals. Successively added proteins in a self-assembly system interact directly by complementary binding sites, so there is much less scope for interpolation or changing one dramatically without influencing its partners than in the higher-order developmental system of animals. Macromolecular assemblies such as the five that constitute the import machinery (Wiedemann et al. 2004) or the three respiratory assemblies in the IM, therefore, may more closely follow the principle that ontogeny recapitulates phylogeny, with a core key molecule such as Tom40 or Tim22 being the primary evolutionary innovation and functionally and geometrically more peripheral molecules being added subsequently. It is striking that one can propose a detailed scenario for the origin of this very complex mitochondrial import machinery that conserves so many intermolecular dependencies and the polarity of the core molecules within the IM and OM throughout evolution (Cavalier-Smith 2006b). This emphasises that such macromolecular assemblies are remarkably conservative in evolution and were assembled over evolutionary time in a logical order, with most major interactions, functions, and selective advantages largely conserved.

The conservatism of multiprotein macromolecular assemblies like ribosomes, signal recognition particles, proteasomes, phycobilisomes, photosynthetic reaction centres, and flagella, makes them more reliable for the cladistic reconstruction of large-scale bacterial phylogeny, and rooting the tree of life, than are typical single-protein metabolic enzymes subject to fewer interactive constraints. Single-gene trees for the latter generally carry too little information for reliable phylogenetic reconstruction of deep branches, even if their evolutionary mode is unbiased across the tree; they often yield mutually contradictory trees. These contradictions sometimes clearly arise because of lateral transfer, which is inherently less likely for whole macromolecular complexes, but more often lack an obvious explanation. Unfortunately, concatenating numerous such proteins, even when obvious lateral transfers are excluded, may amplify rather than avoid systematic biases and thus give a misleading answer. Macromolecular assemblies have particular value in clarifying relationships because they provide 3D geometric/binding information of a similar sort to that which historically was so decisive in establishing correctly the nature of most animal phyla and classes long before sequencing, and which is still often a more reliable guide than gene trees. I therefore gave them strong emphasis in my bacterial megaclassification and phylogeny, in addition to rare genic molecular cladistic characters like indels, gene splits, and fusions; they were particularly valuable in unambiguously rooting the tree of life within negibacteria (Cavalier-Smith 2006c). Intermolecular coevolution is another central feature of megaevolutionary innovation, not only among molecules of one macromolecular assembly such as TOM, but also between those that interact directly with each other (e.g. TOM and the Tim23 complex), and also among complexes and cell components that do not physically directly interact. Such indirect coevolution may occur if they interact independently with a third molecule or just because each forms part of a higher-level system that contributes to the same function and is subject to similar selective forces; thus, both may change in response to the same change in selective forces (Cavalier-Smith 2006a). To understand megaevolution it is therefore as important to consider the selective forces as the physicochemical mechanisms, and it is essential to establish the correct phylogeny and consider interactions among macromolecules, not just evolution of one at a time; all four are needed for transition analysis (Cavalier-Smith 2006c).

## 8.10 Relative Genomic Contributions of the Two Partners

"Each mitochondrial protein will have to be examined in detail for its affinities with purple bacteria on the one hand and posibacteria/archaebacteria on the other to determine whether it came from host or symbiont". (p. 66 in Cavalier-Smith 1987c)

From the topology of Fig. 8.1 it should be clear that if there were a uniform molecular clock (an idea that I refuted for rRNA 25 years ago, Cavalier-Smith 1980, and have never accepted for any protein either), and if one assumes no nuclear contribution from proteobacteria (also contrary to what I have argued), the phagotrophic neomuran theory would expect eukaryotic nuclear genes most to resemble those of *Archaebacteria* rather than actinobacteria; how easy it would be to resolve this similarity would depend on how long stem A is compared with branch B (I suspect stem A is relatively very short, see later, making it hard for any clock-like genes). Given that the proteobacterium certainly contributed many genes to the nucleus, one would predict that many would be closest to *Proteobacteria* and many others closest to *Archaebacteria*, precisely the same prediction as the hydrogen hypothesis (Martin and Müller 1998), which was incorrectly implied to have a closer fit to the similarity data than the neomuran theory (Esser et al. 2004). However, the above prediction is naïve because it ignores both genome size and the non-clock-like nature of protein evolution.

One cannot say whether most genes should be proteobacteria-like or archaebacteria-like without considering genome size – of the original partners, of the cenancestral eukaryote, and of the cenancestral archaebacterium. To their credit, Esser et al. (2004) recognise that genome size matters, but fail to see why. As they show (their Fig. 4c), there is almost no overlap between the genome sizes of archaebacteria and free-living $\alpha$-proteobacteria: the only archaebacterium with a larger genome than the smallest of the *Proteobacteria* is *Methanosarcina mazei*, which the authors correctly contend augmented its genome by secondarily adding numerous eubacterial genes, making it a misleading guide to the size of cenancestral archaebacterial genomes. The much smaller genome size of archaebacteria will bias the search to giving more hits with proteobacteria, irrespective of whether the neomuran or hydrogen hypothesis is correct. The small genome size of archaebacteria is an additional argument to the many previously given (Cavalier-Smith 2002b) against the hydrogen hypothesis (two key ones being the evidence for archaebacterial holophyly and the improbability of an autotrophic anaerobic ancestor to the ancestrally phagotrophic and aerobic eukaryotes). The taxonomic pattern of similarities in their Fig. 4a is dominated by secondary gene loss and genome reduction. Thus, within eubacteria, the deepest dips, which give the impression of intervening peaks, are all caused by secondary gene loss by parasites: the two rickettsias among proteobacteria, two mycoplasmas among endobacteria, *Chlamydia*, and two parasitic spirochaetes. Even the less deep *Helicobacter/Camplylobacter* dip involves parasites. If genes are missing because of loss they will be underrepresented in hits at all levels of similarity. I suggest that the comparably low scores for archaebacteria are caused by their relatively small genome sizes, attributable to massive gene loss in their cenancestor (Cavalier-Smith 2002a), perhaps because many enzymes could not adapt to hyperthermophily.

I suggest that the cenancestral archaebacterium probably had only about 1,500 genes, whereas the proteobacterium that was enslaved probably had about 4,000 if it was photosynthetic. The size of the ancestral neomuran genome is harder to estimate as actinobacteria have an unusually broad range over about fivefold. If such complexities as spores and sterols, ability to make chitin, calmodulin, and phosphatidylinositol were really inherited by eukaryotes from an actinobacterium (Cavalier-Smith 2002b), it is likely that it was a complex cell; large size would also have favoured the origin of phagotrophy, so one might conservatively expect a genome of the order of 6,000–8,000 genes, even allowing for loss of peptidoglycan. If so, the archaebacterial lineage underwent a fourfold to fivefold reduction. By contrast, the cenancestral eukaryote, being an aerobic *Phalansterium*-like cell (Cavalier-Smith 2000a; Cavalier-Smith et al. 2004), with phagotrophy, sex, and cyst formation, probably had a genome of about 14,000 genes. If the enslaved proteobacterium lost about half its genes, and provided 2,000 to the eukaryote, 12,000 genes is a reasonable estimate for the number of protoeukaryote genes ultimately of neomuran (not archaebacterial) origin: about 6 times as many as from the proteobacterium. Thus, conversion of the cenancestral neomuran into the protoeukaryote probably involved a doubling of gene numbers, whether by gene duplications, by duplicating large chromosome segments, or whole genome doubling, or by all three plus loss; contrast this with a fourfold to fivefold reduction by their archaebacterial sisters.

Equally important for interpreting gene similarity data is the falsity of the molecular clock assumption. I have emphasised that this can be violated in two contrasting ways: a branch of the tree may acquire a permanently elevated evolutionary rate, as is true for mitochondrial genes compared with homologues in *Proteobacteria*; or there may be a transiently massive increase in rate in a lineage, followed by dramatic reduction and conservatism: quantum evolution (Cavalier-Smith 2002a). Though extremes of a continuum that actually involves all manner of shifting rates and erratic behaviours that do not obey any statistical model, and should not be expected to, this contrast is important because of its very different consequences for tree proportions. I argued that for the tree in Fig. 8.1 the neomuran revolution entailed dramatic quantum evolution for DNA-handling enzymes (because of the sudden origin of H3/H4-like histones) and for ribosome-related proteins (because of a new emphasis on cotranslational protein secretion associated with the origin of cotranslationally made N-linked glycoproteins instead of murein; Cavalier-Smith 2002a). Though very misleading about elapsed time, this strictly temporary quantum evolution provided so many neomuran synapomorphies never subsequently lost that trees for such molecules robustly recover neomuran monophyly; high hits between eukaryotes and archaebacteria are ensured in such analyses as in Esser et al. (2004).

Even though the changes for these two classes of molecule are very dramatic compared with the much more trivial changes in average metabolic enzymes during eukaryogenesis, they pale into insignificance compared with the revolutionary changes in those thousands of neomuran proteins that generated eukaryote-specific novelties (Fig. 8.1c), e.g. cytoskeletal proteins, nuclear pore complexes, coated vesicles, phagocytosis, Golgi, centriolar, and ciliary structure. Although a few of the most conserved novel eukaryotic proteins can be traced back to specific bacterial ancestors, e.g. tubulins to FtsZ, actin to MreB (Margolin 2005), dynein to an ATPase (Iyer et al. 2004), myosin and kinesin to a GTPase (Leipe et al. 2002), none of these can be traced back to any particular bacterial group. This is expected because change on this scale, often removing most sequence similarity and leaving clues only in conserved 3D structure, necessarily removes the phylum-specific signatures originally present in their ancestors. Most eukaryotic proteins evolve much faster; some can only just be recognised across the whole eukaryote tree and others cannot be. There is no hope of recognising their ancestry by sequence similarity, though tertiary structure might help. This is obvious and inevitable for major innovation. The simplest and biologically most reasonable interpretation of the origin of these divergent proteins, which are the vast majority, is that they evolved from neomuran proteins. To suggest that they came from an entirely unknown kind of organism, whether our ignorance is dignified by pointless names such as chronocyte (Hartman and Fedorov 2002) or not, is science fiction escapism that ignores Occam's razor and Popperian requirements for testability. One might as well suppose that these proteins came from little green men on Mars – or big red women from Venus.

An important corollary of this is that, if, as I have argued, mitochondria were implanted after most of these innovations, then almost all really new eukaryotic proteins would have arisen by the radical transformation of neomuran proteins, rather than proteobacterial ones; by the time mitochondria arrived the party was largely over; they had only a few dreg ends of innovation to finish off. For years the obsession with the symbiogenetic origins of mitochondria and chloroplasts, which are only slightly modified negibacteria, has distracted attention from the real innovation that created phagocytosis, endomembranes, nucleus, mitosis, cilia, and sex. Early critics of symbiogenesis claimed that it smacks of special creation (Uzzell and Spolsky 1974), as it does indeed in the hands of Margulis, who has denied any real innovation in the whole history of life with the absurd claim that every species was made by symbiogenesis (Margulis and Sagan 2002), recombining what was already there. Evolutionary biologists have to accept the reality of genuine innovation and transformation; in the most extreme cases during symbiogenesis, structural information, whether of protein folding or domains or at the level of ultrastructure, is often much more help than sequence bioinformatics. Given this perspective, the presumption that one can extrapolate from the small fraction of slowly evolving sequences that one can align to the large majority that one cannot (e.g. with respect to the fraction that came from proteobacteria rather than neomura; Esser et al. 2004) is mistaken. The original pool of potentially alignable neomuran genes will have been dramatically depleted by the quantum evolution that created novel eukaryotic proteins, which probably happened minimally to the proteobacterial genes.

Therefore my prediction based on the neomuran theory and the classical principles of quantum evolution that I have applied consistently to eukaryogenesis for 30 years (explicitly in Cavalier-Smith 1987b, 2002a, b) is as follows. Most eukaryote-wide nuclear genes should not be readily detectable by BLAST in any bacteria; of those that can be thus detected the largest number should not be clearly attributable to any particular bacterial phylum; the second largest number should be attributable to $\alpha$-proteobacteria (or if faster evolving only to proteobacteria generally); the third and fourth largest number should be most similar to highly conserved archaebacterial genes or to less conserved actinobacterial ones (I see no way of predicting which of these should be more frequent); some will appear to come from some other phylum – most such cases will be the result of the inability of phylogenetic algorithms always to construct the correct tree (a fundamental fact of phylogenetics too often ignored) – a few may be genuine cases of lateral transfer, but this will be exceedingly hard to demonstrate (perhaps impossible) for any universal eukaryotic protein. A very small minority of nuclear genes restricted to one or a few eukaryotic lineages probably came from bacteria after the origin of eukaryotes, especially in phagotrophs, but fewer than often supposed. In deciding between such categories bioinformatics methods as in Esser et al. (2004) are inadequate as they confuse similarity with cladistic relationships; trees are essential, but not infallible even with the very best methods and high taxon sampling.

Given the previous considerations, the results of Esser et al. (2004) seem not to require any participant in eukaryogenesis other than a protoeukaryote host derived from an actinobacterium, from which most genes came, and a proteobacterium that provided a significantly large minority. Possibly even more important than its genome, the proteobacterium provided two novel genetic membranes that needed but slight modification to become the mitochondrial envelope.

## 8.11 Genic Scale, Tempo, and Timing of Mitochondrial Enslavement and Eukaryote Origin

"Mitochondria might therefore have originated and diversified as recently as 800 My ago". (p. 277 in Cavalier-Smith 1983a)

This still seems to me the best lower-bound estimate to the nearest 100 million years. The upper bound is harder to place, but about 900 million years ago is perhaps most reasonable (Cavalier-Smith 2006a). Elsewhere I explained why I think the remarkably well preserved *Bangiomorpha* fossils are probably cyanobacteria not red algae and why I do not accept that any fossils described by Javaux et al. (2001) are assuredly eukaryotic (Cavalier-Smith 2002a, 2006a; Cavalier-Smith and Chao 2003b). None of them show structures that clearly give evidence of either a cytoskeleton or endomembranes; they are probably not eukaryotic. The first indubitably eukaryotic fossils are late Proterozoic vase-shaped testate amoebae assigned to *Melanocyrillium* and four other genera (Mus and Moczydlowska 2000; Porter and Knoll 2000; Porter et al. 2003), which are at least 742 ± 6 million years old (Karlstrom et al. 2000). Because testate amoebae are polyphyletic, having arisen within Amoebozoa (unikonts) and Cercozoa (bikonts, and polyphyletically within them; Bass et al. 2005; Cavalier-Smith and Chao 2003a) we cannot be sure to which phylum they belong. Some genera are so similar to arcellinid lobosean Amoebozoa that it is most likely that they belong to this phylum; as Amoebozoa are sisters to opisthokonts (Fig. 8.1), the conclusion that opisthokonts diverged from amoebozoa more than 742 million years ago (Porter et al. 2003) is probably correct. Because the only earlier divergence in the entire eukaryote tree is that between unikonts and bikonts, the origin of the cenancestral eukaryote need not have been much earlier. Although a recent Bayesian tree based on numerous proteins and six fossil dates puts the date of divergence of Amoebozoa and other eukaryotes somewhat earlier (950–1,259 million years ago: Douzery et al. 2004), the margin of error of such calculations must be considerable, even though they are a dramatic improvement on earlier trees based on the false assumption of a molecular clock that gave greatly inflated estimates. As that tree seems to overestimate bilaterian divergence by 100 million years compared with the fossil record (Douzery et al. 2004), its estimates of the age of opisthokonts and eukaryotes may be even more inflated. A key question is how deep within the amoebozoan tree did the arcellinid phenotype extend? From recent trees it appears that Lobosea diverged from other amoebozoan classes so close to the base of the amoebozoan tree that they need be scarcely any younger than Amoebozoa as a whole (Cavalier-Smith et al. 2004; Kudryavtsev et al. 2005); however, as crown arcellinids appear to branch well within Lobosea (Nikolaev et al. 2005), they are probably distinctly younger than Amoebozoa. If the test-bearing of arcellinids is the ancestral state for Lobosea, as is possible and not unreasonable (Cavalier-Smith 1993b), neither Lobosea nor Amoebozoa, nor even eukaryotes, need be much older, so a reasonable minimum date for the origin of eukaryotes and mitochondria would be about 750 million years ago. But if naked Lobosea never had testate ancestors (also reasonable), Amoebozoa (and therefore eukaryotes) are probably distinctly older than 742 million years, probably at least about 800 million years old, possibly more.

Bikont fossils do not resolve the uncertainty as almost all are restricted to the Phanerozoic, and I am not convinced that any aged over about 600 million years assuredly belong to a modern bikont phylum. The oldest fossils that I accept as rhizarian or plant are very early Cambrian or very late Neoproterozoic, respectively (Cavalier-Smith 2006a). The more than 742 million years old testate amoeba *Melicerion* has been proposed as a euglyphid (Cercozoa) (Porter et al. 2003), but I am unconvinced, as it might be an arcellinid amoebozoan with unusually regular agglutinated mineral particles within an organic test, or even belong to neither phylum (e.g. a stem eukaryote). Although the date of origin of mitochondria is therefore uncertain, it was probably in the Neoproterozoic, 750–1,000 million years ago, very much more recent than the great oxygenation event of the biosphere caused by cyanobacterial oxygenic photosynthesis over a billion years earlier. Mitochondrial origin was therefore not limited by unavailability of atmospheric oxygen, as so often assumed, but by the non-availability of a phagotrophic host until a billion or more years later.

I suggested before that about 100 new genes were needed for originating mitochondria and that about 1,000 old ones were needed to acquire presequences. Assuming that it takes about 500 mutations to make a radically new gene and only two to add a presequence, then about 52,000 new mutations had to spread to convert a proteobacterium into a mitochondrion. How long might this have taken? A protoeukaryote could easily have had over three generations a day. Ten generations gives a population of 1,024 if unconstrained by limited resources, predation, or disease, so under the most favourable conditions a population can multiply $1 \times 10^3$-fold every 3 days, $1 \times 10^6$-fold every 6 days, and $1 \times 10^9$-fold every 9 days. An advantageous new mutant could reach a population of three billion in 10 days. Assuming that a second beneficial mutation occurred with a frequency of only $10^{-8}$ (much less often than typical point mutations), that population (small by microbial standards) would have 30 additional beneficial mutations in every gene. As 98% of beneficial mutations with a 1% selective advantage go extinct by chance while still rare and only about 2% spread through the population (Fisher 1958), only about one extra beneficial mutation likely eventually to become fixed would be generated in 10 days. Its rate of spread depends on its selective advantage and initial frequency; initially very slow when rare, faster when common. In a haploid protoeukaryote with a steady-state population, a mutation with an advantage of only 1% (almost unmeasurable in the laboratory) would take 692 generations to increase tenfold in frequency from $10^{-3}$ to $10^{-2}$, but only 459 generations to increase 50-fold from $10^{-2}$ to 50% (Haldane 1932). However, a third beneficial mutation building on the first two would almost certainly arise in a cell with both before the earlier ones had even reached the $10^{-2}$ level, so in practice for a change needing many successive mutations their spreading would actually overlap in time and their benefit would be at least additive, so cells with more beneficial mutations simultaneously would multiply faster than those with fewer. Assume that about 5 mutations only could spread simultaneously and that each took 10,000 generations to spread; $10^8$ generations would then be needed for making the mitochondrion. As ten generations take 3 days each, a gene would take 3,000 days (8.2 years) to spread and the whole could be complete in 82,000 years. This probably considerably overestimates the time needed to evolve mitochondria, as many key mutations could have tenfold higher selective advantage, 500 mutations may overestimate the number needed to make a new gene, and probably many more than five mutations could spread simultaneously, especially as sex probably evolved in the protoeukaryote and could combine them from different cells (Cavalier-Smith 2002b). Possibly, therefore, the origin of mitochondria could have taken 10,000 years or less. Earlier I suggested that chloroplasts might have been enslaved in as little as 3,000 years (Cavalier-Smith 1982). Many, used to elephant generation times and population numbers, intuitively underestimate the speed with which major evolutionary change can occur in unicells.

I suggested before that about 4,000–6,000 new genes might have arisen during the origin of eukaryotes. Assuming 5,000, that is 50 times those needed for making mitochondria. If the same assumptions are applied, only about 500,000 years needs to have been taken for the evolution of the entire eukaryote cell. This geologically short time for the most radical transformation in the history of life, combined with the improbability of intermediates being able to compete with the fully perfected eukaryotes at the end of this period of quantum evolution, is sufficient explanation of why no intermediates survive. Since then stabilising selection has prevented fundamental changes in mitochondria, cilia, or nuclei, so they are basically the same in the sperm of cycad trees, blue whales, and the malaria parasite. Unique events, quantum evolution, and near stasis are the major features of evolution, not gradual predictable uniform change beloved by mathematical modellers.

*Acknowledgements:* I thank Bill Martin for the stimulus given by the possible but untenable hydrogen hypothesis to developing this improved version of the classical phagotrophic host theory of the symbiogenetic origin of mitochondria by cell enslavement and his scientific generosity in providing a publication niche for ideas that contrast with some of his own. I thank NERC for research grants and NERC and the Canadian Institute for Advanced Research for Fellowship support.

## References

Achleitner G, Gaigg B, Krasser A, Kainersdorfer E, Kohlwein SD, Perktold A, Zellnig G, Daum G (1999) Association between the endoplasmic reticulum and mitochondria of yeast facilitates interorganelle transport of phospholipids through membrane contact. Eur J Biochem 264:545–553

Ahting U, Waizenegger T, Neupert W, Rapaport D (2005) Signal-anchored proteins follow a unique insertion pathway into the outer membrane of mitochondria. J Biol Chem 280: 48–53

Allen JF, Puthiyaveetil S, Strom J, Allen CA (2005) Energy transduction anchors genes in organelles. BioEssays 27:426–435

Amiri H, Karlberg O, Andersson SG (2003) Deep origin of plastid/parasite ATP/ADP translocases. J Mol Evol 56:137–150

Amos LA, van den Ent F, Lowe J (2004) Structural/functional homology between the bacterial and eukaryotic cytoskeletons. Curr Opin Cell Biol 16:24–31

Andersson SG, Karlberg O, Canback B, Kurland CG (2003) On the origin of mitochondria: a genomics perspective. Philos Trans R Soc Lond B Biol Sci 358:165–177; discussion 177–179

Bass D, Moreira D, López-García P, Polet S, Chao EE, von der Heyden S, Pawlowski J, Cavalier-Smith T (2005) Polyubiquitin insertions and the phylogeny of Cercozoa and Rhizaria. Protist 156:149–161

Bhattacharyya SN, Chatterjee S, Goswami S, Tripathi G, Dey SN, Adhya S (2003) "Ping-pong" interactions between mitochondrial tRNA import receptors within a multiprotein complex. Mol Cell Biol 23:5217–5224

Brown JR, Doolittle WF (1997) Archaea and the prokaryote-to-eukaryote transition. Microbiol Mol Biol Rev 61:456–502

Cavalier-Smith T (1975) The origin of nuclei and of eukaryotic cells. Nature 256:463–468

Cavalier-Smith T (1977) Mitocondri e cloroplasti: un problema evolutivo. In: Scienza e Technica. Mondadori, Milan, pp 305–318

Cavalier-Smith T (1980) Cell compartmentation and the origin of eukaryote membranous organelles. In: Schwemmler W, Schenk HEA (eds) Endocytobiology: endosymbiosis and cell biology, a synthesis of recent research. de Gruyter, Berlin, pp 893–916

Cavalier-Smith T (1981) The origin and early evolution of the eukaryotic cell. In: Carlile MJ, Collins JF, Moseley BEB (eds) Molecular and cellular aspects of microbial evolution. Cambridge University Press, Cambridge, pp 33–84

Cavalier-Smith T (1982) The origins of plastids. Biol J Linn Soc 17:289–306

Cavalier-Smith T (1983a) Endosymbiotic origin of the mitochondrial envelope. In: Schwemmler W, Schenk HEA (eds) Endocytobiology II. de Gruyter, Berlin, pp 265–279

Cavalier-Smith T (1983b) A 6-kingdom classification and a unified phylogeny. In: Schwemmler W, Schenk HEA (eds) Endocytobiology II. de Gruyter, Berlin, pp 1027–1034

Cavalier-Smith T (1987a) The origin of cells: a symbiosis between genes, catalysts, and membranes. Cold Spring Harb Symp Quant Biol 52:805–824

Cavalier-Smith T (1987b) The origin of eukaryotic and archaebacterial cells. Ann N Y Acad Sci 503:17–54

Cavalier-Smith T (1987c) The simultaneous symbiotic origin of mitochondria, chloroplasts, and microbodies. Ann N Y Acad Sci 503:55–71
Cavalier-Smith T (1990) Symbiotic origin of peroxisomes. In: Nardon P, Gianinazzi-Pearson V, Grenier AM, Margulis L, Smith DC (eds) Endocytobiology IV. Institut National de la Recherche Agronomique, Paris, pp 515–521
Cavalier-Smith T (1991) The evolution of cells. In: Osawa S, Honjo T (eds) Evolution of life. Springer, Berlin Heidelberg New York, pp 271–304
Cavalier-Smith T (1993a) Evolution of the eukaryotic genome. In: Broda P, Oliver SG, Sims P (eds) The eukaryotic genome. Cambridge University Press, Cambridge, pp 333–385
Cavalier-Smith T (1993b) Kingdom Protozoa and its 18 phyla. Microbiol Rev 57:953–994
Cavalier-Smith T (2000a) Flagellate megaevolution: the basis for eukaryote diversification. In: Green JC, Leadbeater BSC (eds) The flagellates. Taylor and Francis, London, pp 361–390
Cavalier-Smith T (2000b) Membrane heredity and early chloroplast evolution. Trends Plant Sci 5:174–182
Cavalier-Smith T (2002a) The neomuran origin of archaebacteria, the negibacterial root of the universal tree and bacterial megaclassification. Int J Syst Evol Microbiol 52:7–76
Cavalier-Smith T (2002b) The phagotrophic origin of eukaryotes and phylogenetic classification of Protozoa. Int J Syst Evol Microbiol 52:297–354
Cavalier-Smith T (2003a) Genomic reduction and evolution of novel genetic membranes and protein-targeting machinery in eukaryote-eukaryote chimaeras (meta-algae). Philos Trans R Soc Lond B Biol Sci 358:109–134
Cavalier-Smith T (2003b) Protist phylogeny and the high-level classification of Protozoa. Eur J Protistol 39:338–348
Cavalier-Smith T (2004a) Chromalveolate diversity and cell megaevolution: interplay of membranes, genomes and cytoskeleton. In: Hirt RP, Horner DS (eds) Organelles, genomes and eukaryote phylogeny. CRC, London, pp 75–108
Cavalier-Smith T (2004b) The membranome and membrane heredity in development and evolution. In: Hirt RP, Horner DS (eds) Organelles, genomes and eukaryote phylogeny. Taylor & Francis, London, pp 335–351
Cavalier-Smith T (2006a) Cell evolution and Earth history: stasis and revolution. Philos Trans R Soc Lond B Biol Sci 361:969–1006
Cavalier-Smith T (2006b) Origin of mitochondria by intracellular enslavement of a photosynthetic purple bacterium. Proc R Soc Lond B Biol Sci 273:1943–1952
Cavalier-Smith T (2006c) Rooting the tree of life by transition analyses. Biol Direct 1:19
Cavalier-Smith T, Chao EE (2003a) Phylogeny and classification of phylum Cercozoa (Protozoa). Protist 154:341–358
Cavalier-Smith T, Chao EE (2003b) Phylogeny of Choanozoa, Apusozoa, and other Protozoa and early eukaryote megaevolution. J Mol Evol 56:540–563
Cavalier-Smith T, Chao EE, Oates B (2004) Molecular phylogeny of Amoebozoa and the evolutionary significance of the unikont *Phalansterium*. Eur J Protistol 40:21–48
Chew O, Lister R, Qbadou S, Heazlewood JL, Soll J, Schleiff E, Millar AH, Whelan J (2004) A plant outer mitochondrial membrane protein with high amino acid sequence identity to a chloroplast protein import receptor. FEBS Lett 557:109–114
Covello PS, Gray MW (1993) On the evolution of RNA editing. Trends Genet 9:265–268
De Beer G (1940) Embryos and ancestors. Clarendon, Oxford
de Grey A (2005) Forces maintaining organellar genomes: is any as strong as genetic code disparity or hydrophobicity? BioEssays 27:436–446
de Kroon AI, Koorengevel MC, Vromans TA, de Kruijff B (2003) Continuous equilibration of phosphatidylcholine and its precursors between endoplasmic reticulum and mitochondria in yeast. Mol Biol Cell 14:2142–2150
Douzery EJ, Snell EA, Bapteste E, Delsuc F, Philippe H (2004) The timing of eukaryotic evolution: does a relaxed molecular clock reconcile proteins and fossils? Proc Natl Acad Sci USA 101:15386–15391

Embley TM, van der Giezen M, Horner DS, Dyal PL, Foster P (2003) Mitochondria and hydrogenosomes are two forms of the same fundamental organelle. Philos Trans R Soc Lond B Biol Sci 358:191–201; discussion 201–202

Esser C, Ahmadinejad N, Wiegand C, Rotte C, Sebastiani F, Gelius-Dietrich G, Henze K, Kretschmann E, Richly E, Leister D, Bryant D, Steel MA, Lockhart PJ, Penny D, Martin W (2004) A genome phylogeny for mitochondria among alpha-proteobacteria and a predominantly eubacterial ancestry of yeast nuclear genes. Mol Biol Evol 21:1643–1660

Fenchel T, Bernard C (1993) A purple protist. Nature 362:300

Finlay BJ, Esteban GF (2001) Exploring Leeuwenhoek's legacy: the abundance and diversity of protozoa. Int Microbiol 4:125–133

Finlay BJ, Span ASW, Harman JMP (1983) Nitrate respiration in primitive eukaryotes. Nature 333–335

Fisher RA (1958) The genetical theory of natural selection. 2nd edn. Dover, New York

Fox GE, Stackebrandt E, Hespell RB, Gibson J, Maniloff J, Dyer TA, Wolfe RS, Balch WE, Tanner RS, Magrum LJ, Zablen LB, Blakemore R, Gupta R, Bonen L, Lewis BJ, Stahl DA, Luehrsen KR, Chen KN, Woese CR (1980) The phylogeny of prokaryotes. Science 209:457–463

Frazier AE, Chacinska A, Truscott KN, Guiard B, Pfanner N, Rehling P (2003) Mitochondria use different mechanisms for transport of multispanning membrane proteins through the intermembrane space. Mol Cell Biol 23:7818–7828

Frazier AE, Dudek J, Guiard B, Voos W, Li Y, Lind M, Meisinger C, Geissler A, Sickmann A, Meyer HE, Bilanchone V, Cumsky MG, Truscott KN, Pfanner N, Rehling P (2004) Pam16 has an essential role in the mitochondrial protein import motor. Nat Struct Mol Biol 11:226–233

Gentle I, Gabriel K, Beech P, Waller R, Lithgow T (2005) The Omp85 family of proteins is essential for outer membrane biogenesis in mitochondria and bacteria. J Cell Biol 164:19–25

Golding GB, Gupta RS (1995) Protein-based phylogenies support a chimeric origin for the eukaryotic genome. Mol Biol Evol 12:1–6

Gray MW, Lang BF, Burger G (2004) Mitochondria of protists. Annu Rev Genet 38:477–524

Haldane JBS (1932) The causes of evolution. Longmans, Green, London

Hartman H, Fedorov A (2002) The origin of the eukaryotic cell: a genomic investigation. Proc Natl Acad Sci USA 99:1420–1425

Hashimoto T (2003) Plastid division: its origins and evolution. Int Rev Cytol 222:63–98

Iyer LM, Leipe D, Koonin EV, Aravind L (2004) Evolutionary history and higher order classification of AAA+ ATPases. J Struct Biol 146:11–31

Javaux EJ, Knoll AH, Walter MR (2001) Morphological and ecological complexity in early eukaryotic ecosystems. Nature 412:66–69

John P, Whatley FR (1975) *Paracoccus denitrificans* and the evolutionary origin of mitochondria. Nature 254:495–498

John P, Whatley FR (1977) *Paracoccus denitrificans* Davis (*Micrococcus denitrificans* Beijerinck) as a mitochondrion. Adv Bot Res 4:51–115

Kanaji S, Iwahashi J, Kida Y, Sakaguchi M, Mihara K (2000) Characterization of the signal that directs Tom20 to the mitochondrial outer membrane. J Cell Biol 151:277–288

Karlstrom KE, Bowring SA, Dehler CM, Knoll AH, Porter SM, Des Marais DJ, Weil AB, Sharp ZD, Geissman JW, Elrick MB, Timmons JM, Crossey LJ, Davidek KL (2000) Chuar Group of the Grand Canyon: record of breakup of Rodinia, associated change in the global carbon cycle, and ecosystem expansion by 740 Ma. Geology 28:619–622

Kiefel BR, Gilson PR, Beech PL (2004) Diverse eukaryotes have retained mitochondrial homologues of the bacterial division protein FtsZ. Protist 155:105–115

Kudryavtsev AA, Bernhardt D, Schlegel M, Chao EE, Cavalier-Smith T (2005) 18S ribosomal RNA gene sequences of *Cochliopodium* (Himatismenida) and the phylogeny of Amoebozoa. Protist 156:215–224

Leipe DD, Wolf M, Koonin EV, Aravind L (2002) Classification and evolution of P-loop GTPases and related ATPases. J Mol Biol 317:41–72

Lister R, Whelan J (2006) Mitochondrial protein import: convergent solutions for receptor structure. Curr Biol 16:R197–R199

Lucattini R, Likic VA, Lithgow T (2004) Bacterial proteins predisposed for targeting to mitochondria. Mol Biol Evol 21:652–658

Macasev D, Whelan J, Newbigin E, Silva-Filho MC, Mulhern TD, Lithgow T (2004) Tom22′, an 8-kDa trans-site receptor in plants and protozoans, is a conserved feature of the TOM complex that appeared early in the evolution of eukaryotes. Mol Biol Evol 21:1557–1564

Margeot A, Blugeon C, Sylvestre J, Vialette S, Jacq C, Corral-Debrinski M (2002) In *Saccharomyces cerevisiae*, ATP2 mRNA sorting to the vicinity of mitochondria is essential for respiratory function. EMBO J 21:6893–904

Margolin W (2005) Bacterial mitosis: actin in a new role at the origin. Curr Biol 15:R259–R261

Margulis L (1970) Origin of eukaryotic cells. Yale University Press, New Haven

Margulis L (1981) Symbiosis in cell evolution. Freeman, San Francisco

Margulis L, Dolan MF, Guerrero R (2000) The chimeric eukaryote: origin of the nucleus from the karyomastigont in amitochondriate protists. Proc Natl Acad Sci USA 97:6954–6959

Margulis L, Sagan D (2002) Acquiring genomes. Basic Books, New York

Martin W, Müller M (1998) The hydrogen hypothesis for the first eukaryote. Nature 392:37–44

Martin W, Russell MJ (2003) On the origins of cells: a hypothesis for the evolutionary transitions from abiotic geochemistry to chemoautotrophic prokaryotes, and from prokaryotes to nucleated cells. Philos Trans R Soc Lond B Biol Sci 358:59–85

McFadden GI, Ralf SA (2003) Dynamin: the endosymbiosis ring of power. Proc Natl Acad Sci USA 100:3557–3559

Meier S, Neupert W, Herrmann JM (2005) Conserved N-terminal negative charges in the Tim17 subunit of the TIM23 translocase play a critical role in the import of preproteins into mitochondria. J Biol Chem 280:7777–7785

Miyagishima SY, Nozaki H, Nishida K, Matsuzaki M, Kuroiwa T (2004) Two types of FtsZ proteins in mitochondria and red-lineage chloroplasts: the duplication of FtsZ is implicated in endosymbiosis. J Mol Evol 58:291–303

Mukhopadhyay A, Ni L, Weiner H (2004) A co-translational model to explain the in vivo import of proteins into HeLa cell mitochondria. Biochem J 382:385–92

Mus MM, Moczydlowska M (2000) Internal morphology and taphonomic history of the Neoproterozoic vase-shaped microfossils from the Visingsö Group, Sweden. Norsk Geol Tidsskr 80:213–228

Nikolaev SI, Mitchell EA, Petrov NB, Berney C, Fahrni J, Pawlowski J (2005) The testate lobose amoebae (order Arcellinida Kent, 1880) finally find their home within Amoebozoa. Protist 156:191–202

Nowicki M, Muller F, Frentzen M (2005) Cardiolipin synthase of *Arabidopsis thaliana*. FEBS Lett 579:2161–2155

Osteryoung KW, Nunnari J (2003) The division of endosymbiotic organelles. Science 302:1698–1704

Overmann J, Schubert K (2002) Phototrophic consortia: model systems for symbiotic interrelations between prokaryotes. Arch Microbiol 177:201–208

Peretó J, López-García P, Moreira D (2004) Ancestral lipid biosynthesis and early membrane evolution. Trends Biochem Sci 29:469–477

Perry AJ, Hulett JM, Likic VA, Lithgow T. Gooley PR (2006) Convergent evolution of receptors for protein import into mitochondria. Curr Biol 16:221–229

Porter SM, Knoll AH (2000) Testate amoebae of the Chuar Group, Grand Canyon. Paleobiology 27:345–370

Porter SM, Meisterfeld R, Knoll AH (2003) Vase-shaped microfossils from the Neoproterozoic Chuar Group, Grand Canyon: a classification guided by modern testate amoebae. J Paleont 77:409–429

Preuss M, Ott M, Funes S, Luirink J, Herrmann JM (2005) Evolution of mitochondrial Oxa proteins from bacterial YidC: inherited and acquired functions of a conserved protein. J Biol Chem 280:13004–13011

Rehling P, Model K, Brandner K, Kovermann P, Sickmann A, Meyer HE, Kuhlbrandt W, Wagner R, Truscott KN, Pfanner N (2003) Protein insertion into the mitochondrial inner membrane by a twin-pore translocase. Science 299:1747–1751

Richards TA, Cavalier-Smith T (2005) Myosin domain evolution and the primary divergence of eukaryotes. Nature 436:1113–1118

Sagan L (1967) On the origin of mitosing cells. J Theor Biol 14:225–274

Schnepf E (1964) Zur Feinstruktur von *Geosiphon pyriforme*: ein Versuch zur Deutung cytoplasmatischer Membranen und Kompartimente. Archiv Microbiol 49:112–131

Searcy D (1992) Origins of mitochondria and chloroplasts from sulfur based symbioses. In: Hartman H, Matsuno K (eds) The origin and evolution of the cell. World Scientific, Singapore, pp 47–78

Stechmann A, Cavalier Smith T (2003) The root of the eukaryote tree pinpointed. Curr Biol 13:R665–R666

Sylvestre J, Vialette S, Corral Debrinski M, Jacq C (2003) Long mRNAs coding for yeast mitochondrial proteins of prokaryotic origin preferentially localize to the vicinity of mitochondria. Genome Biol 4:R44

Tjalsma H, Bron S, van Dijl JM (2003) Complementary impact of paralogous Oxa1-like proteins of *Bacillus subtilis* on post-translocational stages in protein secretion. J Biol Chem 278:15622–15632

Uzzell T, Spolsky C (1974) Mitochondria and plastids as endosymbionts: a revival of special creation? Am Sci 62:334–343

van Roermund CW, Drissen R, van Den Berg M, Ijlst L, Hettema EH, Tabak HF, Waterham HR, Wanders RJ (2001) Identification of a peroxisomal ATP carrier required for medium-chain fatty acid beta-oxidation and normal peroxisome proliferation in *Saccharomyces cerevisiae*. Mol Cell Biol 21:4321–4329

Van Valen LM, Maiorana VC (1980) The archaebacteria and eukaryotic origins. Nature 287:248–250

Whatley JM, John P, Whatley FR (1979) From extracellular to intracellular: the establishment of mitochondria and chloroplasts. Proc R Soc Lond B Biol Sci 204:165–187

Wiedemann N, Frazier AE, Pfanner N (2004) The protein import machinery of mitochondria. J. Biol Chem 279:14473–14476

Woese CR (1977) Endosymbionts and mitochondrial origin. J Mol Evol 10:93–96

Woese CR, Fox GE (1977) The concept of cellular evolution. J Mol Evol 10:1–6

Wojtkowska M, Szczech N, Stobienia O, Jarmuszkiewicz W, Budzinska M, Kmita H (2005) An inception report on the TOM complex of the amoeba *Acanthamoeba castellanii*, a simple model protozoan in mitochondria studies. J Bioenerg Biomembr 37:261–268

Wolf YI, Aravind L, Koonin EV (1999) Rickettsiae and Chlamydiae: evidence of horizontal gene transfer and gene exchange. Trends Genet 15:173–175

# 9 Constantin Merezhkowsky and the Endokaryotic Hypothesis

VICTOR V. EMELYANOV

## Summary

A century ago, the Russian biologist Constantin Merezhkowsky, regarded as the genuine founder of endosymbiosis theory, logically argued that chloroplasts were free-living cyanobacteria which once entered into intimate relationship with a plastid-free host. In the same work and later on Merezhkowsky extended his view on symbiogenesis, suggesting that the nucleus also originated via endosymbiosis involving two radically different beings, mycoplasm as a symbiont and amoeboid plasm as a host, a view commonly referred to today as the endokaryotic hypothesis. While the endosymbiosis theory for the origin of plastids and mitochondria is now universally accepted, the origin of the nucleus and the eukaryotic cell per se remains a matter of continuing debate. Numerous molecular data show that nucleated cells (Eukarya) are fundamentally chimeric, with information-processing machinery being related to that of *Archaea* and core metabolism more resembling that of *Bacteria*. It is also becoming increasingly clear that eukaryotism is a derived state and that primitively amitochondriate eukaryotes may have never existed. Several models for the origin of eukaryotes have been proposed in the framework of endosymbiosis theory that explicitly incorporate these data. At the extreme, some scientists believe that mitochondria and eukaryotes themselves arose in a single endosymbiotic event involving an archaeal host and a bacterial symbiont; still others argue that eubacterial genes entered an archaeon-derived, yet eukaryotic cell by means of lateral gene transfer both before and during mitochondrial symbiosis. An intermediate view holds that there have been two successive major contributions of *Bacteria* to Eukarya. First, a primitively amitochondriate cell emerged as a chimera created by fusion between an archaebacterium and a eubacterium. Second, this chimeric organism engulfed a eubacterial symbiont which became a mitochondrion. On the basis of a concept called the canonical pattern of mitochondrial ancestry for eukaryotic genes and considering eukaryotic multigene families, I argue here that a pro-eukaryote originated via some fusion event, which started with endosymbiosis of an archaebacterium in a proteobacterial host. Being actually a prokaryote with a pro-nucleus derived from archaea, such a chimeric organism subsequently hosted a mitochondrial endosymbiont – a pivotal event followed by the appearance of the

Origin of Mitochondria and Hydrogenosomes (ed. by William F. Martin and Miklós Müller)

characteristically eukaryotic features, including a true nucleus. Thus, the present view of eukaryogenesis is consistent with the endokaryotic hypothesis, as first proposed by Merezhkowsky. It is also thought to be quite satisfactory when considered from geological, ecological, and biochemical perspectives.

## 9.1 Introduction

The scientific community has recently celebrated the centenary of the outstanding paper by Constantin Merezhkowsky (1855–1921) which introduced the endosymbiosis theory (Martin and Kowallik 1999). In that paper Merezhkowsky convincingly argued that chloroplasts arose as a result of intracellular symbiosis of a free-living photosynthesizing bacterium in a heterotrophic host. To develop this idea, he first reasoned that plastids cannot be regarded as organs. If the capability to form an organ is a heritable trait, it must have arisen in the ancient past and then been passed to offspring from generation to generation; however, chloroplasts never arise de novo by differentiation from the plasm, but always emerge through division of pre-existing chloroplasts. Hence, these structures should be interpreted as organisms, as symbionts. As additional support of endosymbiosis, Merezhkowsky pointed out the apparent similarity of chloroplasts to cyanobacteria and the analogy between chloroplasts and zoochlorellae – algae living within protists and animals. In the same paper, and later on, Merezhkowsky entertained an even bolder assumption that the nucleus of animals and plants was also endosymbiotically derived from mycoplasm which invaded an amoeboplasm host. Being convinced of a diphyletic origin of life, he regarded symbiotic partners as completely unrelated beings that arose at different times and under different conditions (Merezhkowsky 1910). Beyond doubt, Merezhkowsky was a far-sighted scientist and well ahead of his time. His numerous activities, troubled life, and tragic death have been described in an excellent review of Sapp et al. (2002).

The theory of symbiogenesis, proffered by Merezhkowsky in 1905, was however ignored for at least six decades (Martin et al. 2001). In the 1970s this theory was repopularized by Lynn Sagan (Margulis) and further elaborated as a serial endosymbiosis theory to explain not only the origin of organelles, but also complex eukaryotic cells per se (Sagan 1967; Margulis 1996). Since then endosymbiosis has become an explanatory principle in the field of eukaryogenesis. In a new era, the era of sequenced genes and genomes, the endosymbiotic origin of chloroplasts and mitochondria, including mitochondrion-derived genome-free hydrogenosomes and mitosomes, was firmly corroborated (Gray 1999; van der Giezen et al. 2002; Tovar et al. 2003). There is also little doubt that plastids and mitochondria originated once in evolution (Gray 1999; Adams and Palmer 2003). The origin of the nucleus, a defining feature of eukaryotes, remains however a mysterious matter.

Given that mitochondria, powerhouses of most eukaryotic cells, apparently emerged before chloroplasts, the most intriguing questions are centred around the nature of a primary host for the mitochondrial symbiont. These questions are the following:

1. Where did the genes and the genome of this cell come from? As cellular metabolism is fully predetermined by gene content, the origin of metabolic (in particular, energy-generating) pathways is actually addressed by this question.
2. Was the primitively amitochondriate host a eukaryote or a prokaryote? This issue relates to a problem of prokaryote–eukaryote dichotomy, also known as a major evolutionary discontinuity. Indeed, eukaryotes are typified by the presence of a nucleus, an endomembrane system, and a cytoskeleton – salient features distinguishing them from prokaryotes (Cavalier-Smith 1987; Doolittle 1998). In view of this, the question should be rephrased in the following way: In which succession did the mitochondrion, nucleus, endomembrane, and cytoskeleton come into existence? Among these structures, only mitochondria and nuclei contain DNA that might betray their origin. An exhaustive analysis of mitochondrial genomes has made it possible to confirm the endosymbiotic origin of the organelle itself (Lang et al. 1997; Gray 1999). The question then arises of whether the nucleus was also derived endosymbiotically, as envisioned by the endokaryotic hypothesis, or whether it had an autogenous origin.

In this chapter I present phylogenetic data which are consistent with an endosymbiotic origin of what subsequently became the nucleus. 'Subsequently' means here that the nucleus, as well as other typically eukaryotic structures, emerged after the advent of aerobically respiring mitochondria. A concept of canonical pattern of mitochondrial ancestry for eukaryotic genes (Emelyanov 2001b, 2003b) will be shown to be crucial to the issue.

## 9.2 Modern Hypotheses of Eukaryotic Origin

### 9.2.1 Universal Tree and Concept of Archezoa

On the basis of phylogenetic analysis of small-subunit (SSU) ribosomal RNA (rRNA), all living organisms were classified into domains *Archaea*, *Bacteria*, and Eukarya (Olsen et al. 1994). A reciprocal rooting of the tree of life using translation elongation factors and ATPase subunits, which arose via paralogous duplications apparently before diversification of the three domains, revealed that *Archaea* and Eukarya share the last common ancestor to the exclusion of *Bacteria* (Brown and Doolittle 1997). Analysis of paralogously duplicated subunits of the signal recognition particle corroborated this tree topology (Gribaldo and Cammarano 1998). Numerous data (reviewed by

Brown and Doolittle 1997) indicated that *Archaea*, *Bacteria*, and Eukarya possess both unique and shared features. Of importance, archaea are characterized by a unique energy metabolism (Graham et al. 2000) and lipid composition, while their replication, transcription, and translation systems resemble those of eukaryotes (Brown and Doolittle 1997; Woese 2002). On the other hand, bacteria and eukaryotes share energy metabolism and lipid composition (Martin and Russell 2003). The terms 'operational' and 'informational' genes were coined by Jain et al. (1999) for genes specifying metabolism of small molecules and processing of information, respectively. Thus, as a rough approximation, eukaryotes possess archaeal-like informational genes and bacterial-like operational genes.

It is clear that a distribution of the aforementioned characters would be difficult to reconcile with a universal tree. Further complication came from an improved analysis of the signal recognition particle (Brinkmann and Philippe 1999). This study inferred Eukarya as a primary domain of life, from which *Archaea* and *Bacteria* evolved owing to thermoreduction. The authors surmised that the early emergence of *Bacteria* in previous analyses was simply due to a long-branch attraction (LBA) artefact that imposed affiliation of fast-evolving *Bacteria* to an outgroup; however, exhaustive phylogenetic analysis of several paralogs showed that bacterial and eukaryal sequences evolved at nearly equal rates, and slightly faster than archaeal sequences (Kollman and Doolittle 2000). It is worth noting that in all but two cases a sister relationship of *Archaea* and Eukarya was confirmed. In the remaining two cases, *Bacteria* and Eukarya turned out to be the closest relatives. Of particular interest (see later) and in agreement with earlier data (Hashimoto et al. 1998), this relationship was shown by the valyl-transfer RNA (tRNA) synthetase (ValRS) subtree rooted with isoleucyl-tRNA synthetase (IleRS). The shape of the universal tree was recently addressed using a new approach, conditioned reconstruction, applied to complete genomes from representative bacteria, archaea, and eukaryotes. It was shown that the universal tree may in the best way be represented as a ring, on which eukaryotes link archaeal and bacterial branches. Either *Proteobacteria* or a clade comprising *Cyanobacteria* and *Proteobacteria* was offered as a bacterial contributor to the chimeric eukaryotic genome (Rivera and Lake 2004).

Recently, a view of eukaryotes as the deepest-emerging organisms was rigorously criticized by Martin and Russell (2003). They reasoned that such ideas are founded merely upon a consideration of genetic systems, but these can by no means be regarded separately from metabolic processes. By comparing a relatively narrow and virtually eubacterial-like energy metabolism of eukaryotes with a broad spectrum of energy pathways in prokaryotes and assuming an autotrophic origin of life, the authors concluded that eukaryotes must have arisen long after prokaryotes. It should be noted in this regard that a heterotrophic origin of life in Eukarya, from which *Bacteria* inherited a heterotrophic lifestyle vertically, would be incompatible with phylogenetic data. Indeed, rooting of several trees for glycolytic enzymes in eukaryotes would reveal a reverse, compared with accepted relationships, branching order of bacterial phyla, e.g. *Proteobacteria* would emerge first (Sect. 9.3.2). However, the possibility was not considered in the aforementioned work that nucleated organisms might have originated first, and initially possessed archaeal-like metabolism. In effect, this would be reminiscent of a notorious Archezoa concept, with the only difference being that the latter, in a reverse order, recovers Archezoa directly from archaebacteria (Doolittle 1998).

The concept of Archezoa has a long history. Stanier and van Niel (1962) supposed that the endomembrane system and cytoskeleton emerged clonally in a bacterium, making it possible to engulf solid particles. Bacterial precursors of organelles entered a host in this way, but escaped from being digested and gave rise to mitochondria and plastids (Stanier and van Niel 1962). On the basis of a rooted version of the universal tree, this primitively amitochondriate eukaryote, an archezoon, was further regarded as a descendant of archaea (Cavalier-Smith 1987; Doolittle 1998). Three groups of amitochondriate eukaryotes – Parabasalia, Microsporidia, and Diplomonada – prominently figured as candidates for Archezoa (Cavalier-Smith 1987). Primitively amitochondriate status of these unicellular eukaryotes was first challenged by findings that these organisms encode mitochondrial-like heat-shock protein Cpn60 (for a review see Embley and Hirt 1998). As to bacterial-like ValRS in *Trichomonas vaginalis* (a parabasalid) and *Giardia lamblia* (a diplomonad), previously taken as evidence in support of their secondarily amitochondriate nature (Hashimoto et al. 1998), they were proven to be not mitochondrial, but βγ-proteobacterial in origin (Emelyanov 2003a). To rescue the Archezoa concept, Sogin (1997) suspected that Cpn60, an important general chaperone, entered mitochondrion-lacking eukaryotes by means of lateral gene transfer (LGT) from a bacterium related to a mitochondrial progenitor, thereby conferring selective advantage on a host.

Among amitochondriates, parabasalids were known for a long time to have hydrogenosomes (Müller 1993). In recent years, these energy-producing organelles were convincingly argued to be biochemically modified mitochondria (van der Giezen et al. 2002, Hrdy et al. 2004). As for Microsporidia, they appeared to be highly reduced fungi in which a remnant organelle was either firmly suspected (Vivarès et al. 2002) or discovered experimentally (Williams et al. 2002). On the basis of the presence of a mitochondrial-like machinery for FeS cluster assembly, *G. lamblia* was predicted to have originated from ancestors with bona fide mitochondria (Emelyanov 2003b). At the same time a remnant organelle, mitosome, functioning in FeS cluster assembly was reported in this diplomonad (Tovar et al. 2003). As primitively amitochondriate protists are unlikely to lurk somewhere in an anoxic world, it is becoming increasingly clear that they do not presently exist. A possibility cannot be ruled out, however, that truly mitochondria-lacking eukaryotes once existed but became extinct, being outcompeted by more advanced eukaryotes with mitochondria or their derivatives. For this reason, Cavalier-Smith (1998) retained the blank subkingdom Archezoa in his taxonomy of eukaryotes.

Mitochondria are known to have monophyletically originated from within the α subdivision of *Proteobacteria*. During the long-term endosymbiotic relationship of the mitochondrial precursor with the amitochondriate host, most of the bacterial genes were either lost or passed to the host genome (Lang et al. 1999). Proteins functioning in mitochondria are imported into organelles using a mitochondrion-targeting sequence, regardless of whether their genes were derived from a mitochondrial symbiont or otherwise (Kurland and Andersson 2000). Thus, the most parsimonious explanation for the presence of bacterial genes in modern eukaryotes would be that they derived in toto from a mitochondrial ancestor, no matter whether corresponding proteins are targeted to mitochondria or other compartments (Martin and Müller 1998). However, it remains an open question whether the mitochondrial progenitor was the only source of bacterial genes. An advanced version of the Archezoa theory posits that bacterial genes may have entered an archaea-derived eukaryote through LGT following endocytosis (Doolittle 1998). With respect to genes specifying energy metabolism other than mitochondrial oxidative phosphorylation, such a viewpoint should however be disregarded for the following reasons: (1) given that modern archaea possess a unique energy metabolism (Graham et al. 2000), it is unlikely that they needed some bacterial-like energy pathways in the remote past; (2) essential genes specifying housekeeping processes are normally scattered across bacterial genomes; hence, they could not be laterally transferred from *Bacteria* to Archezoa as selfish operons (Lawrence 1999). This conclusion seems to be especially relevant in respect of the glycolytic (Embden–Meyerhof–Parnas) pathway, in light of the observation that all present-day eukaryotes possess bacterial-like glycolysis, while the glycolytic enzymes are either highly deviant or absent among *Archaea* (Martin and Russell 2003). Moreover, the genes for glycolytic enzymes are not contiguous on bacterial genomes (Blattner et al. 1997; Stephens et al. 1998; Wu et al. 2004).

Yet another class of eukaryogenesis models invoke a sort of fusion between an archaebacterium and a eubacterium to explain the chimeric nature of eukaryotes (Gupta 1998). Fusion, or chimera, hypotheses seem attractive for two reasons: (1) unlike the Archezoa theory, invoking ill-defined and perhaps promiscuous LGT, the chimera theories imply complete integration of two different genomes and likely different types of metabolism, provided that selective forces that brought two different entities together would be offered; (2) fhe fusion of an archaea and a bacterium would have somehow predisposed subsequent mitochondrial origin. It should be noted that most versions of chimera theory consider the archaeal partner as an endosymbiont, which later became fully integrated with the host (Gupta 1998; López-García and Moreira 1999; Horiike et al. 2004). As eukaryotes are endowed with archaeal-like genetic apparatus, this is formally consistent with an endosymbiotic origin of the nucleus. In this context, fusion and endosymbiosis should be clearly distinguished. Chimerism must be understood as full integration, i.e. loss of identity of each partner, whereas an outcome of endosymbiosis is host plus endosymbiotically derived organelle (Gupta 1998).

The hypotheses of eukaryogenesis may be summarized in the following way (Emelyanov 2003a):

1. A primitively mitochondrion-lacking eukaryote, an archezoon, was either a direct descendant of an archaebacterium (Doolittle 1998) or a chimera created by fusion between an archaea and a bacterium (Gupta 1998). The first version is often called an archaeal hypothesis (Brown and Doolittle 1997).
2. The host for the mitochondrial ancestor was either an archaebacterium (Martin and Müller 1998) or a chimeric prokaryote (Emelyanov 2003a). Characteristically eukaryotic features emerged after full establishment of the organelle (Martin and Müller 1998; Vellai et al. 1998; Emelyanov 2003a).

### 9.2.2 Phylogenetic Analysis and Lateral Gene Transfer

The origin of eukaryotic cells should be considered from the organizational (organization and composition of cell) and the gene-metabolic perspectives. This distinction is justified by an almost self-evident fact that characteristically eukaryotic structures and compartments usually involve characteristically eukaryotic proteins which have, by definition, no prokaryotic homologs. In contrast, proteins functioning in metabolic networks, e.g. enzymes of energy metabolism, may often be appropriately aligned to apply them for tree-building programs. Phylogenetic analysis is commonly believed to be the only way in attempts to discern the relationships among prokaryotes and the origin of eukaryotes (Brown and Doolittle 1997; Gupta 1998). The evolution of metabolic processes was recently argued to underlie eukaryogenesis, with the major diversification of eukaryotes being largely due to morphological changes (Poole et al. 2003). Thus, phylogenetic analysis of ubiquitous enzymes of core metabolism should be a relevant approach, when the eukaryote origin is concerned.

In recent years, however, it has been suspected by many researchers that tree reconstructions may often be compromised by LGT. LGT was even claimed to be a major evolutionary process among prokaryotes, likely rendering an organismal phylogeny of no sense (Doolittle 1999). In contrast, others concluded that the impact of LGT on evolution of prokaryotic genomes is much less pronounced, as compared with gene loss, gene genesis, and vertical inheritance (Snel et al. 2002). Beyond doubt, LGT plays an important role in evolution (Boucher et al. 2003). It is clear that events of LGT should be subdivided into transfers of essential, i.e. indispensable to the lineages involved, and non-essential genes. It is also clear that indispensable genes may be horizontally transferred with a replacement, no matter whether orthologous or non-orthologous, of resident genes specifying the same function. But the species are believed to exist for a quite long period of time; thus, resident genes are well adapted to function in the workings of an organism. At last, the recipient of the transferred gene is a single organism and not a population, let alone the species as such. In view of this, it may be concluded that this sort of LGT must confer strong selective advantage upon a recipient to be fixed in a population (Brochier et al. 2002). In effect, it means that the recipient has to become a founder of the new species, i.e. such cases of LGT must be (at least part of) the foundational event. For example, recently reported cases of LGT of ribosomal protein Rps14 were argued to be due to its involvement in antibiotic resistance (Brochier et al. 2000). Transfer of ValRS from *Archaea* to rickettsiae was suggested to have shaped evolution of this lineage (Sect. 9.4.2). It was earlier suggested that informational genes are less amenable to LGT than operational genes, because proteins encoded by them normally interact with many other proteins, being involved in macromolecular complexes (Jain et al. 1999). It is however thought that, for example, glycolysis is, in a sense, not much less complex a system than a ribosome. Obviously, glycolysis is subjected to complex regulation and has many ramifications (Dandekar et al. 1999). It is therefore suggested that a replacement of any glycolytic enzyme by its invading partner would be rather disadvantageous to the recipient cell, regardless of whether the foreign copy functions 'better' or 'worse' than the resident copy.

In view of these considerations, aberrant phylogenies of ubiquitous proteins involved in metabolic processes are thought to be largely due to use of an inadequate evolutionary model than to LGT (Daubin et al. 2003). To circumvent this problem, the use of covarion models or exclusion from analysis of highly deviant sequences, which are normally characterized by long branches and biased character composition, are strongly recommended (Horner and Pesole 2004). However, to make a decision about the correctness of phylogenetic constructs, reference trees are required. Despite some shortcomings, such as LBA artefact caused mostly by biased G+C content, the SSU rRNA tree has been proffered as a truly organismal tree (Olsen et al. 1994; Woese 2002). Notably, a tree-independent approach, insertion-deletion (indel) analysis, recovered the branching order of bacterial phyla similar to that observed in rRNA analysis (Gupta 1998, 2000). In these analyses, major eubacterial groups were shown to emanate in the following order: low G+C Gram-positive bacteria (*Firmicutes*), high G+C Gram-positive bacteria (*Actinobacteria*), *Cyanobacteria*, spirochetes, chlamydiae, and the *Cytophaga–Flavobacteria–Bacteroidetes* (CFB) group, and *Proteobacteria*. Recently, the trees inferred from concatenated alignments of rRNAs and proteins involved in translation from 45 eubacteria were shown to be virtually consistent with the previously described trees (Brochier et al. 2002). In this regard, the reliability of Cpn60 as a tracer of bacterial phylogeny and organellar origin was supported on numerous occasions, with overall tree topology being also consistent with the previously given order of emergence of taxa (Gupta 1998; Emelyanov 2001b; Woese 2002; Richards et al. 2003).

### 9.2.3 Canonical Pattern of Mitochondrial Ancestry for Eukaryotic Genes

Earlier phylogenetic data based on SSU rRNA (Olsen et al. 1994) and Cpn60 (Viale and Arakaki 1994) not only pointed to the α subdivision of *Proteobacteria* as a taxonomic source of mitochondria, but also traced the origin of the organelle to the order *Rickettsiales* of obligate intracellular symbionts. Gene complement analysis involving a broad variety of mitochondrial genomes sequenced to date demonstrated that, along with rRNAs, mitochondria unavoidably contain genes encoding apocytochrome *b* and the largest subunit of cytochrome *c* oxidase, as well as three subunits of NADH:ubiquinone oxidoreductase. It was reasonably suggested that these protein-encoding genes always reside in mitochondrial genomes, because intense hydrophobicity of the proteins would be incompatible with their import into mitochondria (Lang et al. 1999; Adams and Palmer 2003). It is clear that the genes for rRNAs and respiratory enzymes, i.e. resident mitochondrial genes, should be regarded as true tracers for the origin of organelle per se. Recent detailed analyses of rRNAs (Emelyanov 2001b) and concatenated subunits of each of five respiratory complexes (Emelyanov 2003c) robustly confirmed a sister relationship of *Rickettsiales* and mitochondria.

The term 'canonical pattern of mitochondrial ancestry' was coined for branching order, in which the order *Rickettsiales* and mitochondria share common ancestor to the immediate exclusion of free-living α-*Proteobacteria* (Emelyanov 2001b, 2003b). It should be noted that rickettsiae have never been shown to emanate from a particular group of free-living α-*Proteobacteria*, being instead a sister group to the latter as a whole (Emelyanov 2003b, c; Boussau et al. 2004). This finding points to an early origin of endosymbiotic bacteria within the α subdivision. A genome of *Wolbachia pipientis* – the species of the *Rickettsiaceae–Anaplasmataceae* assemblage – was recently completed (Wu et al. 2004), thus providing ample information for molecular analysis. Protein-based phylogenies also incorporated the most gene-rich mitochondrial genome of a zooflagellate *Reclinomonas americana* (Lang et al. 1997) and the earlier-sequenced *Rickettsia prowazekii* genome (Andersson et al. 1998). Most of the reported data corroborated a canonical pattern (Wu et al. 2004). However, more recent analysis additionally involving the complete mitochondrial genome of a liverwort *Marchantia polymorpha* showed that in several cases mitochondrial sequences grouped with those of free-living α-*Proteobacteria* exclusive of two rickettsiae (Esser et al. 2004). An apparent shortcoming of these and earlier studies (Kurland and Andersson 2000) is that the species of *Rickettsiales* were underrepresented, whereas the trees obtained were either overloaded with mitochondrial entries or incorporated a small number of eubacterial phyla.

A sister relationship of the *Rickettsiales* group and mitochondria revealed in phylogenetic analysis of resident mitochondrial genes does not necessarily mean that their last common ancestor was an obligate endosymbiont. The order *Rickettsiales* comprises not only true rickettsiae, currently classified into families *Rickettsiaceae* and *Anaplasmataceae* (Dumler et al. 2001), but also a group of rickettsia-like endosymbionts, the RLE group (Emelyanov and Sinitsyn 1999; Emelyanov 2001a) also called *Holosporaceae.* Along with *Paramecium* endosymbionts *Holospora obtusa*, and *Caedibacter caryophilus*, two *Acanthamoeba castellanii* endosymbionts (Horn et al. 1999) and an agent of necrotizing hepatopancreatitis in shrimp (Loy et al. 1996), this growing family now incorporates several uncultured endosymbionts, such as intracellular bacteria of acidophilic protists (Baker et al. 2003). The family *Rickettsiaceae* includes the genera *Rickettsia* and *Orientia*, while the holophyletic family *Anaplasmataceae* includes *Wolbachia* and several ehrlichiae and anaplasmas (Dumler et al. 2001; Emelyanov 2001a). The availability of SSU rRNA sequences from many species of both the *Rickettsiaceae–Anaplasmataceae* and RLE groups allowed the most explicit phylogenetic analysis using extensive species sampling. Unfortunately, Cpn60 has so far been sequenced only from one representative of RLE, *H. obtusa.* Phylogenetic trees based on SSU rRNA and Cpn60 protein sequences showed branching orders that may be called 'full canonical pattern of mitochondrial ancestry'. It appeared that RLE diverged after free-living $\alpha$-*Proteobacteria*, but before sister groups of *Rickettsiaceae–Anaplasmataceae* and mitochondria (Fig. 9.1). In order to minimize the impact of LBA, the trees shown incorporate only the least divergent rickettsiae, the species of the genus *Rickettsia.* Notably, the overall tree topology agrees well with that inferred in other analyses (Sect. 9.2.2). Taken together, these findings point to paraphyletic nature of the order *Rickettsiales* and strongly reinforce the idea that the mitochondrial ancestor was an endosymbiotic bacterium (Emelyanov 2001b).

With respect to the eukaryogenesis models already described, the significance of the canonical pattern seems apparent. Indeed, the canonical pattern of mitochondrial ancestry for resident mitochondrial genes may apply to other bacterial-like genes of Eukarya. If phylogenetic analysis of a particular gene or protein reveals affiliation of rickettsial sequences with eukaryotic sequences to the immediate exclusion of free-living $\alpha$-*Proteobacteria*, it may be taken as evidence in support of archaeal models that predict mitochondrial ancestry for bacterial-like genes. In contrast, affiliation on numerous occasions of eukaryotes to a particular eubacterial phylum other than *Rickettsiales* will support fusion models. In this regard, it was hypothesized that rampant LGT might blur mitochondrial ancestry of eukaryotic genes (Martin et al. 2001); however, this idea undermines a concept of the species and phyla (Sect. 9.2.2). It seems indeed paradoxical that the mitochondrial ancestor is called an $\alpha$-*Proteobacterium* while following the previously given viewpoint it does not actually belong to $\alpha$-*Proteobacteria*, which are still monophyletic in all published phylogenies. After all, the branching order of $\alpha$-*Proteobacteria* on the trees in all probability reflects their evolutionary history, despite the fact that only extant species are always under consideration. It is clear, for instance, that *Rhodobacterales* comprises versatile free-living

(a)

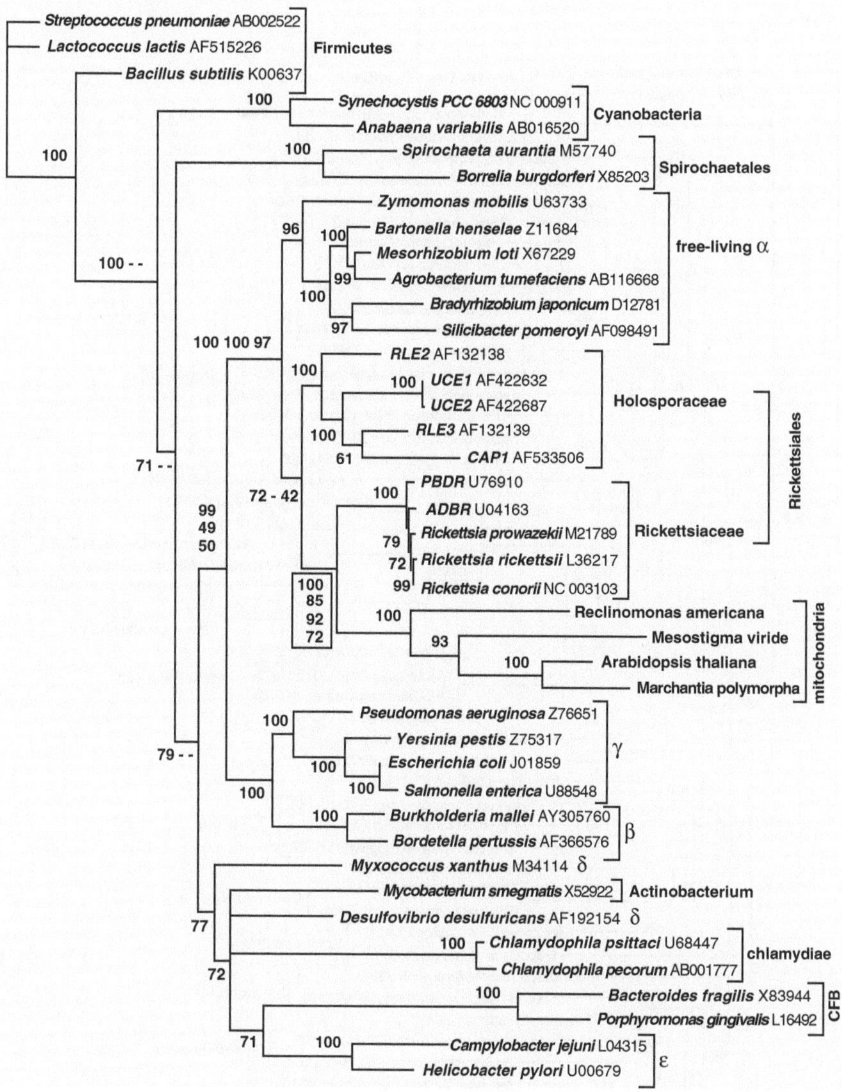


**Fig. 9.1.** Full canonical pattern of mitochondrial ancestry for mitochondrial (eukaryotic) genes. (a) Small-subunit (*SSU*) ribosomal RNA (*rRNA*) tree inferred from Bayesian analysis. Only posterior probabilities are shown at all nodes except for those related to overall branching order. In the latter case, Bayesian posterior probabilities and bootstrap supports in percent for logDet and maximum parsimony (*MP*) consensus trees are placed from *left to right* or from *top to bottom*, respectively, with the fourth value for the critical node being obtained with PUZZLE 4.02 (Strimmer and von Haeseler 1996). Where a single value is shown, support was 100% in all analyses. Supports below 40% are marked with *hyphens*. The *scale bar* corresponds to 0.1 substitutions per site. Bayesian analysis (four chains with 100,000 generations sampled every 100 generations, with the first 200 samples being discarded as burn-in) involved the invgamma model and a GTR substitution matrix (Huelsenbeck and Ronquist 2001). LogDet/paralinear and MP analyses were performed with PAUP 4.0 (Swofford 1998).

(b)

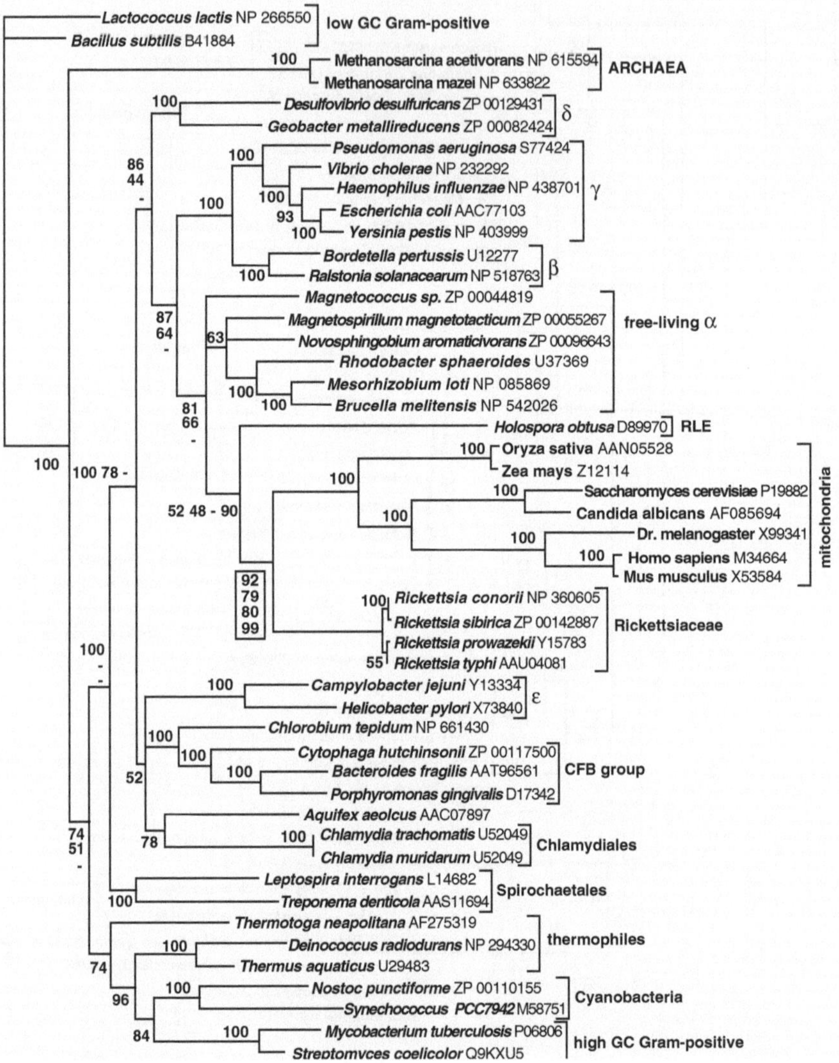


**Fig. 9.1.** (*Continued*) Maximum likelihood (*ML*) analysis with PUZZLE used the HKY85 substitution model and one invariable plus six variable rate categories. Only *Mesostigma viride*, *Reclinomonas americana*, *Mycobacterium smegmatis*, and *Borrelia burgdorferi* did not pass the $\chi^2$ test for compositional homogeneity as implemented in PUZZLE. Similar trees were obtained when other species of *Rickettsiales* (*Holosporaceae* and *Rickettsiaceae–Anaplasmataceae*) were used. Four-cluster analyses were applied to a tree ((mitochondria, *Rickettsiaceae–Anaplasmataceae*), (rickettsia-like endosymbionts, outgroup)). PHYLTEST analysis (Kumar and Rzetsky 1996) supported non-zero internal branch length with CP=0.98, and quartet mapping (Strimmer and von Haeseler 1996) support was 74.5%. *CFB Cytophaga–Flavobacter–Bacteroides*, *RLE* rickettsia-like endosymbionts of *Acanthamoeba castellanii*, *CAP* candidatus captivus acidiprotistae, *UCE* uncultured endosymbionts (Baker et al. 2003),

α-*Proteobacteria*, while a trend to live in intimate association with eukaryotic cells is apparently seen in *Rhizobiales* manifesting in the latest diverging facultative intracellular parasites (genera *Bartonella* and *Brucella*) of animals (Emelyanov 2003a; Boussau et al. 2004). These considerations encourage the use of the canonical pattern in an attempt to pinpoint the eukaryote origin.

## 9.3 Chimeric Nature of a Pro-eukaryote

### 9.3.1 Energy Metabolism of Eukaryotes and the Hydrogen Hypothesis

As noted in Sect. 9.2.1, the energy metabolism of eukaryotes is relatively simple when compared with that of prokaryotes. With the exception of photosynthetic organisms, all eukaryotes are heterotrophs whose energy metabolism relies on oxidative breakdown of reduced compounds. The first steps of carbohydrate utilization proceed along the glycolytic pathway yielding pyruvate. In eukaryotes with mitochondria, pyruvate penetrates the mitochondrial matrix, where it undergoes oxidative decarboxylation via a pyruvate dehydrogenase complex producing acetylcoenzyme A. The latter is fed into the Krebs cycle, yielding most of the NADH. Electrons from NADH enter the inner-membrane-located electron transport chain, in which they

*PBDR* papaya bunchy top disease rickettsia, *ADBR Adalia bipunctata* rickettsia. *Greek letters* denote proteobacterial subdivisions. Species names on the tree are followed by accession numbers. Mitochondrial SSU rRNA sequences were retrieved from the Organellar Genome Megasequencing Project (Lang et al. 1999). (**b**) Bayesian consensus tree based on Cpn60 protein sequences. A few sites of poor homology were removed from alignment leaving 491 positions for phylogenetic analysis. Posterior probabilities and bootstrap supports shown from *left to right* or from *top to bottom* were obtained by Bayesian analysis, and NJ and MP methods, respectively. The fourth number at the node joining genus *Rickettsia* to the mitochondrial cluster represents the bootstrap value obtained with MOLPHY 2.3 using the local rearrangement option (Adachi and Hasegawa 1996). It is worth noting that only the *Holospora obtusa* sequence failed to pass the $\chi^2$ test for amino acid composition homogeneity. Bayesian analysis involved the invgamma model and the JTT substitution matrix. In distance matrix analysis, 400 pseudoreplicates were generated with PHYLIP 3.6 (Felsenstein 1999) and used in PUZZLE-BOOT (http://www.tree-puzzle.de/puzzleboot.sh) to obtain ML distances, with the proportion of invariant sites and the gamma shape parameter $\alpha$ being inferred in PUZZLE using the JTT + Γ + inv model with six variable rate categories. The majority rule consensus tree was then inferred from multiple distances using NEIGHBOR and CONSENSE (PHYLIP package). Unweighted MP analysis was performed using PAUP 4.0 as described earlier (Emelyanov 2003a). Four-cluster tests were applied to the tree ((mitochondria, *Rickettsia*), (free-living α-*Proteobacteria* + *H. obtusa*, outgroup)) as described before (in phylogenetic analyses, *H. obtusa* grouped with free-living α-*Proteobacteria* on numerous occasions). The internal branch of the corresponding tree was of non-zero length with CP=0.96, whereas quartet mapping support was 63%. An involvement of other rickettsial species entirely covering the *Rickettsiaceae*–*Anaplasmataceae* assemblage (*Rickettsia prowazekii, Orientia tsutsugamushi, Ehrlichia chaffeensis, Anaplasma marginale*), did not affect tree topology pass from one respiratory complex to another accompanied by transfer of protons from the matrix into the intermembrane space. Finally, protons are pumped into mitochondria via ATPase coupled with ATP synthesis. ATP efflux from mitochondria to cytosol is mediated by an ATP/ADP carrier (AAC) (Skulachev 1988; Saraste 1999).

The vast majority of mitochondria use oxygen as a terminal acceptor of electrons. Along with aerobically respiring mitochondria, versatile mitochondria exist in which both oxygen and other oxidized compounds, e.g. fumarate and nitrate, serve as electron acceptors. Such sophisticated mitochondria were reported in several ciliates, fungi, and even lower animals (Tielens et al. 2002). The yield of ATP is, however, much lower in the cases of anaerobic respiration, as compared with 32–36 mol per mole of glucose produced by aerobic respiration (Saraste 1999).

A variety of unicellular, mostly microaerophilic eukaryotes lack mitochondria, but instead possess mitochondrion-derived organelles known as hydrogenosomes and mitosomes (Embley et al. 1997; Tovar et al. 2003). Among most well-known hydrogenosome-bearing organisms are the aforementioned parabasalid *T. vaginalis*, several ciliates such as *Nyctotherus ovalis*, and chytrid fungi such as *Neocallimastix frontalis*. It is clear that mitochondria were transformed into hydrogenosomes repeatedly and independently in these phylogenetically unrelated groups (Embley et al. 2003). Mitochondria and hydrogenosomes have much in common both structurally and biochemically, but the latter apparently lack cytochoromes and an electron transport chain. A hallmark function of hydrogenosomes is hydrogen-evolving fermentation mediated by FeS proteins pyruvate:ferredoxin oxidoreductase (PFO) and Fe hydrogenase. The end product of glycolysis, pyruvate, is oxidatively decarboxylated by PFO, yielding reduced ferredoxin, the latter being reoxidized by hydrogenase to produce molecular hydrogen. In mitosome-bearing protists, e.g. *G. lamblia*, reoxidation of ferredoxin proceeds in another way (Müller 1993; Embley et al. 2003). Of interest, *G. lamblia* is known to produce a small amount of $H_2$ (Lloyd and Harris 2002), which is consistent with a recent report on the presence of hydrogenase (Embley et al. 2003). These findings suggest that hydrogen may be generated in tiny remnant organelles.

The consideration of energy pathways of mitochondrion-lacking eukaryotes and well-documented cases of interspecific hydrogen transfer prompted Martin and Müller (1998) to put forth the hydrogen hypothesis for eukaryote origin. In general, the hydrogen hypothesis posits that a versatile $\alpha$-proteobacterium capable of respiration, glycolysis, and $H_2$-evolving fermentation entered into endosymbiotic relationship with a strictly anaerobic archaeon, whose energy metabolism proceeded along $H_2$- and $CO_2$-dependent methanogenesis. Being once deprived of a geological source of hydrogen, the archaeon gradually incorporated into its cytoplasm a bacterium which fuelled methanogenesis with $H_2$ and $CO_2$, waste products of fermentation. As the ultimately engulfed endosymbiont could not import organic compounds from outside in order to feed glycolysis and fermentation, its genes for carrier proteins were transferred to the host genome, and carriers were relocated to the archaeal membrane. Finally, glycolysis itself was relocated to the host cytoplasm, while methanogenesis disappeared. This syntrophy-based endosymbiosis resulted in the emergence of a pro-eukaryote endowed with bacterial-like energy metabolism and a mixed genome. Aerobic respiration has been retained in such an anaerobic microorganism with a hydrogenosome-like organelle to rid the host of toxic oxygen. It was also stated that characteristically eukaryotic structures and compartments emerged subsequent to mitochondrial origin (Martin and Müller 1998).

The hydrogen hypothesis was recently criticized as being mechanistically problematic (Emelyanov 2003a). Indeed, several unique evolutionary commitments suggested by the authors must have occurred, in effect, simultaneously to give rise to a viable chimeric entity – methanogen-derived host with a bacterium-derived organelle. Several other points are also disputable. For instance, why was eubacterial-type/eukaryotic-type phospholipid biosynthesis retained and did it later even replace an unrelated lipid fabric of host, if bacterial-derived membrane transporters from the start functioned in a heterologous archaeal membrane? It is also unclear, how an internalized bacterium could efficiently scavenge the host from oxygen? At last, the 'hydrogenosome first' concept implicated in the hydrogen hypothesis is at odds with molecular dating. With the divergence time of higher eukaryotes being set to 1.2 billion years ago (Feng et al. 1997), *Rickettsiaceae* and mitochondria were estimated using clock-like Cpn60 protein to have diverged 1.8 billion years ago. In the same way, Protozoa groups were shown to have diversified 1.45 billion years ago (Emelyanov 2003a), which is consistent with fossil data (Javaux et al. 2001). It is notable that 1.2 billion years ago for divergence of animals, plants, and fungi was recently supported using improved molecular dating techniques (Rodríguez-Trelles et al. 2002; Douzery et al. 2004). These data strongly argue for an aerobic origin of mitochondria. However, following the hydrogen hypothesis, such a syntrophy-based endosymbiosis might have occurred much earlier, i.e. under anoxic conditions, making respiration unnecessary.

### 9.3.2 Non-mitochondrial Origin of Eukaryotic Glycolysis

The hydrogen hypothesis would predict the closest relationship between eukaryotes and $\alpha$-*Proteobacteria* in the trees based on glycolytic and fermentation enzymes, when denying promiscuity of LGT (Sect. 9.2.2). Whereas key enzymes of fermentation, PFO and Fe hydrogenase, have so far been described only in amitochondriate protists, glycolytic enzymes are ubiquitous among eukaryotes. The origin of eukaryotic glycolysis was addressed by numerous investigations. Most of the works were concerned with the origin of glycolytic enzymes in amitochondriate protists. In several cases, tree reconstructions placed different eukaryotes at different tree positions.

These data were interpreted as a tolerance of glycolysis to incorporate components of different origins. Eubacterial ancestry of glycolytic enzymes was supported in most cases, although tree topologies had little in common with accepted bacterial relationships. It should be stressed that in no case eukaryotes grouped with *α-Proteobacteria* (reviewed by Richards et al. 2003). It is notable that triose phosphate isomerase (TPI) and glyceraldehyde-3-phosphate dehydrogenase (GAPDH) trees placed eukaryotes close to *γ-Proteobacteria* (Liaud et al. 2000; Emelyanov 2003a). Recently, a variety of glycolytic enzymes were analysed phylogenetically, and the data obtained were interpreted as supporting a distinctness of Eukarya from two prokaryotic domains (Canback et al. 2002). However, aberrant trees dominated by long branches cast doubt on this inference.

The aforementioned data raise two questions. May conventional phylogenetic methods still apply to glycolytic enzymes? If yes, which bacterial group, if not an α subdivision, was a taxonomic source of eukaryotic glycolysis? To address these questions, phylogenetic analysis of several enzymes was carried out in the present work based on the following premises:

1. Among eukaryotes, only 'crown' groups were involved. As far as possible, Protozoa were omitted for several reasons:
   (a) As unicellular eukaryotes, protists may be still amenable to gene acquisition by means of LGT.
   (b) Most of the well-characterized protists are parasites, whose sequences are often highly deviant.
   (c) It might occur that some gene under study was sequenced not from a eukaryotic organism, but from its endosymbiotic or engulfed bacteria.
2. Eukaryotic sequences were suitably aligned, then one of them was used as a query to search for finished and unfinished prokaryotic genomes via a BLAST server. If several homologs were found in a single species, only the most similar one was chosen for alignment. In the basis of the BLAST hit, the annotated complete sequence was retrieved from the finished genome when available.
3. Neighbour-joining (NJ) analysis with approximate correction for among-sites rate variation was applied to an alignment, which did not incorporate weakly homologous sequences. Species or even phyla characterized by long branches were additionally excluded.
4. Finally, bootstrapped analyses were performed with an alignment refined in the above ways.

Of importance, genome sequencing revealed that *W. pipientis* possesses an incomplete glycolytic pathway (Wu et al. 2004). BLAST search showed that the genes for glycolytic enzymes found in *W. pipientis* are also present in all publicly available unfinished genomes of *Anaplasmataceae.* This finding provided an excellent opportunity to examine the canonical pattern of mitochondrial ancestry for glycolytic enzymes.

Whereas the species of the genus *Rickettsia* lack glycolytic genes, all species of *Rickettsiaceae* and *Anaplasmataceae* possess pyruvate pyrophosphate dikinase (PPDK). PPDK, an auxiliary enzyme of glycolysis, mediates interconversion of phosphoenolpyruvate and pyruvate, and may compensate for the absence of pyruvate kinase (PYK) (Dandekar et al. 1999). Among eukaryotes, PPDK has been reported only in plants and protists (Bringaud et al. 1998). It is also very scarcely distributed among prokaryotes, being however amply present in α-*Proteobacteria*. For this reason, PPDK phylogeny was expected to be consistent with a canonical pattern. It appears, however, to be not the case – the closest eukaryotic relatives are bacteria of the CFB group (Fig. 9.2a). It is worth mentioning that proteobacterial subdivisions other than α-*Proteobacteria* lack this enzyme. As expected, a sister group relationship between *Rickettsiales* and free-living α-*Proteobacteria* was strongly supported, and was not affected by species sampling.

TPI once was a milestone of archaeal theory (Keeling and Doolittle 1997) positing that all bacterial-like genes of Eukarya derive from an α-proteobacterial progenitor of mitochondria. Later on, more careful phylogenetic analyses revealed that eukaryotic TPI groups not with α-*Proteobacteria*, but with γ-*Proteobacteria* (Liaud et al. 2000, Emelyanov 2003a). Compared with previous work (Emelyanov 2003a), several aberrant sequences were now removed while many other bacterial groups were incorporated in the analysis. Again, the canonical pattern was not observed, although rickettsiae and free-living α-*Proteobacteria* formed a clade. Instead, eukaryotes were shown to emanate from the βγ-proteobacterial subdivision (Fig. 9.2b).

Bacteria of the β and γ subdivisions are known to contain up to three paralogs of GAPDH (Dandekar et al. 1999, Liaud et al. 2000), one of which robustly grouped with Eukarya (Liaud et al. 2000, Emelyanov 2003a). Unlike previous work (Emelyanov 2003a), the present study involved only that enzyme from these groups which exhibited the greatest similarity to eukaryotic sequences, and was therefore regarded as an ortholog. It should be also noted that previous analysis involved only one species of *Anaplasmataceae*, *W. pipientis*, the sequence of which was known at that time. In that analysis, *W. pipientis* occupied a root position apparently due to LBA (Emelyanov 2003a). In the present trees, *Anaplasmataceae* expectedly grouped with free-living α-*Proteobacteria* (Fig. 9.3a). The affinity of Eukaryota to the βγ-*Proteobacteria* was convincingly supported, but curiously when CFB bacteria were incorporated, these formed a sister group with βγ-*Proteobacteria* (data not shown). This observation may be explained by the inability of tree-building methods to handle tightly related sequences in a proper way. It is clear from this analysis that paralogous duplication of GAPDH occurred very early in bacterial evolution, i.e. after divergence of *Firmicutes*.

The pyruvate kinase (PYK) tree was rooted with *Archaea*, which is also consistent with midpoint rooting (Fig. 9.3b). Importantly, *Firmicutes* branches off first in this case, in agreement with Gupta's proposal (1998).

As with the GAPDH tree, the PYK tree consists of two subtrees corresponding to two paralogs. One of the two γ-proteobacterial paralogs, fructose-stimulated PykF, was shown to group with eukaryotes (Fig. 9.3b). It should be emphasized that the overall topology of TPI-, GAPDH-, and PYK-based trees and subtrees conforms to the accepted order of emergence of bacterial phyla (Sect. 9.2.2).

Taken together, the present phylogenetic data support a hypothesis that eukaryotic glycolysis derived from a common ancestor of βγ-*Proteobacteria* (Emelyanov 2003a), albeit other glycolytic enzymes need to be examined using the previously described approach. It is notable that in most cases the closest relationship of eukaryotic and βγ-proteobacterial enzymes was sup-

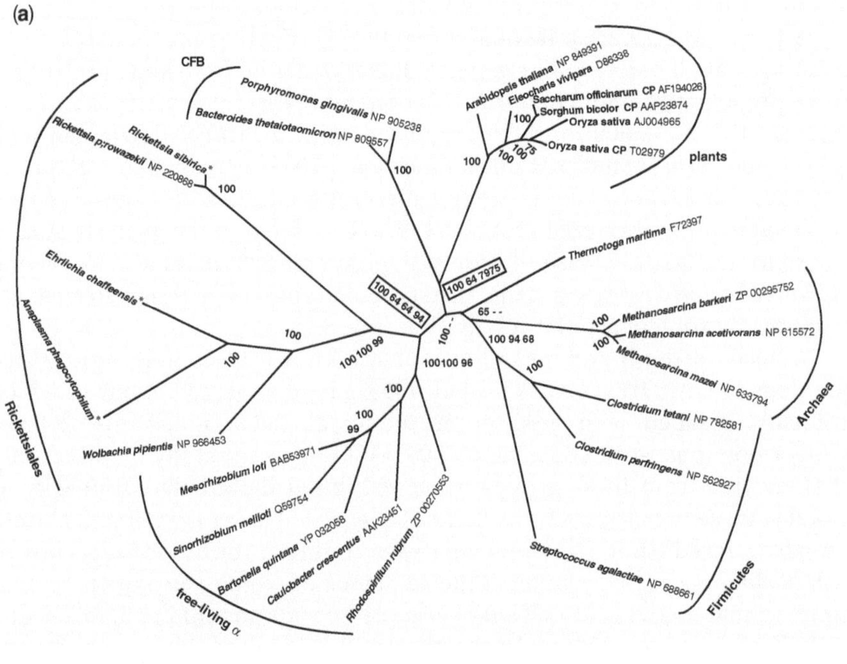


**Fig. 9.2.** Bayesian consensus trees based on pyruvate pyrophosphate dikinase (**a**) and triose phosphate isomerase (*TPI*) (**b**) protein sequences. (**a**) Deviant sequences of protists, *Neorickettsia sennetsu* and *Magnetococcus* sp. were not incorporated. No orthologs were found in *Archaea* except for Methanosarcinales. *Thermotoga maritima* and all rickettsiae except *Wolbachia pipientis* failed to pass the $\chi^2$ test for compositional homogeneity. Semiconstrained trees generated in MOLPHY 2.3, on which the plant sequences were grouped with *Rickettsiales*, i.e. topology consistent with a canonical pattern of mitochondrial ancestry for eukaryotic (plant) proteins, were rejected at 0.05 confidence limit using CONSEL 0.1d (Shimodaira and Hasegawa 2001) statistical tests. The same tree topology was recovered upon species sampling that involved four *Rickettsia* and *Wolbachia* spp. (**b**) Deviant sequences excluded from analysis were those from *Archaea*, *Actinobacteria*, *Mollicutes*, and *Proteobacteria* of ε and δ subdivisions. Only the *Rickettsiales* sequences did not pass the $\chi^2$ test. The trees on which Eukarya were constrained to *Rickettsiales* were not rejected by CONSEL statistical tests at 0.05 confidence level.

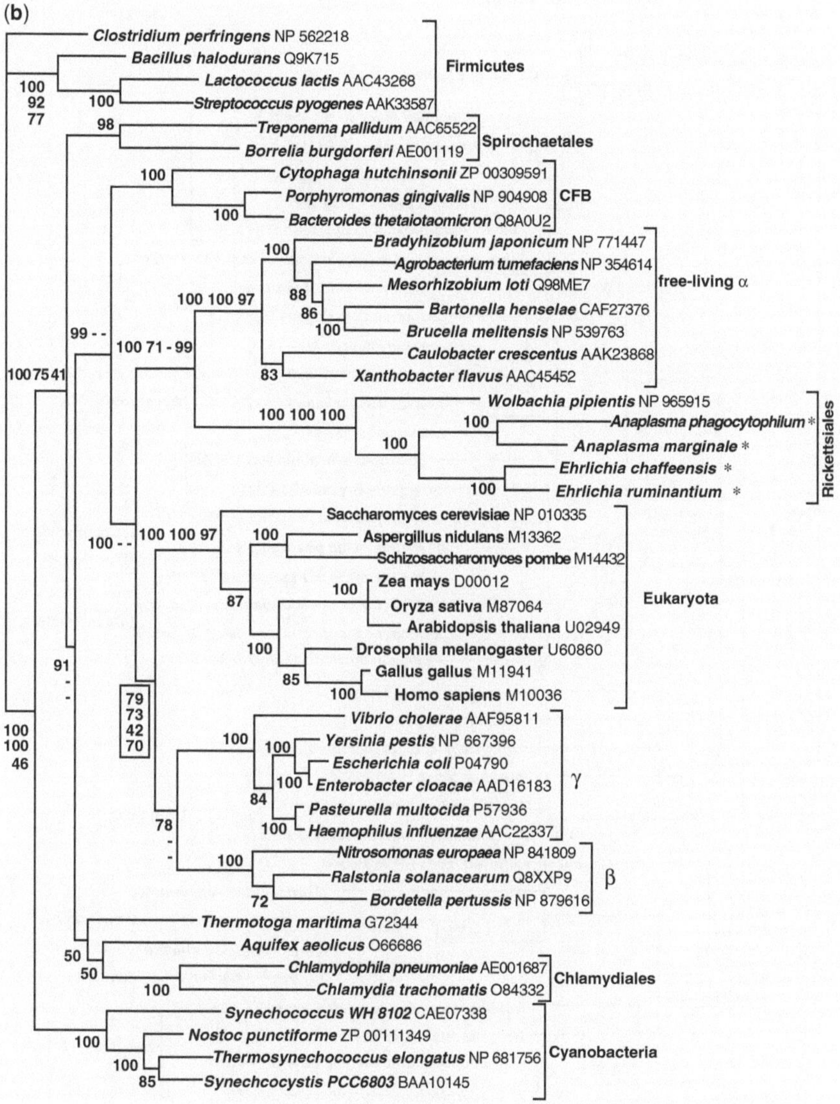


Fig. 9.2. (*Continued*) Supports for tree ((Eukarya, βγ-*Proteobacteria*), (α-*Proteobacteria*, outgroup)) obtained in quartet mapping and internal branch length analysis (see legend to Fig. 9.1) were 50% and 0.98, respectively. For other details see the legend to Fig. 9.1

ported by statistical tests (see the legends to Figs. 9.2, 9.3). The data presented are consistent with the observation of Gupta (2000) that neither the β nor the γ subdivision alone might contribute to eukaryotes. It is clear that the distribution of some glycolytic enzymes among both the bacteria and the eukaryotes may well be accounted for by a differential gene loss/retention model, e.g. one of several paralogs of βγ-proteobacterial GAPDH and PYK has been

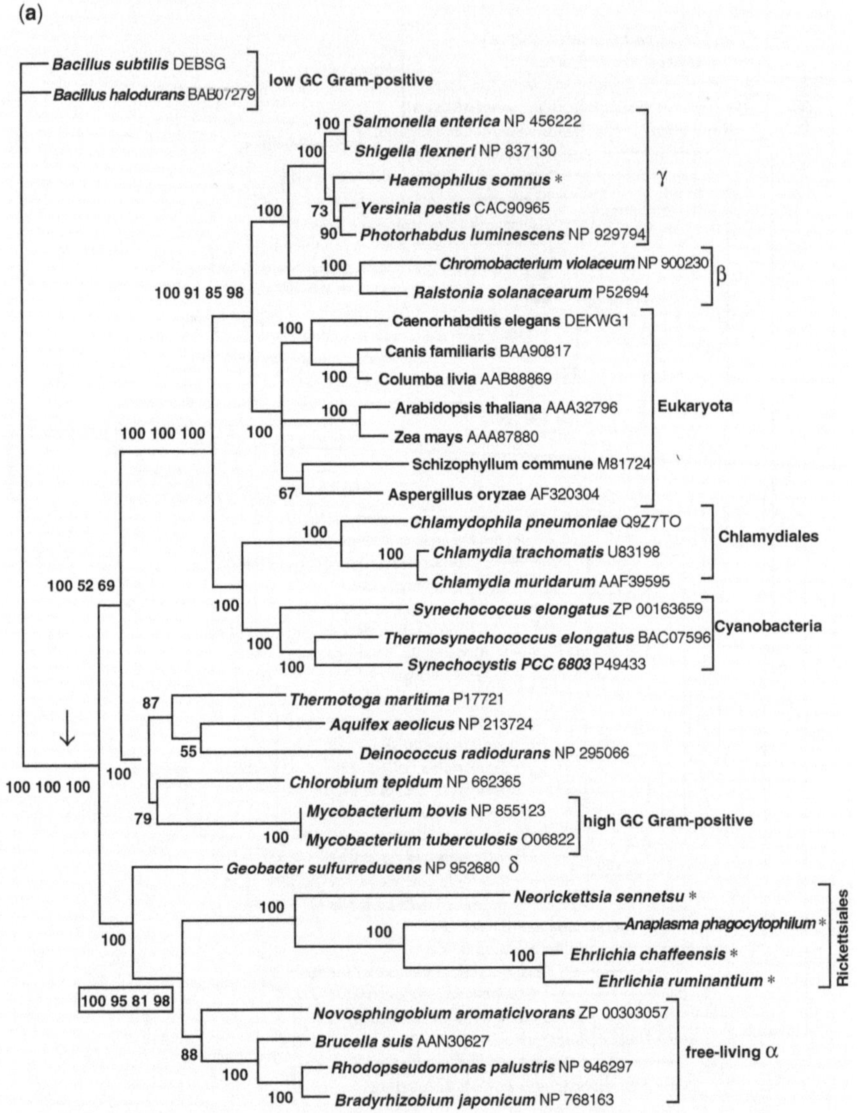


**Fig. 9.3.** Phylogenetic trees of glyceraldehyde-3-phosphate dehydrogenase (**a**) and pyruvate kinase (**b**) protein sequences. (**a**) The sequences from *Archaea*, spirochetes and the ε subdivision exhibited low homology to those of eukaryotes, and were not used. Only *E. ruminantium*, *E. chaffeensis*, and *Synechococcus elongatus* failed to pass the $\chi^2$ test. Semiconstrained trees consistent with a canonical pattern of mitochondrial ancestry were strongly rejected ($P$=0.00) by all statistical tests used. (**b**) The sequences from methanogens, *Chlamydiales*, *Mollicutes*, *Spirochaetales*, the CFB group, and *Proteobacteria* of β, δ, and ε subdivisions were excluded from phylogenetic analysis as highly aberrant. Only the sequences from *Clostridium perfringens* (NP 56065), *Caulobacter crescentus*, and *Leishmania mexicana* failed to pass the $\chi^2$ test.

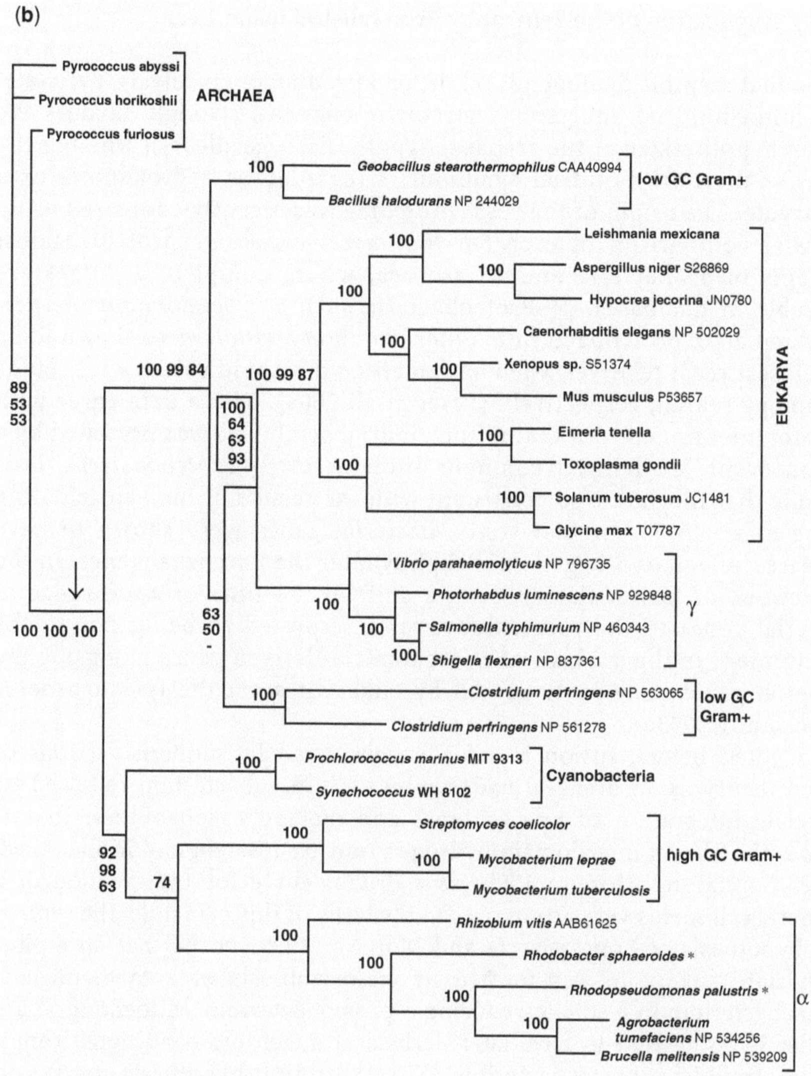


**Fig. 9.3.** (*Continued*) Semiconstrained trees consistent with a canonical pattern were strongly rejected ($P$=0.00) by statistical tests. Paralogous duplications are marked by an *arrow*. See the legend to Fig. 9.1 for other details

retained in eukaryotes. An affinity of the CFB group to βγ-*Proteobacteria* merits attention. It is suggested that some genes may have changed minimally during vertical transfer from CFB bacteria to the common ancestor of βγ-*Proteobacteria*. With respect to PPDK, that common ancestor may have possessed the enzyme which is characteristic of CFB but is lacking in β and γ subdivisions.

### 9.3.3 Application of the Syntrophy Principle to Fusion Event

Given that respiration depends on the end product of glycolysis, pyruvate, non-mitochondrial ancestry of glycolytic enzymes strongly favours the fusion hypothesis over the archaeal hypothesis, regardless of whether the host for the mitochondrial symbiont is regarded as a prokaryote or a eukaryote. The origin of the yeast proteome was recently addressed using pairwise comparison of each *Saccharomyces cerevisiae* protein against proteins of 45 bacteria and 15 archaea, whose complete genomes are available in databases. βγ-Proteobacteria such as *Pseudomonas aeruginosa* and an α-proteobacterium *Sinorhizobium meliloti* were shown to be the closest yeast relatives when low and high threshold values were set for homology search, respectively (Esser et al. 2004). These data agree with the aforementioned idea that mitochondrial symbiosis was preceded by a fusion event involving a common ancestor of βγ-*Proteobacteria*. It is notable that they are also consistent with the results of molecular dating (Feng et al. 1997). In that study, bacterial genes were shown to have appeared in eukaryotic genomes slightly later than archaeal genes. In the framework of fusion theory, such a shift in the time of appearance of bacterial genes to the present day was interpreted as being due to the involvement in the analysis of mitochondria-derived genes alongside the bacterial genes previously gained by eukaryotes via the fusion process (Emelyanov 2003a).

As noted before, fusion is actually considered by modern versions of fusion theory as an archaeal endosymbiosis in a eubacterium followed by complete integration of two genomes and metabolic capabilities. In this regard, the idea of interdomain hydrogen transfer (Martin and Müller 1998; López-García and Moreira 1999) as a driving force for the creation of a chimeric cell seems very attractive. On the basis of this principle, the syntrophy hypothesis of López-García and Moreira (1999) posits that an amitochondriate eukaryote was formed by endosymbiosis of a methanogenic archaebacterium in a sulfate-reducing δ-proteobacterium. Mitochondria are further hypothesized to have later derived at anaerobic conditions from a methylotrophic α-proteobacterium. The syntrophy hypothesis for fusion should clearly be preferred over the hydrogen hypothesis for two reasons: (1) it directly accounts for bacterial-like phospholipids in eukaryotic membranes; (2) a eubacterial host would effectively protect an archaeal symbiont against oxygen penetrating from outside. The view of a syntrophic association is not, however, sustained by two aspects of the data. First, methanogens and sulfate reducers are known to compete for hydrogen in sulfate-containing environments, so methanogenesis runs backwards to produce hydrogen; thus, oxidation of methane in the absence of oxygen is carried out by methanogenic archaea but not methylotrophic bacteria (Fenchel and Finlay 1995). Second, there are no molecular data in support of a δ-proteobacterial contribution to eukaryotes.

The data presented here are consistent with hypothesis that the host for the archaeal symbiont was a fermenting proteobacterium of the $\beta\gamma$ subdivision capable of respiration (Emelyanov 2003a). At relatively low oxygen concentration, the latter may have served mostly to maintain redox balance. A syntrophic association would thus be mutually advantageous to the methanogenic endosymbiont and fermenting host. Indeed, while the methanogen strictly depended on hydrogen and carbon dioxide produced by the host, it supplied the latter with a sink for reduced equivalents, making energetically inefficient respiration of no value. The loss of respiration would entail the loss of the outer membrane, thereby facilitating import of organic substances for glycolysis and fermentation. Such a view of fusion raises, however, the question why archaeal genetic apparatus has been retained in a chimeric organism. As suggested by Gupta (1998), this may have been due to strong pressure of antibiotics. In view of their important role in competition between microorganisms, antibiotics might be much more widespread in an ancient, prokaryotic world than today. Gupta (1998) convincingly argued that *Archaea* diverged from *Bacteria* owing to selection for multiple resistance to antibiotics.

It seems indeed very likely that bacterial genetic machinery has been radically remodelled in archaea to make an irreversible departure from the action of antibiotics. Following this logic, transfer of proteobacterial genes to an archaeal nucleoid might be driven by antibiotic pressure. It is also clear that in the cases of functional redundancy the archaeal genes must have been preserved in a chimeric organism. In this regard, it is worth mentioning a different fate of ubiquitous chaperones Cpn60 and Hsp70. It may be clear that two distantly related chaperonins, bacterial GroEL also known as Cpn60 and archaeal chaperonin-containing T-complex polypeptide (CCT), entered chimeric organisms via a fusion event. Among them, only the archaeal chaperonin has been retained, giving rise to the eukaryotic CCT family (Archibald et al. 2000), while the bacterial Cpn60 was lost and later regained from the mitochondrial ancestor to serve in organelles. In contrast, the bacterial Hsp70 was suggested to have been preserved in eukaryotes, whose paralogous duplication resulted in endoplasmic reticulum (ER) and cytosolic isoforms (Gupta 1998). Otherwise, Hsp70 may even have been lost during formation of a chimeric cell and later acquired from a mitochondrial progenitor (Sect. 9.4.3). It should be noted in this regard that several methanogens lack Hsp70 (Bult et al. 1996). Like the situation for chaperonin, a system for FeS cluster assembly of a chimeric organism might well be of archaeal heritage (Emelyanov 2003b).

When considered in this way, the chimeric organism would be a fermenting aerotolerant prokaryote, surrounded by a single bacterial membrane and endowed with compartmentalized methanogenesis and a mosaic genome dominated by archaeal genes. Such an eccentric prokaryote is further suggested to have once become a host for a mitochondrial symbiont (Sect. 9.4.2).

## 9.4 Mitochondrial Origin and Eukaryogenesis

### 9.4.1 Common Ancestry of Rickettsiae and Mitochondria

As stated before, most detailed phylogenetic data based on Cpn60 and SSU rRNA sequences (Fig. 9.1) point to the common origin of mitochondria and *Rickettsiaceae–Anaplasmataceae* families from within endosymbiotic α-*Proteobacteria*, whose closest extant relatives are the RLE group of endosymbionts. Within the *Rickettsiaceae–Anaplasmataceae* assemblage, the genus *Rickettsia* is the least divergent group (Emelyanov and Sinitsyn 1999, Emelyanov 2001b). A view of *Anaplasmataceae* as a diversification of the genus *Rickettsia* is consistent with biological data (Emelyanov 2001a). Whereas most rickettsiae are endosymbionts of arthropods, with only few of them occasionally infecting mammals (Hackstadt 1996), two species were reported in the protist *A. castellanii* (Horn et al. 1999). All but one species of the RLE group known so far are protist endosymbionts. Given that *Anaplasmataceae* possess glycolytic enzymes of α-proteobacterial origin (Figs. 9.2, 9.3), the species of RLE should also possess a glycolytic pathway. Thriving in very rich niche, host cytosol, rickettsiae are known to have dispensed with many metabolic capacities (Andersson et al. 1998; Wu et al. 2004). A reductive mode of evolution may have started in the last common ancestor of *Rickettsiaceae* and mitochondria. In particular, the absence of some biosynthetic pathways may well be shared trait derived by rickettsiae and organelles from their common endosymbiotic progenitor. Like most mitochondria, all known rickettsiae are obligate oxygen respirers. This is not surprising, given that modern rickettsiae inhabit strictly aerobic hosts (Hackstadt 1996). In contrast, the common ancestor may have been a sophisticated endosymbiont capable of both aerobic and anaerobic respiration. It was suggested that such an endosymbiotic α-proteobacterium might invade the previously described chimeric prokaryote by virtue of tightly membrane bound phospholipase activity, as modern rickettsiae do (Emelyanov 2001b, 2003a). This idea is supported by observation of an endosymbiotic relationship involving two different proteobacteria (von Dolen et al. 2001). Curiously, a true rickettsia was recently reported to invade even mitochondria (Beninati et al. 2004).

### 9.4.2 First Steps Towards Organelle

The previously described chimeric organism and rickettsia-like endosymbiont may have met in a microaerobic environment rich in antibiotics. Finding that all species of *Rickettsiaceae–Anaplasmataceae* characterized to date possess ValRS of an archaeal genre, unlike free-living α-*Proteobacteria* which have typically proteobacterial enzyme (Emelyanov 2003a), may partially support this idea. The most parsimonious explanation of this finding is that archaeal ValRS, which may have been insensitive to some antibiotic, was acquired by a common ancestor of rickettsiae and mitochondria from a chimeric host. The latter may have been endowed with both the characteristically βγ-proteobacterial (Emelyanov 2003a) and the archaeal enzyme inherited from bacterial and archaeal fusion partners, respectively. This would be reminiscent of the recent observation that some pathogens possess two different forms of IleRS, with one of them being a eukaryote-derived enzyme resistant to pseudomonic acid (Brown et al. 2003). An alternative explanation that ancient rickettsia might have acquired ValRS from archaea is unlikely, given that endosymbiotic bacteria rarely exchange genes through LGT (Lawrence 1999). The biological context of such an LGT would also be unclear.

During the long-term relationship, the rickettsia-like endosymbiont might ultimately pull the chimeric host out of the anaerobic world. This process would be accompanied by loss of methanogenesis and the archaeal membrane. Since then, the endosymbiont might have helped maintain redox balance via respiration. Natural selection would have favoured the endosymbiont–host pairs, in which a relationship between partners has been mutualistic. This might be achieved in two ways. First, the symbiont invented AAC to share respiration-derived ATP with the host, thereby fuelling its metabolism which supplied the symbiont with pyruvate. Some researchers argue that living organisms can neither produce ATP in an amount exceeding their need, nor they can suffer from a shortage in ATP (Martin and Müller 1998; López-García and Moreira 1999). Perhaps so, if symbiont and host are treated separately. But endosymbiosis is an entirely different matter – both partners might gain a great advantage from complete oxidation of carbohydrates via combined energy metabolism (John and Whatley 1975), i.e. host glycolysis and symbiont aerobic respiration. Second, some indispensable genes might be functionally transferred from the endosymbiont to the genome of the prokaryotic host in a relatively easy manner. Such a gene transfer would irreversibly tie the symbiont to its host, and would also allow the latter to take partial control over the invader – a prerequisite of endosymbiont taming. This is consistent with the idea that some primitive capability for protein import must have arisen prior to the complex import system. These two events could provide a firm basis for coevolution of rickettsia-like endosymbiont and chimeric host, first of all, invention of a sophisticated protein import machinery. Obviously, this would require a considerable amount of time. It was hypothesized that rickettsiae and mitochondria diverged from one another just during this period; thus, rickettsiae remained as endosymbiotic bacteria, whose obligate dependence upon the eukaryotic host rests on the need to import some mitochondrial proteins which are essential to their survival (Emelyanov 2001b).

In contrast, the immediate ancestor of the organelle lost the ability to escape from pro-eukaryotic host, with its proliferation having been forced to keep pace with host cell division. The establishment of highly precise and complex protein-import machinery, which emerged during coevolution of endosymbiont and host, made possible a massive transfer of endosymbiont genes to the host genome. It is clear that symbiont-derived genes had a different fate. Given that only a limited subset of bacterial genes must have been retained during primary fusion, i.e. mostly those supporting methanogenesis (Sect. 9.3.3), some endosymbiont proteins may have been recruited to serve in cytosol (Gabaldon and Huynen 2003). In turn, some pre-existing bacterial proteins may have been readily attracted to an emerging organelle merely upon addition of a targeting sequence, assuming that the βγ-proteobacterial fusion partner might be similar to the α-proteobacterial progenitor of the mitochondrion in many respects (Emelyanov 2003a). This idea may easily account for observations that only a minority of mitochondrial proteins are α-proteobacterial in origin (Kurland and Andersson 2000; Gabaldon and Huynen 2003). These are mostly proteins directly or indirectly supporting oxidative phosphorylation, a hallmark function of mitochondria. It was argued that the respiratory chain is the only true novelty brought into an anaerobic host by a mitochondrial symbiont (Emelyanov 2003c). Given that most of the FeS cluster containing proteins function in mitochondria (Lill and Kispal 2000), the endosymbiont system for FeS cluster assembly was retained in place of the pre-existing system (Emelyanov 2003b). Findings that mitochondria of some stramenopiles import TPI and GAPDH (Liaud et al. 2000) and that *Euglena* mitochondria import pyruvate:NADP+ oxidoreductase, an enzyme related to PFO (Hoffmeister et al. 2004), were interpreted as supporting the mitochondrial origin of glycolytic and fermentation pathways; however, these data can by no means be taken as proof of mitochondrial ancestry, given that any potentially importable protein may be targeted to an organelle when supplemented with targeting presequence (Neupert 1997). Finally, the vast majority of mitochondrial proteins were argued to have been a host invention (Kurland and Andersson 2000). Special emphasis should be placed on AAC – carrier protein which has played a key role in the transformation of the symbiont into an energy-generating organelle (see before). Bacterial-type AAC – inner-membrane protein with 12 transmembrane segments – has been described in *Rickettsia*, *Chlamydiales*, and plastids (Winkler and Neuhaus 1999), and recently in a microsporidian *Encephalitozoon cuniculi* (Vivarès et al. 2002). As predicted in a version of classic endosymbiont theory (Emelyanov 2001b), it was recently sequenced from species of the RLE group (Linka et al. 2003).

It was earlier hypothesized that the AAC had been invented either by chlamydiae with subsequent LGT to rickettsiae or by rickettsiae, and was physically but not functionally transferred from the latter to the genome of an amitochondriate host. Similarly to Cox1, also containing 12 transmembrane domains incompatible with import, rickettsial AAC could not be imported into mitochondria, but was later recruited to chloroplasts in a plant lineage (Emelyanov 2001b). Therefore, unrelated mitochondrial AAC – a member of the mitochondrial carrier family proteins containing six transmembrane domains – was instead recruited by the host to function in mitochondria (Emelyanov 2003a). Recent phylogenetic data (Amiri et al. 2003; Schmitz-Esser et al. 2004) fully corroborated this hypothesis. As shown here in parenthetic notation, an unrooted tree ((*Rickettsiales*, *E. cuniculi*), (plants, *Chlamydiales*)) was inferred from these analyses. It is seen that eukaryotes are paraphyletic on this tree. In order to keep eukaryotic holophyly, one should accept that a root of the tree lies either in chalmydiae or in *Rickettsiales*, with LGT having occurred either from Microsporidia to *Rickettsiales* or from plants to *Chlamydiales*, respectively. But only the latter possibility is logically correct. Indeed, if placing the root in chlamydiae, one would have to postulate that AAC was transferred from secondarily amitochondriate *E. cuniculi* to RLE which emerged before mitochondrial origin (Fig. 9.1). To rescue this latter scenario, one should accept that multiple LGT events involving the *Rickettsiales* have occurred subsequent to mitochondrial origin. The origin of the non-mitochondrial AAC in the RLE group is consistent with molecular data showing that the order *Chlamydiales* diversified approximately 0.7 billion years ago (Schmitz-Esser et al. 2004) while plants arose 1.2 billion years ago (Sect. 9.3.1). Thus, the acquisition of bacterial-type AAC from plant lineage, as earlier suggested (Stephens et al. 1998), may have played a crucial role in the diversification of chlamydiae.

In general, it is clear that the origin of mitochondria was accompanied by drastic reorganization involving both the host and the endosymbiont. In the next section, it will be argued that these dramatic changes entailed the emergence of characteristically eukaryotic features such as the nucleus, endomembrane system, and cytoskeleton, i.e. the origin of a full-fledged eukaryote.

### 9.4.3 Origin of the True Eukaryote

As pointed out in Sect. 9.2.1, it may not be excluded that primitively amitochondriate eukaryotes once existed. The question was therefore addressed of whether characteristically eukaryotic compartments emerged before or subsequent to the origin of mitochondria. Along with specifically eukaryotic proteins, eukaryotic structures and compartments are often characterized by paralogous (multigene) families of proteins which are well conserved among prokaryotes and eukaryotes (Gupta 1998; Lawrence 1999). Thus, phylogenetic tools are thought to be useful in the attempt to answer the question. The following chain of arguments demonstrate that the canonical pattern of mitochondrial ancestry may be instrumental in this regard It may be suggested that some eukaryotic structure/compartment is specified by a paralogous family, so a prokaryote is characterized by a single protein, while a eukaryote is typified by just a family. A hypothetical tree is considered in which a eukaryotic paralogous family forms a monophyletic group with *Rickettsiales* to the immediate exclusion of free-living $\alpha$-*Proteobacteria*, i.e.

a canonical pattern. It follows from this observation that the eukaryotic family derives from a mitochondrial homolog; hence, the respective eukaryotic compartment typified by this family originated subsequent to the establishment of a mitochondrion. Indeed, if the eukaryotic compartment were characterized not by a family but by a single protein, whose phylogeny conforms to a canonical pattern, it might be argued that this protein (hence, compartment itself) appeared before mitochondrial origin but was later substituted by a mitochondrial partner. This would be not the case for the paralogous family. One may indeed suggest that some paralogous family characteristic of a eukaryotic compartment would, say, derive from a βγ-proteobacterial homolog. This would mean that the family originated after the fusion event but before mitochondrial origin; hence, the compartment as such already existed before mitochondrial origin. However, to reconcile this view with phylogenetic result, showing the canonical pattern for a paralogous family, one would have to make an unrealistic assumption that the mitochondrial homolog gave rise to the observed family, then the pre-existing family that originated during fusion and accommodated to the compartment in which it functions have totally disappeared. The absurdity of such a scenario, invoking a replacement of the multigene family, seems apparent.

Application of this model to the MutS homolog (MSH) family of DNA mismatch repair enzymes was recently published (Emelyanov 2003a). Given that archaea lack MutS, bacteria possess single MutS, while eukaryotes are characterized by a multigene MSH family including a mitochondrial homolog (Culligan et al. 2000) and that phylogenetic analysis revealed a canonical pattern of mitochondrial ancestry for the eukaryotic family, it was concluded that the nucleus as such emerged after establishment of mitochondria (Emelyanov 2003a and unpublished data).

Along with organellar homologs, all known eukaryotes are characterized by nucleocytosolic homologs of heat-shock proteins Hsp40, Hsp70, and Hsp90 whose ancient duplications resulted in ER and cytosolic isoforms. As convincingly argued by Gupta (1998), this paralogous duplication must have accompanied the origin of the ER. The phylogeny of Hsp70 and Hsp90 clearly points to eubacterial origin of eukaryotic proteins, represented by sister groups of ER and cytosolic isoforms, but their affinity to any bacterial group is uncertain (Gupta 1998, Emelyanov 2002, 2003a). On the basis of indel analysis, Gupta (1998) concluded that Hsp70 paralogs trace their descent to α, δ, or ε but not β or γ subdivision of *Proteobacteria*. These data may be interpreted in such a way that nucleocytosolic Hsp70 derived from Hsp70 of a mitochondrial ancestor; hence, the ER itself originated following mitochondrial origin. An alternative view holds that a source of the eukaryotic chaperone might have been the common ancestor of βγ-*Proteobacteria*, i.e. the fusion partner. It may be logically argued that the first possibility should still be preferred. The arguments here are as follows. Were the host for the mitochondrial ancestor a chimeric prokaryote with single βγ-proteobacterial

Hsp70, the latter might have been easily recruited to the emerging organelle. In this case, phylogenetic analysis would reveal an affiliation of the organellar chaperone to that of βγ-*Proteobacteria*. Alternatively, were the host a eukaryote endowed with both cytosolic and ER isoforms, the latter could have been easily targeted to the organelle via modification of an aminoterminal presequence and removal of the ER retention signal. Thus, phylogeny would recover ER and mitochondrial homologs as a clade. Such a situation was recently reported for plastid Hsp90 which was been derived from ER homolog (Emelyanov 2002). In any case, the original endosymbiont copy might have been lost. However, mitochondrial Hsp70 always groups with α-proteobacterial homologs and never with the ER isoform in phylogenetic trees (e.g. Emelyanov 2003a). This reasoning leads to the following scenario for the origin of nucleocytosolic Hsp70. Hsp70 was brought into the pro-eukaryote with the mitochondrial ancestor followed by paralogous duplication and rapid divergence of the non-organellar, i.e. the nucleocytosolic, copy. This is why the phylogenetic position of the nucleocytosolic isoforms is uncertain. During concomitant origin of the ER, the non-mitochondrial copy underwent an additional duplication resulting in ER and cytoplasmic copies. As to the ER and cytosolic forms of Hsp90, they were reported to share a specific insert of two to three amino acids mostly with α-*Proteobacteria* including *Rickettsiaceae–Anaplasmataceae* (Emelyanov 2002, 2003a, and unpublished data).

Selection for more efficient respiration must have entailed an enlargement of the mitochondrial inner membrane via formation of cristae (John and Whatley 1975). Recently, it was shown that a dynamin-related protein (DRP) carries out a remodelling of the inner mitochondrial membrane in yeast (Wong et al. 2000). One may suggest that DRP originated first from some large GTPase to serve this function, followed by its paralogous duplication, resulting in a dynamin family. The latter may have carried out plasma membrane fission that led to the formation of sheaths and vesicles. These events may well explain the origin of the ER, the Golgi system, and, finally, endocytosis.

It was previously thought that the eukaryotic cytoskeletal proteins tubulin and actin trace their descent to prokaryotic FtsZ and FtsA or MreB, respectively, in spite of a very low similarity at the level of primary structure (Faguy and Doolittle 1998; van den Ent et al. 2001). Curiously, perfect homologs of tubulins α and β were recently reported in *Prosthecobacter*, phylum *Verrucomicrobia* (Jenkins et al. 2002). On the basis of this finding, one may speculate that genes for tubulins were acquired by the pro-eukaryote via LGT. These two events – origin of dynamin via paralogous duplication of DRP, initially involved in cristae formation, and acquisition of tubulins from *Verrucomicrobia* – might form the basis for endocytosis. This scenario is thought to be more parsimonious than that proposed by the classic Archezoa concept (Sect. 9.2.1).

Taken together, the data presented here suggest that the emergence of typically eukaryotic structures – the nucleus, the endomembrane system, and the cytoskeleton -may have occurred during a relatively short period of time following mitochondrial origin. In this regard, it seems unlikely that the origins of these compartments and structures were much more dramatic events than those which accompanied the establishment of the mitochondrion.

### 9.4.4 Secondarily Amitochondriate Eukaryotes

The most compelling evidence for the absence of primitively amitochondriate eukaryotes came from observation that hydrogenosome-containing *T. vaginalis* and mitosome-containing *G. lamblia* possess a mitochondrial-type organelle-associated system for FeS cluster assembly (Sect. 9.2.1). Evidently, retention of the mitosome was driven by the need for FeS proteins which are present in all organisms regardless of whether their energy metabolism rests on respiration or hydrogen-evolving fermentation. It is however unclear why the latter is localized to the hydrogenosome in hydrogenosome-containing eukaryotes and, possibly, to the mitosome in *G. lamblia*. Given that biochemical repertoires of mitochondria and hydrogenosomes essentially overlap (Embley et al. 2003, Hrdy et al. 2004), it is conceivable that a refitting of mitochondria with hydrogen-evolving fermentation might be a much easier evolutionary event than loss of the organelle and establishment of fermentation in cytosol, let alone that loss of the organelle would entail irreversible loss of FeS cluster assembly apparatus.

Long ago, the mitochondrion was regarded by Embley et al. (1997) as a 'bag' to which different sorts of energy metabolism may be targeted(. It was also suggested that both aerobic and anaerobic respiration originated with the mitochondrial ancestor (Emelyanov 2001b). A possibility cannot, however, be excluded that some or even all types of anaerobic respiration are later acquisitions in various lineages as adaptation to life with little oxygen (Tielens et al. 2002). At any rate, in view of the much higher ATP yield in aerobic respiration compared with that in anaerobic processes, it is suggested that the first eukaryotes were aerobically respiring organisms. Given that aerobically respiring higher eukaryotes, including yeast, are in all probability incapable of hydrogen-producing fermentation (Embley et al. 2003), the question arises as to a source of fermentation enzymes in mitochondria-lacking eukaryotes. Reported monophyly of PFO and, possibly, hydrogenase from diverse organisms argues for the presence of these enzymes in a primitively amitochondriate cell (Embley et al. 2003). If so, they were retained in various eukaryotes, so as to be easily recruitable during transition to an anoxic world (Embley et al. 1997). This means that degradation of aerobic mitochondria might occur only in those eukaryotes in which fermentation enzymes have been preserved. Yet another possibility remains that transformation of mitochondria into hydrogenosomes or mitosomes occurred concomitantly with acquisition of PFO and hydrogenase from anaerobic prokaryotes inhabiting the same oxygen-poor niche. Recent genome analysis of diplomonads *G. lamblia* and *Spironucleus barkhanus* revealed that many of their genes, supposedly acquired via LGT, are involved in anaerobic processes (Andersson et al. 2003). At any rate, once resident mitochondrial genes encoding respiratory subunits (Sect. 9.2.3) had been lost, conversion of mitochondria into hydrogenosomes and remnant organelles would become an irreversible event.

### 9.4.5 A View of Eukaryogenesis From Geological, Ecological, and Bioenergetic Perspectives

After the initial inventory of abiotic organic material had been depleted, methanogenesis could have been the main source of primary production at the reduced conditions of primitive Earth (Catling et al. 2001; Kasting and Seifert 2002). Hydrogen may not only have escaped from the atmosphere to space, but may for the most part have been consumed by methanogenic archaea. Depletion of the atmosphere with hydrogen must have entailed mass extinction of methanogens. Via methanogenesis, geological hydrogen and carbon dioxide would have been converted into reduced organic material. The latter could have fed glycolysis and fermentation, thereby providing conditions for enormous diversification of heterotrophic bacteria. Owing to the high biodiversity of both the archaea and the fermenting bacteria, large numbers of pairs of methanogenic endosymbionts and fermenting hosts must have been formed during exhaustion of geological hydrogen. Thus, methanogenesis might eventually survive owing to syntrophic association with hydrogen-producing bacteria. As argued here, the latter may have been a representative of the lineage ancestral to βγ-*Proteobacteria*. This view of a proteobacterial fusion partner is consistent with phylogenetic data suggesting that the latest diverged β and γ subdivisions should be the most-species-rich phyla (Gupta 2000). It is thought that a great variety of the endosymbiont-host consortia may have been subjected to natural selection driven by the pressure of antibiotics. This evolutionary process could have resulted in a single-membraned chimeric organism endowed with archaea-derived genetic apparatus, in which the compartmentalized methanogenesis and host's hydrogen-evolving fermentation have been successfully united. The cell may have rapidly propagated, having once become a target for an aerobically respiring endosymbiotic α-proteobacterium. This may have occurred around 1.8 billion years ago (Sect. 9.3.1), when atmospheric oxygen concentration had reached a level at which aerobic respiration would have been a highly efficient ATP-generating pathway. Oxidative phosphorylation would thus have been successfully integrated with eubacterial-type glycolysis of the chimeric host with concomitant loss of extremely oxygen sensitive methanogenesis, which so far served a host to maintain redox balance.

High yield of ATP from aerobic respiration made possible the development of typically eukaryotic features (Vellai et al. 1998). It is suggested that the last common ancestor of all eukaryotes was an aerobically respiring organism capable of complete oxidation of carbohydrates to carbon dioxide and water. Some unicellular eukaryotes have either retained or secondarily acquired the ability for anaerobic respiration and hydrogen-evolving fermentation, which has allowed their adaptation to life under microaerophilic or anaerobic conditions.

## 9.5 Conclusions

Accumulating molecular data suggest that there are no representatives of the premitochondrial stage of eukaryogenesis among the species living today. However, it may not be excluded that amitochondriate eukaryotes once existed but became extinct upon the origin of mitochondria. There is also no room for doubt that present-day eukaryotes are fundamentally chimeric, being endowed with both bacterial and archaeal features. In view of these data, the variety of eukaryogenesis models (Martin et al. 2001) may be reduced to four, when considered from standpoints of genome-metabolism and cellular organization. The classic Archezoa concept posits that primitively amitochondriate eukaryotes existed and that they were direct descendants of archaea (Doolittle 1998). Classic fusion theory (Gupta 1998) and its modern version (Lopez-Garcia and Moreira 1999) also posit that a primitively amitochondriate organism was a true eukaryote, but it is inferred as a sort of fusion between an archaebacterium and a eubacterium. According to the hydrogen hypothesis (Martin and Müller 1998), the primary host for the mitochondrial symbiont was a methanogenic archaebacterium. The fourth hypothesis for eukaryote origin advocated in this review holds that the primitively amitochondriate organism was a prokaryote, which emerged during endosymbiosis of a methanogenic archaeon in a fermenting proteobacterium, and that eukaryotic characters evolved after establishment of an aerobically respiring mitochondrion. A canonical pattern of mitochondrial ancestry for eukaryotic genes and proteins was developed (Emelyanov 2001b, 2003b,c) and used here in an attempt to distinguish between the models. It appears that the data obtained fit the last model in the best way. Thus, it may well be that it was the prophecy of Merezhkowsky that a nucleated cell originated during endosymbiotic relationship of the two entirely different beings and that just the endosymbiont was precursor of the nucleus.

*Acknowledgements:* I am grateful to Radhey S. Gupta and T. Martin Embley for helpful comments on the manuscript. Thanks are also due to the chief and staff of Joint Supercomputer Centre of the Russian Academy of Sciences (JSCC, Moscow) for providing the computer resources.

## References

Adachi J, Hasegawa M (1996) Computer science monograph 28. Institute of Statistical Mathematics, Tokyo

Adams KL, Palmer JD (2003) Evolution of mitochondrial gene content: gene loss and transfer to the nucleus. Mol Phylogenet Evol 29:380–395

Amiri H, Karlberg O, Andersson SGE (2003) Deep origin of plastid/parasite ATP/ADP translocases. J Mol Evol 56:137–150

Andersson SGE, Zomorodipour A, Andersson JO, Sicheritz-Pontén T, Alsmark UCM, Podowski RM, Näslund AK, Eriksson A-S, Winkler HH, Curland CG (1998) The genome sequence of *Rickettsia prowazekii* and the origin of mitochondria. Nature 396:133–140

Andersson JO, Sjögren AM, Davis LA, Embley TM, Roger AJ (2003) Phylogenetic analyses of diplomonad genes reveal frequent lateral gene transfers affecting eukaryotes. Curr Biol 13:94–104

Archibald JM, Logsdon JM jr, Doolittle WF (2000) Origin and evolution of eukaryotic chaperonins: phylogenetic evidence for ancient duplications in CCT genes. Mol Biol Evol 17:1456–1466

Baker BJ, Hugenholtz P, Dawson SC, Banfield JF (2003) Extremely acidophilic protists from acid mine drainage host Rickettsiales-lineage endosymbionts that have intervening sequences in their 16S rRNA genes. Appl Environ Microbiol 69:5512–5518

Beninati T, Lo N, Sacchi L, Genchi C, Noda H, Bandi C (2004) A novel alpha-Proteobacterium resides in the mitochondria of ovarian cells of the tick *Ixodes ricinus*. Appl Environ Microbiol 70:2596–2602

Blattner FR, Plunkett G III, Block CA, Perna NT, Burland V, Riley M et al (1997) The complete genome sequence of *Escherichia coli* K-12. Science 277:1453–1462

Boucher Y, Douady CJ, Papke RT, Walsh DA, Boudreau MER, Nesbo CL, Case RJ, Doolittle WF (2003) Lateral gene transfer and the origins of prokaryotic groups. Annu Rev Genet 37:283–328

Boussau B, Karlberg EO, Frank AC, Legault B-A, Andersson SGE (2004) Computational inference of scenarios for alpha-proteobacterial genome evolution. Proc Natl Acad Sci USA 101:9722–9727

Bringaud F, Baltz D, Baltz T (1998) Functional and molecular characterization of a glycosomal $PP_i$-dependent enzyme in trypanosomatids: pyruvate, phosphate dikinase. Proc Natl Acad Sci USA 95:7963–7968

Brinkmann H, Philippe H (1999) Archaea sister-group of Bacteria? Implications from tree reconstruction artifacts in ancient phylogenies. Mol Biol Evol 16:817–825

Brochier C, Philippe H, Moreira D (2000) The evolutionary history of ribosomal protein RpS14: horizontal gene transfer at the heart of the ribosome. Trends Genet 16:529–533

Brochier C, Bapteste E, Moreira D, Philippe H (2002) Eubacterial phylogeny based on translational apparatus proteins. Trends Genet 18:1–5

Brown JR, Doolittle WF (1997) Archaea and prokaryote-to-eukaryote transition. Microbiol Mol Biol Rev 61:456–502

Brown JR, Gentry D, Becker JA, Ingraham K, Holmes DJ, Stanhope MJ (2003) Horizontal transfer of drug-resistant aminoacyl-transfer-RNA synthetases of anthrax and Gram-positive pathogens. EMBO Rep 4:692–698

Bult CJ, White O, Olsen GJ, Zhou L, Fleischmann RD, Sutton GG et al (1996) Complete genome sequence of the methanogenic archaeon *Methanococcus jannaschii*. Science 273:1058–1072

Canback B, Andersson SGE, Kurland CG (2002) The global phylogeny of glycolytic enzymes. Proc Natl Acad Sci USA 99:6097–6102

Catling DC, Zahnle KJ, McKay CP (2001) Biogenic methane, hydrogen escape, and the irreversible oxidation of early Earth. Science 293:839–843

Cavalier-Smith T (1987) Eukaryots with no mitochondria. Nature 326:332–333

Cavalier-Smith T (1998) A revised six-kingdom system of life. Biol Rev Camb Philos Soc 73:203–266

Culligan KM, Meyer-Gauen G, Lyons-Weiler J, Hays JB (2000) Evolutionary origin, diversification and specialization of eukaryotic MutS homolog mismatch repair proteins. Nucleic Acids Res 28:463–471
Dandekar T, Schuster S, Snel B, Huynen M, Bork P (1999) Pathway alignment: application to comparative analysis of glycolytic enzymes. Biochem J 343:115–124
Daubin V, Moran NA, Ochman H (2003) Phylogenetics and the cohesion of bacterial genomes. Science 301:829–832
Doolittle WF (1998) You are what you eat: a gene transfer ratchet could account for bacterial genes in eukaryotic nuclear genomes. Trends Genet 14:307–311
Doolittle WF (1999) Phylogenetic classification and the universal tree. Science 284:2124–2129
Douzery EJ, Snell EA, Bapteste E, Delsuc F, Philippe H (2004) The timing of eukaryotic evolution: does a relaxed molecular clock reconcile proteins and fossils? Proc Natl Acad Sci USA 101:15386–15391
Dumler JS, Barbet AF, Bekker CPJ, Dasch GA, Palmer GH, Ray SC, Rikihisa Y, Rurangirwa FR (2001) Reorganization of genera in the families Rickettsiaceae and Anaplasmataceae in the order Rickettsiales: unification of some species of *Ehrlichia* with *Anaplasma*, *Cowdria* with *Ehrlichia* and *Ehrlichia* with *Neorickettsia*, descriptions of six new species combinations and designation of *Ehrlichia equi* and 'HGE agent' as subjective synonyms of *Ehrlichia phagocytophila*. Int J Syst Evol Microbiol 51:2145–2165
Embley TM, Hirt RP (1998) Early branching eukaryotes? Curr Opin Genet Dev 8:624–629
Embley TM, Horner DS, Hirt RP (1997) Anaerobic eukaryote evolution: hydrogenosomes as biochemically modified mitochondria. Trends Ecol Evol 12:437–441
Embley TM, van der Giezen M, Horner DS, Dyal PL, Bell S, Foster PG (2003) Hydrogenosomes, mitochondria and early eukaryotic evolution. IUBMB Life 55:387–395
Emelyanov VV, Sinitsyn BV (1999) A groE-based phylogenetic analysis shows the closest evolutionary relationship of mitochondria to obligate intracytoplasmic bacterium *Rickettsia prowazekii*. In: Raoult D, Brouqui P (eds) Rickettsiae and rickettsial diseases at the turn of the third millennium. Elsevier, Paris, pp 31–37
Emelyanov VV (2001a) Rickettsiaceae, rickettsia-like endosymbionts, and the origin of mitochondria. Biosci Rep 21:1–17
Emelyanov VV (2001b) Evolutionary relationship of rickettsiae and mitochondria. FEBS Lett 501:11–18
Emelyanov VV (2002) Phylogenetic relationships of organellar Hsp90 homologs reveal fundamental difference to organellar Hsp70 and Hsp60 evolution. Gene 299:125–133
Emelyanov VV (2003a) Mitochondrial connection to the origin of the eukaryotic cell. Eur J Biochem 270:1599–1618
Emelyanov VV (2003b) Phylogenetic affinity of a *Giardia lamblia* cysteine desulfurase conforms to canonical pattern of mitochondrial ancestry. FEMS Microbiol Lett 226:257–266
Emelyanov VV (2003c) Common evolutionary origin of mitochondrial and rickettsial respiratory chains. Arch Biochem Biophys 420:130–141
Esser C, Ahmadinejad N, Wiegand C, Rotte C, Sebastiani F, Gelius-Dietrich G, Henze K, Kretschmann E, Richly E, Leister D, Bryant D, Steel MA, Lockhart PJ, Penny D, Martin W (2004) A genome phylogeny for mitochondria among α-Proteobacteria and a predominantly eubacterial ancestry of yeast nuclear genes. Mol Biol Evol 21:1643–1660
Faguy DM, Doolittle WF (1998) Cytoskeletal proteins: the evolution of cell division. Curr Biol 8:R338–R341
Felsenstein J (1999) PHYLIP – Phylogeny inference package, version 3.6. University of Washington, Seattle
Fenchel T, Finlay BJ (1995) Ecology and evolution in anoxic worlds. Oxford University Press, Oxford
Feng D-F, Cho G, Doolittle RF (1997) Determining divergence times with a protein clock: update and reevaluation. Proc Natl Acad Sci USA 94:13028–13033
Gabaldon T, Huynen MA (2003) Reconstruction of the protomitochondrial metabolism. Science 301:609

Graham DE, Overbeek R, Olsen GJ, Woese CR (2000) An archaeal genomic signature. Proc Natl Acad Sci USA 97:3304–3308

Gray MW (1999) Evolution of organellar genomes. Curr Opin Genet Dev 9:678–687

Gribaldo S, Cammarano P (1998) The root of the universal tree of life inferred from anciently duplicated genes encoding components of the protein-targeting machinery. J Mol Evol 47:508–516

Gupta RS (1998) Protein phylogenies and signature sequences: a reappraisal of evolutionary relationships among Archaebacteria, Eubacteria, and Eukaryotes. Microbiol Mol Biol Rev 62:1435–1491

Gupta RS (2000) The phylogeny of Proteobacteria: relationships to other eubacterial phyla and eukaryotes. FEMS Microbiol Rev 24:367–402

Hackstadt T (1996) The biology of rickettsiae. Infect Agents Dis 5:127–143

Hashimoto T, Sánchez LB, Shirakura T, Müller M, Hasegawa M (1998) Secondary absence of mitochondria in *Giardia lamblia* and *Trichomonas vaginalis* revealed by valyl-tRNA synthetase phylogeny. Proc Natl Acad Sci USA 95:6860–6865

Hoffmeister M, van der Klei A, Rotte C, van Grinsven KWA, van Hellemond JJ, Henze K, Tielens AGM, Martin W (2004) *Euglena gracilis* rhodoquinone:ubiquinone ratio and mitochondrial proteome differ under aerobic and anaerobic conditions. J Biol Chem 279:22422–22429

Horiike T, Hamada K, Miyata D, Shinozawa T (2004) The origin of eukaryotes is suggested as the symbiosis of pyrococcus into gamma-Proteobacteria by phylogenetic tree based on gene content. J Mol Evol 59:606–619

Horn M, Fritsche TR, Gautom RK, Schleifer KH, Wagner M (1999) Novel bacterial endosymbionts of *Acanthamoeba* spp. related to the *Paramecium caudatum* symbiont *Caedibacter caryophilus*. Env Microbiol 1:357–367

Horner DS, Pesole G (2004) Phylogenetic analyses: a brief introduction to methods and their application. Expert Rev Mol Diagn 4:339–350

Hrdy I, Hirt RP, Dolezal P, Bardonova L, Foster PG, Tachezy J, Embley TM (2004) *Trichomonas* hydrogenosomes contain the NADH dehydrogenase module of mitochondrial complex I. Nature 432:618–622

Huelsenbeck JP, Ronquist F (2001) MRBAYES: Bayesian inference of phylogenetic trees. Bioinformatics 17:754–755

Jain R, Rivera MC, Lake JA (1999) Horizontal gene transfer among genomes: the complexity hypothesis. Proc Natl Acad Sci USA 96:3801–3806

Javaux EJ, Knoll AH, Walter MR (2001) Morphological and ecological complexity in early eukaryotic ecosystems. Nature 412:66–69

Jenkins C, Samudrala R, Anderson I, Hedlund BP, Petroni G, Michailova N, Pinel N, Overbeek R, Rosati G, Staley JT (2002) Genes for the cytoskeletal protein tubulin in the bacterial genus *Prosthecobacter*. Proc Natl Acad Sci USA 99:17049–17054

John P, Whatley FR (1975) *Paracoccus denitrificans* and the evolutionary origin of the mitochondriaon. Nature 254:495–498

Kasting JF, Seifert JL (2002) Life and the evolution of Earth's atmosphere. Science 296:1066–1068

Keeling PJ, Doolittle WF (1997) Evidence that eukaryotic triose-phosphate isomerase is of α-proteobacterial origin. Proc Natl Acad Sci USA 94:1270–1275

Kollman JM, Doolittle RF (2000) Determining the relative rates of change for prokaryotic and eukaryotic proteins with anciently duplicated paralogs. J Mol Evol 51:173–181

Kumar S, Rzhetsky A (1996) Evolutionary relationships of eukaryotic kingdoms. J Mol Evol 42:183–193

Kurland CG, Andersson SG (2000) Origin and evolution of the mitochondrial proteome. Microbiol Mol Biol Rev 64:786–820

Lang BF, Burger G, O'Kelly CJ, Cedergren R, Golding GB, Lemieux C, Sankoff D, Turmel M, Gray MW (1997) An ancestral mitochondrial DNA resembling a eubacterial genome in miniature. Nature 387:493–497

Lang BF, Gray MW, Burger G (1999) Mitochondrial genome evolution and the origin of eukaryotes. Annu Rev Genet 33:351–397
Lawrence JG (1999) Selfish operons: the evolutionary impact of gene clustering in prokaryotes and eukaryotes. Curr Opin Genet Dev 9:642–648
Liaud MF, Lichtlé C, Apt K, Martin W, Cerff R (2000) Compartment-specific isoforms of TPI and GAPDH are imported into diatom mitochondria as a fusion protein: evidence in favor of a mitochondrial origin of the eukaryotic glycolytic pathway. Mol Biol Evol 17:213–223
Lill R, Kispal G (2000) Maturation of cellular Fe-S proteins: an essential function of mitochondria. Trends Biochem Sci 25:352–356
Linka N, Hurka H, Lang BF, Burger G, Winkler HH, Stamme C, Urbani C, Seil I, Kusch J, Neuhaus HE (2003) Phylogenetic relationships of non-mitochondrial nucleotide transport proteins in bacteria and eukaryotes. Gene 306:27–35
Lloyd D, Harris JC (2002) *Giardia*: highly evolved parasite or early branching eukaryote? Trends Microbiol 10:122–127
López-García P, Moreira D (1999) Metabolic symbiosis at the origin of eukaryotes. Trends Biochem Sci 24:88–93
Loy JK, Dewhirst FE, Weber W, Frelier PF, Garbar TL, Tasca SI, Templeton JW (1996) Molecular phylogeny and in situ detection of the etiological agent of necrotizing hepatopancreatitis in shrimp. Appl Env Microbiol 62:3439–2445
Margulis L (1996) Archaeal-eubacterial mergers in the origin of Eukarya: phylogenetic classification of life. Proc Natl Acad Sci USA 93:1071–1076
Martin W, Müller M (1998) The hydrogen hypothesis for the first eukaryote. Nature 392:37–41
Martin W, Kowallik KV (1999) Annotated English translation of Mereschkowsky's 1905 paper 'Über Natur und Ursprung der Chromatophoren im Pflanzenreiche'. Eur J Phycol 34:287–295
Martin W, Russell MJ (2003) On the origins of cells: a hypothesis for the evolutionary transitions from abiotic geochemistry to chemoautotrophic prokaryotes, and from prokaryotes to nucleated cells. Philos Trans R Soc Lond B Biol Sci 358:59–85
Martin W, Hoffmeister M, Rotte C, Henze K (2001) An overview of endosymbiotic models for the origin of eukaryotes, their ATP-producing organelles (mitochondria and hydsrogenosomes), and their heterotrophic lifestyle. Biol Chem 382:1521–1539
Merezhkowsky C (1910) Theorie der zwei Plasmaarten als Grundlage der Symbiogenese, einer neuen Lehre von der Entstehung der Organismen. Biol Zentralbl 30:277–303, 321–347, 353–367
Müller M (1993) The hydrogenosome. J Gen Microbiol 139:2879–2889
Neupert W (1997) Protein import into mitochondria. Annu Rev Biochem 66:863–917
Olsen GJ, Woese CR, Overbeek R (1994) The winds of (evolutionary) change: breathing new life into microbiology. J Bacteriol 176:1–6
Poole AM, Phillips MJ, Penny D (2003) Prokaryote and eukaryote evolvability. Biosystems 69:163–185
Richards TA, Hirt RP, Williams BAP, Embley TM (2003) Horizontal gene transfer and the evolution of parasitic Protozoa. Protist 154:17–32
Rivera MC, Lake JA (2004) The ring of life provides evidence for a genome fusion origin of eukaryotes. Nature 431:152–155
Rodríguez-Trelles F, Tarrío R, Ayala FJ (2002) A methodological bias toward overestimation of molecular evolutionary time scales. Proc Natl Acad Sci USA 99:8112–8115
Sagan L (1967) On the origin of mitosing cells. J Theor Biol 14:225–274
Sapp J, Carrapiço F, Zolotonosov M (2002) Symbiogenesis: the hidden face of Constantin Merezhkowsky. Hist Philos Life Sci 24:413–440
Saraste M (1999) Oxidative phosphorylation at the fin de siècle. Science 283:1488–1493
Schmitz-Esser S, Linka N, Collingro A, Beier CL, Neuhaus HE, Wagner M, Horn M (2004) ATP/ADP translocases: a common feature of obligate intracellular amoebal symbionts related to chlamydiae and rickettsiae. J Bacteriol 186:683–691
Shimodaira H, Hasegawa M (2001) CONSEL: for assessing the confidence of phylogenetic tree selection. Bioinformatics 17:1246–1247

Skulachev VP (1988) Membrane bioenergetics. Springer, Berlin Heidelberg New York
Snel B, Bork P, Huynen MA (2002) Genomes in flux: the evolution archaeal and proteobacterial gene content. Genome Res 12:17–25
Sogin ML (1997) History assignment: when was the mitochondrion founded? Curr Opin Genet Dev 7:792–799
Stanier RY, van Niel CB (1962) The concept of a bacterium. Arch Mikrobiol 42:17–35
Stephens RS, Kalman S, Lammel C, Fan J, Marathe R, Aravind L et al (1998) Genome sequence of an obligate intracellular pathogen of humans: *Chlamydia trachomatis*. Science 282:754–759
Strimmer K, von Haeseler A (1997) Likelihood mapping: a simple method to visualize phylogenetic content of a sequence alignment. Proc Natl Acad Sci USA 94:6815–6819
Swofford DL (1998) PAUP* – phylogenetic analysis using parsimony (* and other methods), version 4.0. Sinauer, Sunderland
Tielens AGM, Rotte C, van Hellemond JJ, Martin W (2002) Mitochondria as we don't know them. Trends in Biochem Sci 27:564–572
Tovar J, León-Avila G, Sánchez LB, Sutak R, Tachezy J, van der Giezen M, Hernández M, Müller M, Lucocq JM (2003) Mitochondrial remnant organelles of *Giardia* function in iron-sulfur protein maturation. Nature 426:172–176
van den Ent F, Amos LA, Löwe J (2001) Prokaryotic origin of the actin cytoskeleton. Nature 413:39–44
van der Giezen M, Slotboom DJ, Horner DS, Dyal PL, Harding M, Xue GP, Embley TM, Kunji ER (2002) Conserved properties of hydrogenosomal and mitochondrial ADP/ATP carriers: a common origin for both organelles. EMBO J 21:572–579
Vellai T, Takács K, Vida G (1998) A new aspect to the origin and evolution of eukaryotes. J Mol Evol 46:499–507
Viale AM, Arakaki AK (1994) The chaperone connection to the origins of the eukaryotic organelles. FEBS Lett 341:146–151
Vivarès CP, Gouy M, Thomarat FT, Méténier G (2002) Functional and evolutionary analysis of a eukaryotic parasitic genome. Curr Opin Microbiol 5:499–505
von Dohlen CD, Kohler S, Alsop ST, McManus WR (2001) Mealybug $\beta$-proteobacterial endosymbionts contain $\gamma$-proteobacterial symbionts. Nature 412:433–435
Williams BAP, Hirt RP, Lucocq JM, Embley TM (2002) A mitochondrial remnant in the microsporidian *Trachipleistophora hominis*. Nature 418:865–869
Winkler HH, Neuhaus HE (1999) Non-mitochondrial ATP transport. Trends Biochem Sci 24:64–66
Woese CR (2002) On the evolution of cells. Proc Natl Acad Sci USA 99:8742–8747
Wong ED, Wagner JA, Gorsich SW, McCaffery JM, Shaw JM, Nunnari J (2000) The dynamin-related GTPase, Mgm 1p, is an intermembrane space protein required for maintenance of fusion competent mitochondria. J Cell Biol 151:341–352
Wu M, Sun LV, Vamathevan J, Riegler M, Deboy R, Brownlie JC et al (2004) Phylogenomics of the reproductive parasite *Wolbachia pipientis* wMel: a streamlined genome overrun by mobile genetic elements. PLoS Biol 2:327–341

# 10 The Diversity of Mitochondrion-Related Organelles Amongst Eukaryotic Microbes

MARIA JOSÉ BARBERÀ, IÑAKI RUIZ-TRILLO, JESSICA LEIGH, LAURA A. HUG, ANDREW J. ROGER

## Summary

The origin of mitochondria was a fundamental step in the evolution of eukaryotes. A popular view in the 1980s and 1990s held that the eukaryotic cell evolved prior to the origin of mitochondria and that contemporary mitochondrion-lacking protists such as diplomonads, parabasalids, pelobionts and microsporidia were Archezoa – living descendants of organisms from this early "pre-organellar" stage of eukaryote evolution. However, over the last decade, intensive phylogenetic and cell biological studies of these and other anaerobic protists lacking classical mitochondria indicate that they all harbour organelles and/or genes derived from mitochondria. Thus, all currently known extant eukaryotes derive from a common ancestor that harboured the mitochondrial endosymbiont. Anaerobic protists contain a wide variety of mitochondrion-derived organelles that are functionally diverse, ranging from "hydrogenosomes" that carry out anaerobic energy metabolic reactions to extremely reduced "mitosomes" of unknown function. There are even some organisms, such as the ciliate *Nyctotheus ovalis*, that contain organelles that are appear to be transitional stages between mitochondria and hydrogenosomes, displaying properties of both organelles. Despite the functional diversity of mitochondrion-derived organelles in diverse anaerobic eukaryote lineages, common evolutionary themes can be discerned that yield insights into the selective forces involved in biochemical remodelling of the mitochondrial organelle in eukaryote evolution.

## 10.1 Introduction

The transition from prokaryotic to eukaryotic cells and the subsequent events that took place during the origin of the eukaryotic cell remain some of the greatest unresolved, and hotly debated, biological puzzles. One of the few widely accepted facts of early eukaryotic evolution is that eukaryotic cells are fundamentally chimaeric: organelles such as mitochondria and chloroplasts are derived from endosymbiotic precursors that are

Origin of Mitochondria and Hydrogenosomes (ed. by William F. Martin and Miklós Müller)

phylogenetically distant from the ancestor of the nucleocytoplasmic lineage (Gray et al. 1999). Just one decade ago, we would have also argued that the genes from the nucleocytoplasmic lineage (and by default, the majority of the non-mitochondrial and non-chloroplast genes) were most similar to archaebacterial orthologs (reviewed in Brown and Doolittle 1997). Thus, the "host" of those early endosymbiotic events was probably an organism related to the lineage that gave rise to *Archaebacteria*. However, this view has been challenged by recent genomic analyses purporting to show that the nucleocytoplasmic lineage is also chimaeric, containing a mix of genes that are either most similar to archaebacterial or to eubacterial homologs (Rivera et al. 1998; Esser et al. 2004). The exact delineation of that mix remains unknown, but a general pattern arises from the analysis of current complete genome sequences: "operational genes" (genes involved in metabolic and biosynthetic processes) seem to be most similar to eubacterial orthologs; while "informational genes" (genes involved in transcription, translation and DNA maintenance) are closest to archaebacterial orthologs. Although it is well accepted that many of the original mitochondrion-encoded genes were gradually transferred to the nucleus during eukaryote evolution, it is not clear whether this process is responsible for the quantity or the quality (e.g. genes unrelated to mitochondrial biogenesis and function) of the eubacterial-like genes in the nuclear genome (Markos et al. 1993; Keeling and Doolittle 1997).

Whether this arising pattern of eukaryotic chimaeric composition is consistent with the different hypotheses of eukaryotic origins and evolution will be touched on in this chapter, including the debate regarding potential hosts and symbionts. However, the main goal of this chapter is to explore the events that occurred subsequent to the origin of mitochondria in eukaryotes. Thus, we will pay special attention to the sometimes unfairly underestimated biochemical and structural diversity of mysterious organelles found within anaerobic eukaryotes that appear to be of mitochondrial origin. Until recently, information about the structure and function of these organelles was scarce. New data, however, are being gathered from a wide variety of anaerobic eukaryotes, providing some important insights into this biological issue. Our two main evolutionary concerns here are the origin(s) of these organelles in anaerobic protists and the degree to which mitochondrion-derived organellar diversity has already been revealed. We will discuss new data on organelle origins, present a glimpse of the vast diversity of these organelles, and argue that all extant eukaryotes derive from a unique ancestor that contained an endosymbiont-derived organelle that, amongst extant eukaryotes, can take the form of mitochondria, mitosomes or hydrogenosomes. Finally, we will show that mapping the presence and absence of these organelles on the "new" hypothetical tree of eukaryotes requires a surprisingly strange and complex series of gene losses, gains and biochemical remodelling.

### 10.1.1 The Origin of Mitochondria – the Symbiont

The evidence that modern mitochondria are derived from a eubacterial ancestor is sufficient that no reasonable argument can be made to the contrary; however, the identity of the closest extant relatives to mitochondria is a more contentious issue. Phylogenetic analyses indicate that mitochondria likely emerged from among the α-proteobacteria, a diverse group with lifestyles ranging from autotrophic to parasitic. Most studies have suggested that, within the α-proteobacteria, the sister group to mitochondria is likely the order *Rickettsiales* (Gray et al. 1999; Fitzpatrick et al. 2005), although some studies indicate that mitochondria emerged from within the *Rickettsiales* (Emelyanov 2001). Other authors suggest that mitochondria may branch elsewhere among the α-proteobacteria (e.g. Esser et al. 2004 conclude that mitochondria are more likely to be sisters of the *Rhodospirillales*).

In order to shed some light on the position of mitochondria within the α-proteobacteria, we have analysed 37 mitochondrion-encoded amino acid sequences (11,231 characters) from three eukaryotes, along with their homologues in 16 α-proteobacteria and three γ-proteobacteria, as an outgroup. We concatenated all 37 sequences and inferred a maximum likelihood (ML) tree using PHYML (Guindon and Gascuel 2003) (under WAG + Γ) and determined bootstrap support from 100 replicates. In addition, we used MrBayes (Ronquist and Huelsenbeck 2003) (under WAG + Γ) to infer a Bayesian phylogeny from this same data set. The topology inferred by both methods was identical (Fig. 10.1).

From these analyses, the closest α-proteobacterial relatives of the mitochondria appear to be the *Rickettsiales*, with 93% bootstrap support (ML) and a posterior probability of 1 (Bayesian analysis). However, phylogenetic analysis of this group is problematic for a few reasons. First, as can be clearly seen in Fig. 10.1, the branches leading to both the *Rickettsiales* and the mitochondria are very long, suggesting that their apparent monophyly may be the result of long-branch attraction if the assumptions of the evolutionary model employed are violated (model misspecification). Indeed, both mitochondria and *Rickettsiales* genomes are very rich in adenine and thymine bases, whereas other α- and γ-proteobacterial genomes are rich in guanine and cytosine (Table 10.1). In this case, the extreme base composition bias has affected amino acid usage. To see this effect we calculated the ratio of glycine, alanine, arginine and proline (GARP) amino acid usage to phenylalanine, tyrosine, methionine, isoleucine, asparagine and lysine (FYMINK) usage as a measure of amino acid level bias resulting from nucleotide bias (GARP amino acids are all encoded by codons containing only glycine and cysteine in the first two positions, whereas FYMINK amino acids codons contain only alanine and threonine in the first two positions) (Foster et al. 1997; Foster and Hickey 1999). *Rickettsiales* and mitochondria have a GARP to FYMINK ratio less than 1, whereas all other taxa studied here have a GARP to FYMINK

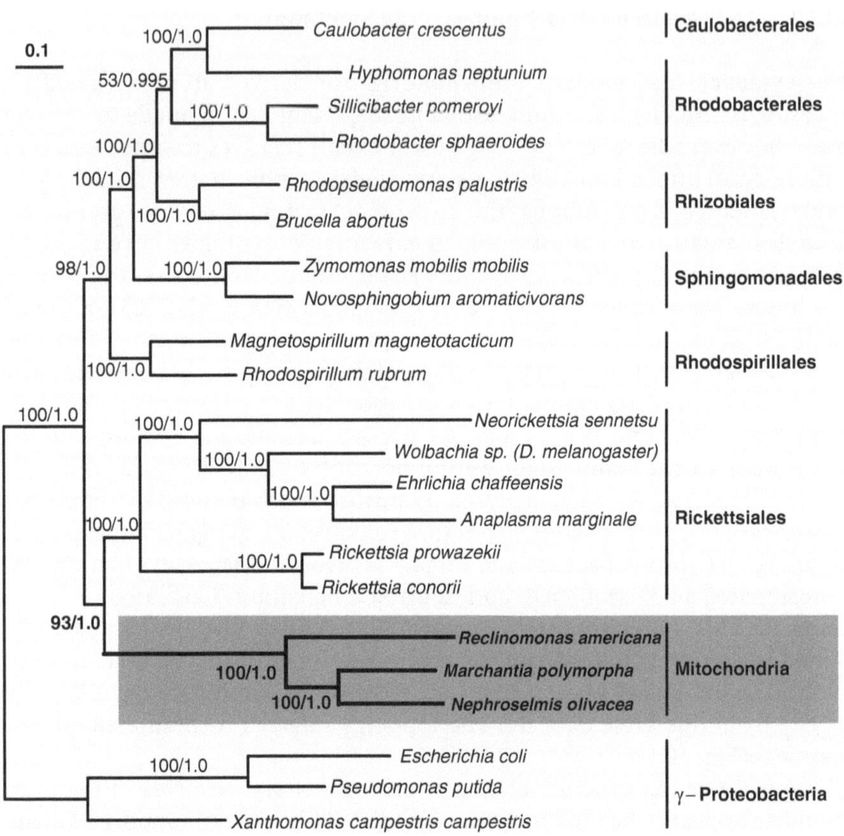


**Fig. 10.1.** Phylogeny of the α-proteobacteria and mitochondria. The phylogeny shown here was inferred from 37 concatenated mitochondrial proteins using PHYML. The same topology was recovered using MrBayes. Maximum likelihood (*ML*) bootstrap support (*left*) and Bayesian posterior probability (*right*) values are shown for nodes where ML bootstrap support was less than 100%. These 37 proteins analysed included three ATP synthase subunits, three cytochrome *c* oxidase subunits, 11 NADH dehydrogenase subunits, two succinate dehydrogenase subunits, 13 ribosomal proteins, three RNA polymerase subunits, apocytochrome *b* and the elongation factor Tu. All sequences were aligned separately using Muscle (Edgar 2004) and alignments were further refined manually

ratio greater than 1 (Table 10.1). The evolutionary models employed here assume that the equilibrium amino acid composition does not vary in different sequences. Consequently, the apparent monophyly of *Rickettsiales* and mitochondria may be nothing more than a phylogenetic reconstruction artefact resulting from a shared bias in amino acid usage.

In any case, our results concur with most other published phylogenies of α-proteobacteria, suggesting that the closest eubacterial relatives of mitochondria are the *Rickettsiales*; however, until a phylogenetic analysis method

**Table 10.1.** Nucleotide and amino acid composition for sequences examined

| Species | %G+C | GARP:FYMINK |
|---|---|---|
| ***Anaplasma*** | 43.86 | 0.9084 |
| *Caulobacter* | 64.25 | 1.1954 |
| ***Ehrlichia*** | 32.55 | 0.6511 |
| *Escherichia* | 53.79 | 1.123 |
| *Hyphomonas* | 60.07 | 1.1688 |
| *Magnetospirillum* | 62.96 | 1.1436 |
| ***Marchantia*** | 36.98 | 0.7142 |
| ***Neorickettsia*** | 43.06 | 0.8691 |
| ***Nephroselmis*** | 32.78 | 0.7119 |
| *Novosphingobium* | 63.28 | 1.2076 |
| *Pseudomonas* | 59.99 | 1.1579 |
| ***Reclinomonas*** | 27.52 | 0.5351 |
| *Rhizobium* | 60.68 | 1.0844 |
| *Rhodobacter* | 64.81 | 1.2126 |
| *Rhodopseudomonas* | 63.58 | 1.1319 |
| *Rhodospirillum* | 60.45 | 1.1816 |
| ***Rickettsia conorii*** | 35.63 | 0.6691 |
| ***Rickettsia prowazekii*** | 32.91 | 0.6283 |
| *Silicibacter* | 61.68 | 1.0959 |
| ***Wolbachia*** | 36.34 | 0.6839 |
| *Xanthomonas* | 62.31 | 1.2099 |
| *Zymomonas* | 49.80 | 1.1718 |

Species in *boldface* are mitochondria and *Rickettsiales* α-proteobacteria with lowered %G+C and GARP:FYMINK content

is developed that accommodates changing amino acid composition over the tree, we remain skeptical that *Rickettsiales* are truly the sister group of mitochondria.

### 10.1.2 The Host: the Rise and Decline of the Archezoa Hypothesis

Another hotly debated question regarding the origin of eukaryotes concerns the nature of the host cell that acquired the mitochondrial endosymbiont. The main point of contention is whether the host that engulfed the organelle ancestral to mitochondria was a prokaryote, a eukaryote or something in-between. If the host were a prokaryote, then the origin of the eukaryotic cell was contemporaneous with, or occurred after, the endosymbiotic event. Alternatively, the host cell was already a eukaryote (with a nucleus) or a proto-eukaryote (lacking a nucleus, but displaying other key eukaryote features) that engulfed the ancestral mitochondrial endosymbiont. Although the original serial endosymbiotic model implied that there was no pre-endosymbiotic phase of eukaryote evolution (Sagan 1967), some authors at the time suggested that phagocytosis (which requires distinctive eukaryotic features such as the cytoskeleton and the endomembrane system) had to evolve prior to the engulfment of the symbiont that gave rise to the mitochondria (de Duve 1969; Stanier 1970). In the early 1980s, Cavalier-Smith elaborated on the view that the host was a true eukaryote and argued that some extant protist groups represented mitochondrion-lacking and chloroplast-lacking eukaryotes that had diverged before the endosymbiotic events that gave rise to these organelles. He proposed that these organisms should be classified in a eukaryotic subkingdom, the Archezoa, to emphasize the evolutionary importance of their "primitive" cellular architecture (Cavalier-Smith 1983, 1987). The first ribosomal RNA (rRNA) phylogenies appeared to validate the Archezoa hypothesis, consistently showing amitochondriate archezoan taxa, such as microsporidians, parabasalids (*Trichomonas vaginalis*) and diplomonads (*Giardia intestinalis*) branching prior to mitochondriate groups at the base of the eukaryotic tree (Sogin 1989).

However, subsequent evolutionary, genetic and cell biological investigations of diverse amitochondriate eukaryotes have completely overturned the Archezoa hypothesis. Ironically, the first data to challenge this view were small-subunit (SSU) rRNA phylogenies that showed "amitochondriate" amoebae such as *Entamoeba* and the pelobionts as branching after other mitochondriate protist lineages within the eukaryotic tree (Sogin 1989, 1997; Hinkle et al. 1994; Cavalier-Smith and Chao 1996). Current multiple gene phylogenies place these amoebae in a group containing mitochondriate lobose amoebae and slime moulds (Arisue et al. 2002a; Bapteste et al. 2002). Similarly, some protein phylogenies showed Microsporidia to be highly derived *Fungi* rather than early diverging eukaryotes (Keeling and Doolittle 1996; Hashimoto et al. 1997, 1998; Roger et al. 1999), a position that has now been confirmed by several phylogenomic analyses (Thomarat et al. 2004; Brinkmann et al. 2005). Although an alternative phylogenetic position for amitochondriates such as diplomonads and parabasalids has not been resolved in molecular phylogenies, their basal position in the EF1-$\alpha$ and rRNA trees appears to be artefactual, resulting from a highly increased evolutionary rate in their sequences and inadequate phylogenetic models (Embley and Hirt 1998; Philippe and Laurent 1998). Indeed, a variety of ultrastructural characters strongly suggest that, rather than being "basal eukaryotes", the diplomonads and parabasalids belong in a "supergroup", the Excavata, with mitochondriate lineages such as the Euglenozoa and the jakobids (Fig. 10.2), a position that requires that they had a mitochondriate ancestry (Simpson and Roger 2004).

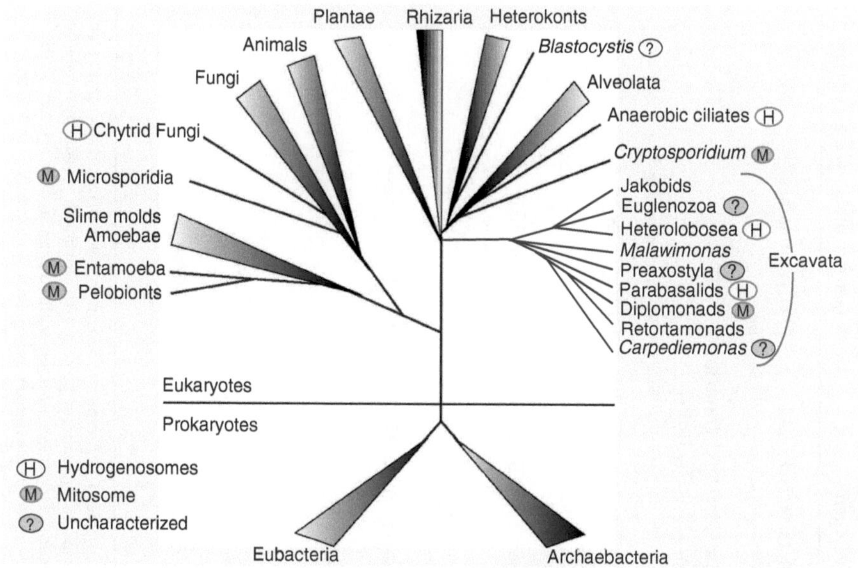


**Fig. 10.2.** The phylogenetic distribution of mitochondrion-related organelles in eukaryotes. Organelle function is indicated as *H* hydrogenosome, *M* mitosome or ? unknown

Yet, by far the most conclusive evidence against the Archezoa hypothesis is the presence in many putative archezoans (e.g. diplomonads, parabasalids, *Entamoeba* and Microsporidia) of mitochondrion-derived organelles or genes that encode mitochondrion-targeted products (such as chaperonins) that trace their history to mitochondria and α-proteobacteria (Clark and Roger 1995; Germot et al. 1996, 1997; Roger et al. 1996, 1998; Hirt et al. 1997; Roger 1999; Table 10.2). It seems reasonable, therefore, to assume that all extant eukaryotes either contain mitochondria or are descended from organisms that once did; however, the unlikely possibility that some true undiscovered archezoan may still lurk in a poorly studied anaerobic environment cannot be ruled out.

It has been suggested that the apparent absence of extant Archezoa supports the idea that the "host" was most probably a prokaryote and that endosymbiosis was an early event in eukaryotic history, if not contemporaneous with eukaryotic origins. However, we advocate a more cautious interpretation. The facts indicate that no lineages descending from transitional forms between prokaryotes and fully formed mitochondriate eukaryotes are currently alive. Yet in order for the evolutionary transformation between prokaryote to eukaryote to take place, such transitional organisms *must* once have existed, but left no descendants. Therefore, any hypothesis that posits the existence of such organisms (including extinct Archezoa or mitochondrion-containing prokaryotes, for example) cannot be rejected out of hand. The difficulty will be in evaluating the relative merits of these hypotheses since the

**Table 10.2.** Amitochondriate taxa with our current knowledge of the organelle composition and metabolism

| Protistan group | Losses | Organelle | Species | Mitochondrial derived genes | Anaerobic metabolism | Genome |
|---|---|---|---|---|---|---|
| Entamoebae | 1 | M | *Entamoeba histolytica* | PNT, Cpn60, Hsp70, IscU, IscS, AAC | PFO, hydrogenase | No |
| Pelobionts | 1 | M (?) | *Mastigamoeba balamuthi* | – | PFO | – |
| Microsporidia | 1 | M | *Encephalitozoon cuniculi* | PDH E1, ATM 1p, IscS, IscU, Fd, frataxin, Hsp70, ERV1p, Tim22, Tom70, IMP-2 | – | No |
| Chytrid fungi | 1 | H | *Neocallimastix* | Hsp70, Cpn60, AAC | PFL, hydrogenase | No |
| Ciliophora (7 orders) | 4–7 (?) | H | *Nyctotheus ovalis* | AAC, Cpn60, Hsp70, Tom34, PDH E1, PDH E2, CI, CII | Hydrogenase | Yes |
| Heterokonts | 1 (?) | H (?) | *Blastocystis hominis* | – | – | Yes |
| Apicomplexa | 1 (?) | M | *Cryptosporidium parvum* | AOX, PNT, IscU, IscS, Fd, MDH, Hsp60, Hsp70 | PNO, Narf-like hydrogenase | No |
| Parabasalids | 1 | H | *Trichomonas vaginalis* | Hmp31, Hsp60, Cpn10, Hsp70, IscS, IscU, 51 kDa, 24 kDa, MPP, Pam18 | PFO, hydrogenase | No |
| Diplomonads | 1 (?) | M | *Giardia intestinalis* | IscS, IscU, Cpn60, mHsp70, Fd, MPP, Pam18 | PFO, hydrogenase | No |
| Retortamonads | | – | | – | – | – |
| Carpediemonas | | H (?) | *Carpediemonas membranifera* | – | – | – |
| Oxymonads | 1 | – | | – | – | – |
| *Trimastix* | | H (?) | *Trimastix convexa* | – | – | – |
| Heterolobosea | 1 | H | *Psalteriomonas lanterna* | Fd | Hydrogenase | – |
| Euglenozoa | (?) | H (?) | *Postgaardi mariagerensis* | – | – | – |

*AAC* ATP/ADP carrier, *AOX* alternative oxidase, *ATM* mitochondrial ABC tranporter, *CI* respiratory chain complex I, *CII* respiratory chain complex II, *Cpn* chaperonin, *ERV* flavin adenine dinucleotide dependent sulphydryl oxidase, *Fd* ferredoxin, *H* hydrogenosome, *Hmp* hydrogenosomal membrane protein, *Hsp* heat shock protein, *hydrogenase* Fe-hydrogenase, *IMP* inner-membrane peptidase, *IscS* pyridoxal phosphate-dependent cysteine desulfurase, *IscU* FeS cluster assembly scaffold protein, *M* mitosome, *MDH* malate dehydrogenase, *MPP* matrix processing peptidase, *Narf* nuclear prelaminin A recognition factor, *Pam* presequence translocase-associated motor, *PDH* pyruvate dehydrogenase, *PFL* pyruvate:formate lyase, *PFO* pyruvate:ferredoxin oxidoreductase, *PNO* pyruvate:$NADP^+$ oxidoreductase, *PNT* pyridine nucleotide transhydrogenase, *Tim* translocator inner membrane, *Tom* translocator outer membrane, *51 kDa* subunit of NADH dehydrogenase, *24 kDa* subunit of NADH dehydrogenase order in which the mitochondrion, the cytoskeleton, the endomembrane system and the nucleus evolved cannot be deduced by evolutionary comparisons amongst extant organisms. Nevertheless, certain orders for these transitions seem more likely than others. Almost every major protist group harbours members with living bacterial symbionts in their cytosol, whereas only a few examples of endosymbiosis have been described for prokaryotes (von Dohlen et al. 2001). Thus, all other things being equal, it seems likely that the presence of the machinery for phagocytosis available to present-day eukaryotes (at least an actin–myosin cytoskeleton and a rudimentary endomembrane system) would have made the uptake and maintenance of the mitochondrial symbiont much easier. This a priori feasibility argument, albeit weak, would suggest that whether or not the host possessed a nucleus, it had evolved at least the cellular machinery that permits phagocytosis.

Regardless of the nature of the ancestral host, the presence of an endosymbiont-derived mitochondrial organelle in the common ancestor of eukaryotes raises intriguing evolutionary questions regarding why, when and how the peculiar transformations of these organelles in diverse anaerobic eukaryote lineages occurred.

## 10.2 Diversity of Anaerobic Protists with Mitochondrion-Related Organelles

Although the vast majority of unicellular and multicellular eukaryotes are aerobes that contain classical mitochondria, a vast diversity of anaerobic eukaryotic lineages have been uncovered in the last few years. Indeed, there are as many as 16 distinct lineages of protists that contain mitochondrion-related organelles that apparently function under low oxygen conditions (Fig. 10.2, Table 10.2). More than half of these lineages have yet to be intensively studied and the biochemical functions of their mitochondrion-derived organelles are largely unknown. For those lineages that have been studied, a rich diversity of form and function has been uncovered. Some possess hydrogen-producing organelles that function in anaerobic energy metabolism and iron–sulfur (Fe-S) cluster biogenesis (hydrogenosomes), while others possess organelles that have only FeS cluster biogenesis functions. Still others contain organelles with a variety of proteins apparently sufficient only for their own biogenesis, making their biochemical role a complete mystery. In this review we will refer to all hydrogen-producing organelles involved in energy metabolism as “hydrogenosomes” and organelles with no obvious energy metabolic role as “mitosomes”. These terms are for expediency only and using them risks masking their underlying diversity in form and function. It is important to keep in mind that hydrogenosomes or mitosomes from different evolutionary lineages may be quite biochemically distinct and some of their specific properties may have different evolutionary origins.

In the following sections we discuss each of the anaerobic groups containing mitochondrion-derived organelles in turn and outline our current understanding of their evolutionary affinities, as well as their biochemical and cell biological properties.

### 10.2.1 Parabasalids

The phylum Parabasalia is a diverse group of parasitic, symbiotic and free-living protists. Parabasalids, together with Diplomonads and some other groups, are proposed to belong to the new eukaryote "supergroup" Excavata, on the basis of ultrastructural synapomorphies (Simpson and Roger 2004; Fig. 10.2). The human sexually transmitted parasite *T. vaginalis* is by far the best studied parabasalid.

The hydrogenosome of *T. vaginalis* is well characterized, as many of its enzymatic functions have been extensively studied and the corresponding genes sequenced. Although this organelle was discovered long ago (Lindmark and Muller 1973), the "amitochondriate" nature of *Trichomonas* was not challenged until very recently, as the hydrogenosome of parabasalids was considered be completely unrelated to mitochondria, perhaps deriving from a separate endosymbiosis. This view was challenged by the discovery of nuclear genes encoding for three mitochondrial chaperones, Cpn60, mtHsp70 and Cpn10 (Bui et al. 1996; Germot et al. 1996; Horner et al. 1996; Roger et al. 1998). Phylogenetic analyses of these genes indicated that they were clearly related to mitochondrial isoforms (Fig. 10.3a, b) nested within α-proteobacterial homologs. Moreover, Cpn60 and mtHsp70 proteins appeared to be targeted to, and located within, the hydrogenosomes of *T. vaginalis* (Bui et al. 1996; Bozner 1997). These data argue very strongly for a mitochondriate ancestry of Parabasalids and were the first results to indicate that mitochondria and hydrogenosomes share a common eubacterial ancestor. Later, these results were further corroborated by the characterization of the IscS enzyme in *Trichomonas*, a protein that in eukaryotes is typically found in mitochondria (Tachezy et al. 2001) and clusters with mitochondrial homologs in phylogenetic analyses (Fig. 10.3c). Studies of the physiology and biochemistry of *Trichomonas* hydrogenosomes have yielded additional, and overwhelming, evidence supporting their mitochondrial origin (discussed later).

*Trichomonas* hydrogenosomes, like mitochondria, are double-membrane-bounded organelles that compartmentalize postglycolytic carbohydrate metabolism to produce ATP and molecular hydrogen. However, they differ from mitochondria in that they utilize both pyruvate and malate to produce ATP anaerobically via substrate-level phosphorylation and they completely lack DNA, cytochromes and the citric acid cycle (van der Giezen et al. 2005). In these organelles, pyruvate is decarboxylated into acetylcoenzyme A (acetyl-CoA) by an enzyme called pyruvate:ferredoxin oxidoreductase (PFO), instead of using the pyruvate dehydrogenase (PDH) complex found in

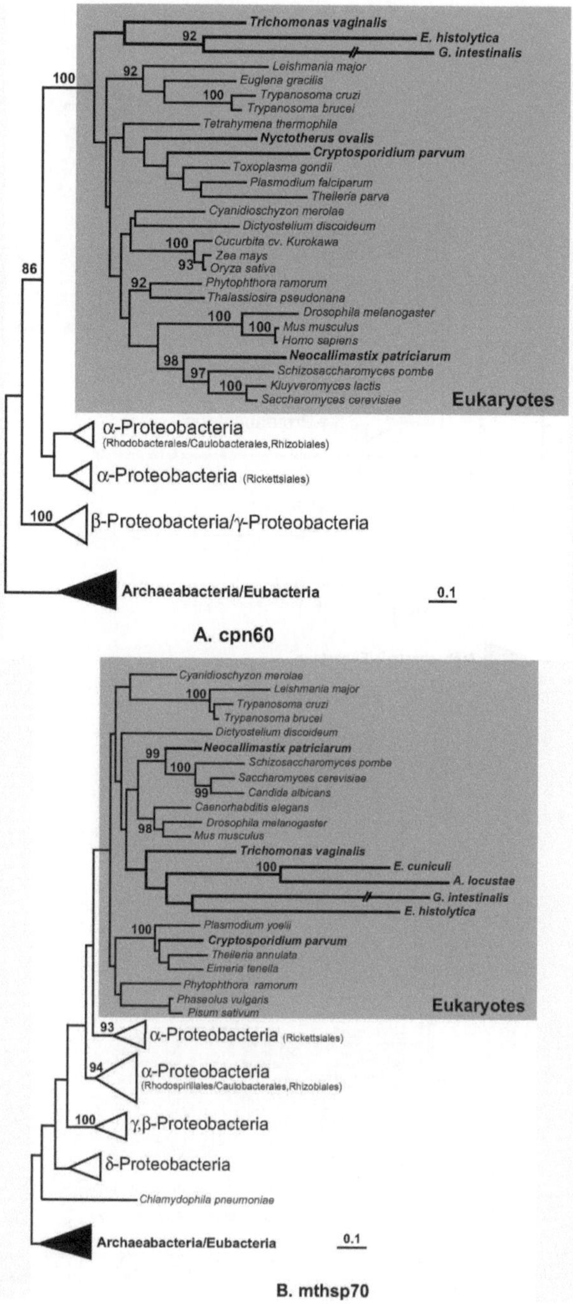


**Fig. 10.3.** ML phylogenetic trees of the mitochondrial markers cpn60, mthsp70, IscS and IscU. Trees were obtained with the Iqpnni program (Vinh le and von Haeseler 2004), using a JTT + Γ model of evolution, with eight gamma rate categories and at least a 100 iterations. Branch support was evaluated by a 100-bootstrap replicates in phyml (http://atgc.lirmm.fr/phyml/) using a JTT + Γ model of evolution with four gamma rate categories. Branch support above branches is shown only for those nodes with greater than 85% bootstrap support. Anaerobic eukaryotes are shown in *bold*

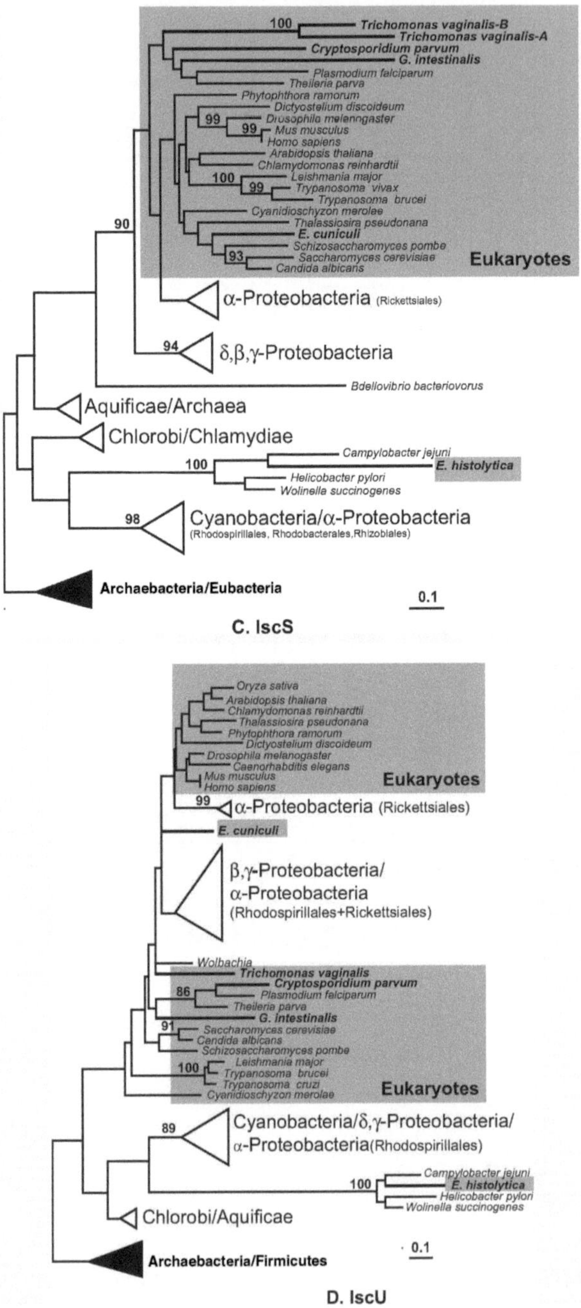


Fig. 10.3. Continued

canonical mitochondria (Hrdy and Muller 1995). PFO transfers the electrons generated in the process to the ferredoxin protein, and an Fe-hydrogenase enzyme is responsible for reoxidizing the reduced ferredoxin with the concomitant production of $H_2$. The acetate:succinate CoA transferase (ASCT) together with succinate thiokinase recovers the CoA moiety and synthesizes ATP by substrate-level phosphorylation (Martin et al. 2001). In *T. vaginalis*, it appears that at least three different Fe-hydrogenases exist: two short forms known to be targeted to the hydrogenosome (Bui and Johnson 1996) and a much larger third form that contains an N-terminal extension that suggests it has a hydrogenosomal location (Horner et al. 2000). The three cluster together in phylogenetic trees, suggesting they diversified by gene duplication within the lineage leading to *Trichomonas* (Horner et al. 2000; Fig. 10.4); however, the specific function of each hydrogenase isoform remains unclear.

A common function for mitochondria and hydrogenosomes, therefore, is the synthesis and export of ATP to the cytosol of the cell. Although the presence of a canonical ATP/ADP carrier (AAC) in the hydrogenosome of *Trichomonas* has yet to be reported, a different member of the mitochondrial carrier family (MCF), the Hmp31 protein, appears to be very abundant in *T. vaginalis* hydrogenosome membranes (Dyall et al. 2000). *Trichomonas* Hmp31 is phylogenetically distantly related to mitochondrial AAC, possibly indicating a mitochondrial origin of the protein. This was further corroborated by the finding that Hmp31 can be targeted to the inner membrane of yeast mitochondria using the same import pathway used for the assembly of mitochondrial carrier proteins and, conversely, that yeast mitochondrial AAC was directed into the hydrogenosome of *T. vaginalis* (Dyall et al. 2000). These results suggest the conservation of the import pathways for inner-membrane proteins between hydrogenosomes and mitochondria. Moreover, the Hmp31 homolog of *Trichomonas gallinae* has been demonstrated to specifically transport ADP and ATP very efficiently, supporting the initial evidence that Hmp31 is a nucleotide carrier in the hydrogenosome (Tjaden et al. 2004).

Another canonical mitochondrial function is the synthesis of Fe-S clusters, a process that was recently demonstrated to take place within *Trichomonas* hydrogenosomes (Tachezy et al. 2001; Sutak et al. 2004). The synthesis of Fe-S clusters seems to be an essential function for many of the organelles derived from the original mitochondrial endosymbiont.

Although hydrogenosomes do not contain a respiratory chain, two different groups have reported that the *T. vaginalis* genome contains genes encoding for the 51-kDa (NuoF) and 24-kDa (NuoE) subunits of the NADH dehydrogenase module in complex I (NADH:ubiquinone oxidoreductase), the first step in the mitochondrial respiratory chain (Dyall et al. 2004; Hrdy et al. 2004). Phylogenetic analyses by Dyall and co-workers did not support the mitochondrial origin of *Trichomonas* NADH dehydrogenase; they proposed instead an independent endosymbiotic origin for these components of the *Trichomonas* hydrogenosome. However, the *Trichomonas* NuoF homolog

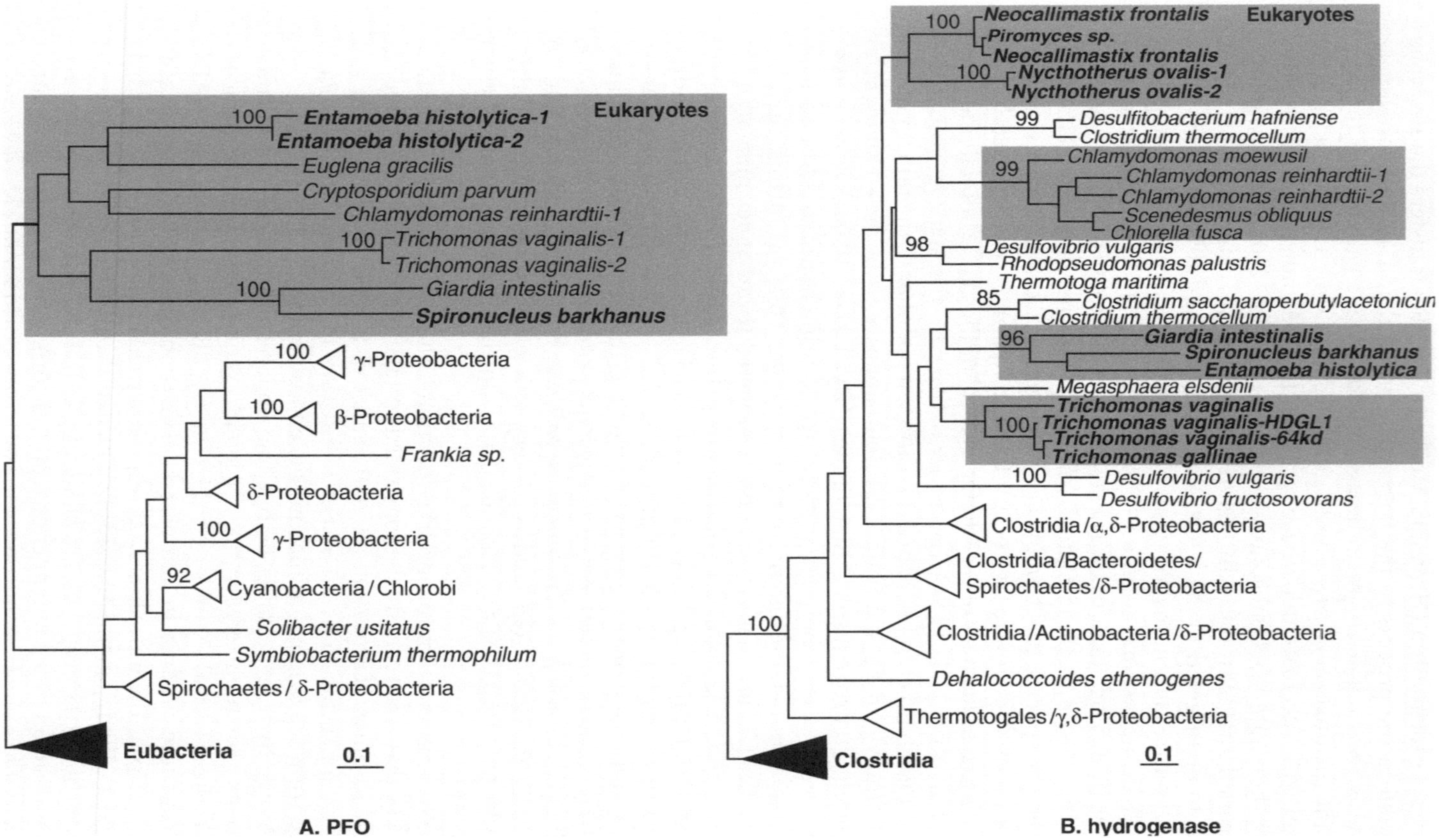


**Fig. 10.4.** Maximum likelihood phylogenetic trees of (**a**) pyruvate:ferredoxin oxidoreductase (*PFO*) and (**b**) Fe-hydrogenase genes. Trees and branch support were obtained as in Fig. 10.3. Branch support above branches is shown only for those nodes with greater than 85% bootstrap support. Anaerobic eukaryotes are shown in *bold*

has an extremely divergent sequence with a distinctive amino acid composition and its position outside the mitochondrial clade is likely a phylogenetic artefact. Consistent with this, more sophisticated analyses intended to overcome this artefact indicate that it does share a common ancestry with mitochondrial proteins (Hrdy et al. 2004). Regardless of their phylogenetic positions, all authors agree that both subunits of NADH dehydrogenase are hydrogenosomal, active, and may have a role oxidizing the NADH produced by the hydrogenosomal malic enzyme (Dyall et al. 2004; Hrdy et al. 2004).

The biogenesis and protein-import systems of the *Trichomonas* hydrogenosomes have been extensively studied, providing key evidence corroborating their common ancestry with mitochondria. The first evidence was the presence of mitochondrial chaperonins in the hydrogenosome, essential enzymes in the import system (Bui et al. 1996; Germot et al. 1996; Horner et al. 1996; Roger et al. 1998). A second piece of evidence was the presence in the hydrogenosomal proteins of N-terminal targeting peptides with sequence features similar to those of mitochondrial N-terminal targeting peptides (Dyall and Johnson 2000). Translocation experiments using Hmp31 and mitochondrial AAC (see before) showed that the targeting signals in hydrogenosomal Hmp31 and the yeast mitochondrial AAC are conserved, and that the import systems in both organelles are completely compatible (Dyall et al. 2000). In addition, the import of Hmp31 in yeast mitochondria uses the same Tim9/Tim10 import pathway as the AAC protein, suggesting the presence of these proteins in the *Trichomonas* hydrogenosome (Dyall et al. 2000). Furthermore, the partial genome sequence of *T. vaginalis* revealed the presence of a gene coding for the β subunit of the mitochondrial-processing peptidase (MPP), the enzyme responsible for the cleavage of the N-terminal targeting peptides in mitochondria (Dolezal et al. 2005). A gene for the mitochondrial inner-membrane import protein Tim14/PAM18 was also found in the *Trichomonas* genome and was shown to localize in the hydrogenosome (Dolezal et al. 2005). The genome project of *T. vaginalis* will surely provide more insights on the machinery of the import system in hydrogenosomes. Recently, the conservation of the protein targeting mechanisms between hydrogenosomes and mitosomes of the diplomonad *G. lamblia* was also reported (Dolezal et al. 2005), indicating deep-level conservation in the protein-import mechanisms of mitochondrion-derived organelles.

### 10.2.2 Chytrid Fungi

There has never been any doubt that anaerobic chytridiomycetes (fungal symbionts such as *Neocallimastix* or *Piromyces* that colonize the gastrointestinal tract of herbivorous mammals) evolved from mitochondrion-bearing ancestors, as phylogenetic analyses demonstrated that they form a monophyletic group with the aerobic chytridiomycetes and all other fungi (Akhmanova et al. 1998a; van der Giezen et al. 2002; Voncken et al. 2002a;

Fig. 10.3). Thus, anaerobic chytrid fungi have secondarily lost their aerobic mitochondrial function in adaptation to their anaerobic life style.

Although some early evidence suggested a peroxisomal origin for chytrid hydrogenosomes, current data indicate that they share a common ancestry with mitochondria (van der Giezen et al. 2002, 2003; Voncken et al. 2002a; Fig. 10.4). These data include the presence of a double membrane surrounding the organelle (Benchimol et al. 1997; van der Giezen et al. 1997a) and phylogenetic analyses of Cpn60, mtHsp70 and AAC, which showed that sequences of chytrid homologs cluster with mitochondrial sequences (van der Giezen et al. 2002, 2003). It has recently been reported that chytrid hydrogenosomes differ from the hydrogenosomes of *Trichomonas* or ciliates, relying on mixed-acid fermentation (Akhmanova et al. 1999; Boxma et al. 2004), as in facultative anaerobic bacteria. Indeed, despite early reports that anaerobic chytrids contain PFO activity (Yarlett et al. 1986), it has been argued that the function of this enzyme is replaced by pyruvate:formate lyase (PFL), which splits pyruvate into acetyl-CoA and formate. These chytrid PFLs function in their hydrogenosomes and cluster with a homolog found in chlorophyte chloroplasts and mitochondria, which are in turn related to eubacterial homologs (Gelius-Dietrich and Henze 2004). However, like other hydrogenosomes, a highly active "long-form" Fe-hydrogenase has been detected and localized to the hydrogenosome of *Neocallimastix* (Davidson et al. 2002; Voncken et al. 2002b).

An important mitochondrial function that is also common to the hydrogenosomes is the export of ATP in exchange for ADP. In the hydrogenosomes of *Neocallimastix* spp. a mitochondrial-like ACC has been identified and functional analyses have shown that it works as an adenine nucleotide exchanger (van der Giezen et al. 2002; Voncken et al. 2002a). Moreover, this hydrogenosomal AAC has similar susceptibility to classical inhibitors of the mitochondrial AAC function and phylogenetic analyses place the *Neocallimastix* AAC sequence in a clade with the mitochondrial AAC of aerobic fungi.

Little is known about the protein-import systems of the hydrogenosomes of chytrid fungi. So far, no proteins belonging to the membrane complexes of the import system have been identified, although some preliminary evidence shows that the import system of the *Neocallimastix* hydrogenosome might share many features with mitochondrial import pathways. For example, in functional analyses, *Neocallimastix* hydrogenosomal AAC was targeted into yeast mitochondria and restored the growth of a yeast strain lacking a functional AAC. Interestingly, mitochondrial AAC import depends on internal targeting sequences and the system responsible for this is different from the import system for mitochondrial matrix proteins. Hence, the targeting signals and import system for the AAC appear to be conserved between yeast mitochondria and chytrid hydrogenosomes (van der Giezen et al. 2002). It is also known that, like the *T. vaginalis* hydrogenosome, the *Neocallimastix* organelle contains Cpn60 and mtHsp70, two proteins that function in mitochondrial import and protein folding (van der Giezen et al. 2003). Translocation experiments in mammalian cells showed that both proteins are targeted into the mitochondrion and that the N-terminal presequences are necessary for the localization of the proteins into the organelle, confirming conservation of the mitochondrial-like targeting signal. Similar experiments with the *Neocallimastix frontalis* hydrogenosomal malic enzyme confirm the existence of a cleavable N-terminal targeting sequence (van der Giezen et al. 1997b, 1998).

### 10.2.3 Anaerobic Ciliates

Ciliates comprise a diverse group of unicellular protists containing both aerobic and anaerobic lineages. Within the anaerobic ciliates we can find free-living organisms inhabiting anoxic freshwater and marine sediments along with species that live in intestinal tracts. Many anaerobic ciliates have hydrogenosomes as well as intracellular methanogen (archaebacterial) symbionts that transform the hydrogen produced by the organelle into methane (Akhmanova et al. 1998b). Like anaerobic chytrids, anaerobic ciliates have never been considered primitive organisms as abundant ultrastructural and phylogenetic evidence has supported the idea that anaerobic ciliates evolved secondarily from mitochondrion-bearing ciliates. In fact, nucleus-encoded SSU rRNA phylogenies show that hydrogenosome-bearing ciliates are not a monophyletic group, but have adapted to anoxia at least four times (Embley et al. 1995; Table 10.2), and possibly as many as seven times (Fenchel and Finlay 1995).

Until very recently, most of the information about ciliate hydrogenosomes came from biochemical studies related to the metabolism of the organelle (see van der Giezen et al. 2005 for a review). However, some years ago the presence of a genome was reported in the hydrogenosome of the ciliate *Nyctotherus ovalis*, an inhabitant of the cockroach hindgut. In this section, we will focus on the hydrogenosome of *N. ovalis* since it is one of the best studied and because the presence of a genome in its organelle and its peculiar biochemistry make it a perfect example for understanding the evolution of anaerobic organelles. The *N. ovalis* hydrogenosome is quite unique in that it is the only hydrogenosome described to date that harbours a functional genome (Akhmanova et al. 1998b; Boxma et al. 2005). The genome of the *Nyctotherus* hydrogenosome was first identified by immunocytochemistry and was shown to express SSU rRNA (Akhmanova et al. 1998b). Phylogenetic analyses indicate that these hydrogenosomal SSU rRNA sequences are closely related to the mitochondrial sequences from aerobic ciliates, corroborating the hypothesis that ciliate hydrogenosomes evolved from ciliate mitochondria (Akhmanova et al. 1998b; van Hoek et al. 2000). Recently, the hydrogenosomal genome has been partly sequenced, allowing the identification of its direct ancestor. The genome encodes four genes from respiratory complex I, two genes encoding mitochondrial ribosomal proteins, and a transfer RNA (Boxma et al. 2005); phylogenetic analyses show that they group with mitochondrial homologs from aerobic ciliates. Since we only have a partial sequence of the genome, it will be interesting to see which other genes have been retained in it.

*N. ovalis* hydrogenosomes are double-membraned organelles with cristae and putative ribosomes (Akhmanova et al. 1998b). *Nyctotherus* hydrogenosomes are closely associated with archaebacterial methanogenic endosymbionts, which absorb the hydrogen produced by the hydrogenosomes and transform it into methane, thereby driving increased hydrogenosomal metabolism. A gene encoding for an Fe-hydrogenase was identified in the nuclear genome of *N. ovalis* and is targeted to the hydrogenosome (Akhmanova et al. 1998b). However, this Fe-hydrogenase is quite peculiar compared with any other hydrogenase described in eukaryotes so far, as it is fused with two proteins that bear strong resemblance to NuoE and NuoF (24 and 51 kDa of mitochondria NADH:ubiquinone oxidoreductase, respectively). The unique structure of the hydrogenase–NuoE/NuoF protein suggests that it may couple the production of hydrogen to the reoxidation of NADH (Akhmanova et al. 1998b).

Carbohydrate metabolism in the hydrogenosome of *N. ovalis* also seems to differ from the metabolism described for other anaerobes. So far, no gene encoding for PFO has been identified, although an expressed sequence tag (EST) survey identified subunits E1 and E2 of PDH (Boxma et al. 2005). It is possible that in the *Nyctotherus* hydrogenosome, PDH is the enzyme responsible for the decarboxylation of pyruvate, instead of PFO. Moreover, as in some anaerobic mitochondria, *N. ovalis* has an incomplete tricarboxylic acid (TCA) cycle and appears to use fumarate as a terminal electron acceptor. The end products of carbohydrate metabolism are thus acetate and succinate, the same end products known in the anaerobic mitochondria of parasitic worms (Boxma et al. 2005; Tielens et al. 2002). *Nyctotherus* contains an incomplete respiratory chain; complexes I and II are functional as shown by assays of specific inhibitors and complex I is believed to be responsible for the proton gradient of the organelle. Complexes III, IV and alternative oxidase (AOX) were not found in the EST survey, but a low amount of rhodoquinone, typical of anaerobic mitochondria producing succinate, was detected (Boxma et al. 2005). All in all, both the genome and the biochemistry of the *N. ovalis* hydrogenosome suggest that this organelle represents a transitional stage between mitochondria and hydrogenosomes since it has features of both.

### 10.2.4 Diplomonads

Diplomonads are a group of microaerophilic unicellular protists once considered to be an ancient protist group owing to their ultrastructural features and their early branching position in a number of gene phylogenies (Sogin 1991; Hashimoto et al. 1997; Rensing et al. 1997; Stiller et al. 1998).

The best-studied member is *G. intestinalis*, a human parasite and the causative agent of the waterborne disease giardiasis. However, the early-branching position of diplomonads in molecular phylogenies is almost certainly a long-branch attraction artefact resulting from their extremely divergent sequences (Embley and Hirt 1998; Hashimoto et al. 1998; Roger 1999; Simpson et al. 2002; Figs. 10.3, 10.4). Ultrastructural data strongly suggest they instead belong in the Excavata (Simpson and Roger 2004; Fig. 10.2).

Although diplomonads are not considered basal eukaryotes anymore (Embley and Hirt 1998; Hashimoto et al. 1998; Roger 1999), *G. intestinalis* remains a hallmark of both the analysis of eukaryotic cell evolution and the study of anaerobic protists. It was not until the late 1990s that evidence indicating that *Giardia* secondarily lacked mitochondrial functions first surfaced. Genes encoding proteins of probable mitochondrial origin, such as Cpn60, mtHsp70 and IscS were found in the nuclear genome of *Giardia* (Hashimoto et al. 1998; Roger et al. 1998; Tachezy et al. 2001). The key evidence for a mitochondriate ancestry was furnished by the discovery of a double-membrane-bounded mitochondrial relic in *Giardia* to which the enzymes IscS and IscU were targeted (Tovar et al. 2003). This finding was decisive because IscS and IscU are known to be essential enzymes of the FeS cluster biosynthesis in mitochondria and *Trichomonas* hydrogenosomes (Johnson et al. 2005; Lill and Muhlenhoff 2005). *Giardia* mitosomes are very small, approximately 140 mm × 60 nm, and have no cristae, but they are abundant in the trophozoite stage, consistent with an important physiological role. More recently it has been reported that the mitochondrial proteins Cpn60, mtHsp70 and ferredoxin are also located in this organelle (Regoes et al. 2005).

Like the distantly related amoebozoan *Entamoeba histolytica*, *Giardia* carries out a fermentative-type metabolism whereby extended glycolysis and substrate-level phosphorylation generate ATP exclusively in the cytosol (Lloyd et al. 2002; Müller 1998), and which completely lacks the aerobic respiratory components of mitochondria. Some phylogenetic studies suggest that most of the genes encoding fermentation enzymes and other proteins of *Entamoeba*, *Giardia* and *T. vaginalis* may actually have been laterally transferred from prokaryotes (Rosenthal et al. 1997; Field et al. 2000; Nixon et al. 2002; Loftus et al. 2005; Andersson et al. 2003). One of the fates of pyruvate in *Giardia* is to be oxidized via a cytosolic PFO, as in the *Trichomonas* hydrogenosome. There is also some evidence for hydrogenase activity in *Giardia*. This includes the expression of a "short-form" of Fe-hydrogenase in the *Giardia* cell (Nixon et al. 2003), as previously seen in the diplomonad *Spironucleus barkhanus* (Horner et al. 2000), and the production of small amounts of molecular hydrogen (Lloyd et al. 2002). The importance of hydrogenase in anaerobic metabolism in *Giardia* is still unknown.

It has recently been reported that *Giardia* mitosomes, *Trichomonas* hydrogenosomes and mitochondria share a similar mode of protein targeting and translocation (Dolezal et al. 2005; Regoes et al. 2005). The import of *Giardia* IscU and ferredoxin into the organelle was shown to be dependent on an N-terminal extension (Dolezal et al. 2005; Regoes et al. 2005), whereas no leader peptides are apparently required for IscS, mtHsp70 or Cpn60 transport, which instead possibly rely on internal targeting sequences (Dolezal et al. 2005; Regoes et al. 2005). Curiously, the mtHsp70 in the microsporidian *Trachipleistophora hominis* also seems to be imported into the mitosomes of that organism using internal targeting sequences (Williams et al. 2002). Other experiments also confirm the conservation of the protein targeting signals between mitosomes, hydrogenosomes and mitochondria. For example, the overexpression of *Giardia* IscU and ferredoxin in *T. vaginalis* results in targeting of these mitosomal proteins to the hydrogenosome (Dolezal et al. 2005). A similar experiment in mammalian cells, overexpressing a fusion protein with the ferredoxin leader sequence and green fluorescent protein (GFP), had the same result, directing the fusion protein into mitochondria. Components of the protein targeting pathway have been found in the *Giardia* genome: two genes encoding the β subunit of MPP, the enzyme responsible for the cleavage of the N-terminal presequence, and a protein related to a subunit of the presequence translocase-associated motor (PAM) complex, a component of the translocase of the mitochondrial inner membrane. Both proteins were shown to be located in the mitosome of *Giardia* and the *T. vaginalis* hydrogenosome (Dolezal et al. 2005). Collectively, these results suggest that *Giardia* mitosomes possess presequence-dependent and presequence-independent protein import pathways that are similar to those found in mitochondria and hydrogenosomes, providing further evidence for a common origin for these organelles.

### 10.2.5 *Entamoeba* and Pelobionts (Archamoebae)

The Archamoebae is a group comprising the intestinal parasitic amoebae of the genus *Entamoeba* and the free-living pelobiont amoeboflagellates. Although they were initally assumed to be primitive on the basis of both their peculiarities in ultrastructure and their metabolism (Reeves 1984; Cavalier-Smith 1991; Bakker-Grunwald and Wostmann 1993), recent analyses robustly place them in the Amoebozoa with aerobic amoebae and slime moulds (Bapteste et al. 2002; Arisue et al. 2002b).

The microaerophilic parasite *E. histolytica*, the cause of human amebiasis, is the best-studied member of Archamoebae and was the first amitochondriate protist in which genes of mitochondrial origin were discovered (Clark and Roger 1995). These genes encoded the enzyme pyridine nucleotide transhydrogenase (PNT) and the mitochondrial chaperonin Cpn60 (Clark and Roger 1995). Additionally, both proteins were shown to have an N-terminal extension similar to a mitochondrial targeting peptide, suggesting the presence of a relic mitochondrial organelle (Clark and Roger 1995; Roger et al. 1998; see Fig. 10.3 for an updated phylogenetic analysis). Indeed, later studies showed that the Cpn60 protein was located in a cellular compartment of possible mitochondrial origin (Mai et al. 1999; Tovar et al. 1999). The fact that *Entamoeba* carries out fermentation in the cytosol (Müller 1998) led these authors to infer that the organelle did not participate in energy metabolism. The names "crypton" (Mai et al. 1999) or "mitosome" (Tovar et al. 1999) were proposed for the organelle; the first of its kind to be identified in an amitochondriate eukaryote. The import of Cpn60 into the mitosome is dependent on a targeting peptide (Tovar et al. 1999) as removal of the first few N-terminal amino acids disrupted the targeting of the protein to the organelle; however, targeting was restored by the addition of a mitochondrial presequence from *Trypanosoma cruzi*, providing evidence for the conservation of the protein-import pathways between mitochondria and these organelles.

Recent evidence suggests that mitosomes are more abundant and smaller in the trophozoite of *Entamoeba* than previously thought, and do not seem to contain a genome, contrary to previous reports (Ghosh et al. 2000; Leon-Avila and Tovar 2004; Loftus et al. 2005). Interestingly, a very peculiar ATP/ADP transporter has also been recently described in the *Entamoeba* mitosome (Chan et al. 2005) and is the only member of the MCF detected in the genome of this organism, underlining the extreme reduced nature of this mitosome. The function of the *Entamoeba* mitosome remains unknown, as only Cpn60 and the ATP/ADP transporter have been clearly shown to be located within the organelle; although mtHsp70 and the PNT are also predicted to be mitosomal proteins (Clark and Roger 1995; Bakatselou and Clark 2000). These proteins seem to be only sufficient for maintenance of the organelle itself and do not betray its raison d'être.

The fermentative metabolism of *Entamoeba*, as previously mentioned, is considered to be cytosolic and the immunolocalization of some fermentative enzymes like alcohol dehydrogenase E (ADHE) as well as ferredoxin support the lack of core energy metabolism compartamentalization (Mai et al. 1999). In both, ADHE and ferredoxin, a targeting peptide responsible for localization to the mitosome is missing, This is also seen with PFO, which is found to be associated with the plasma membrane and cytoplasmic structures not related to mitosomes (Rodriguez et al. 1998; Leon-Avila and Tovar 2004). Like *T. vaginalis*, *Entamoeba* has more than one form of Fe-hydrogenase with unclear roles: a short-form clusters with *Giardia* (which is probably the result of a lateral gene transfer, LGT, from a diplomonad-related organism) and a long-form with a long branch that appears to have a prokaryotic origin (Embley et al. 2003; Nixon et al. 2003; Fig. 10.4)

*E. histolytica*, as with other amitochondriates, has also reduced or eliminated most mitochondrial metabolic pathways, including some pathways thought to be essential for a eukaryotic cell (Anderson and Loftus 2005; Loftus et al. 2005). Indeed, phylogenomic analyses of *Entamoeba* show that its metabolism has been shaped by secondary gene loss and LGT (Loftus et al. 2005). An interesting potential LGT event in *Entamoeba* is two proteins of the Nif system, NifS and NifU, a distinct biosynthetic pathway to generate FeS proteins typically present in nitrogen-fixing bacteria (Ali et al. 2004; van der

Giezen et al. 2004; Lill and Muhlenhoff 2005). These Nif proteins are related to, and share the same functions as, IscS and IscU from the ISC system (Johnson et al. 2005). Phylogenetic analysis indicates that *E. histolytica* Nif proteins form a well-supported clade with ε-proteobacteria, clearly indicating a horizontal transfer of these genes from ε-proteobacteria to *Entamoeba* (Ali et al. 2004; van der Giezen et al. 2004; Fig. 10.3c d). An LGT for these proteins is further supported by insertion/deletion characters in the alignments (van der Giezen et al. 2004). Since no ISC proteins have been identified in *Entamoeba* and its Nif system has been shown to replace both the ISC and SUF systems in *Escherichia coli* under anaerobic conditions, we may assume that the Nif system is the only pathway responsible for FeS cluster synthesis in *Entamoeba* (Ali et al. 2004). Intriguingly, *Entamoeba* differs in this respect from all the other anaerobic protists studied to date, which harbour proteins of the ISC system within their mitochondrion-derived organelles (Katinka et al. 2001; LaGier et al. 2003; Tovar et al. 2003).

Pelobionts are mostly free-living anaerobic protists. Early on, the presence of densely stained membrane-bounded structures was reported in *Pelomyxa palustris* (Seravin and Goodkov 1987) and *Mastigamoeba balamuthi* (Chávez et al. 1986), suggesting that pelobionts may also contain vestigial mitochondrion-derived organelles. This has been confirmed in our laboratory with the discovery of genes encoding mitochondrial proteins in the genome of *M. balamuthi* (E. Gill, M.J. Barberà, J.D. Silberman, A. Stechmann and A.J. Roger, unpublished results). These findings, together with the presence of small double-membrane-bounded organelles abundant in the cytoplasm of several species of *Mastigamoeba* (Simpson and Roger 2004), strongly support the existence of mitochondrion-derived organelles in these organisms. Investigations into the function of pelobiont mitosomes and characterization of genes of mitochondrial origin will surely provide insights into the similarities and differences between the organelles of pelobionts and Entamoebae.

### 10.2.6 Microsporidia

Microsporidia is a group of obligate intracellular parasites that include more than a thousand species of unicellular eukaryotes (Keeling and Fast 2002). Like parabasalids, diplomonads and Archamoebae, they were also once considered to be early-branching Archezoa (Cavalier-Smith 1983); however, it is now well established that they are highly derived *Fungi* (Keeling and Doolittle 1996; Fast et al. 1999; Hirt et al. 1999; Keeling 2003).

Like *Giardia* and *Entamoeba*, the presence of a mitochondrial relic in Microsporidia was first inferred from the existence of genes encoding proteins of mitochondrial origin in their genomes (Germot et al. 1997; Hirt et al. 1997; Fast and Keeling 2001; Arisue et al. 2002a). Indeed, the complete genome sequence of *Encephalitozoon cuniculi* appears to contain 22 genes coding for proteins of mitochondrial function and/or origin (Katinka et al. 2001). Phylogenetic analysis showed that six of these proteins are closely related to homologs from α-proteobacteria. Moreover, in some of these proteins a mitochondrial targeting peptide was predicted.

The first evidence of vestigial mitochondrial organelles in the microsporidian *T. hominis* was uncovered by immunolocalization experiments using a specific antibody against mtHsp70 (Williams et al. 2002). The cytoplasm of the meront (intracellular stage) appears to contain numerous tiny organelles with a double membrane that lack cristae. Although the specific function of the organelle and its contribution to general microsporidian metabolism is still unknown, the large number of these organelles could indicate an active role. The mitosomal function has been partially reconstructed on the basis of the genes found in the *E. cuniculi* genome and from some data obtained from *Antanospora locustae*, but the nature and roles of the pathways are not completely clear (Fast and Keeling 2001; Katinka et al. 2001; Keeling and Fast 2002). However, the presence of genes for the FeS cluster assembly machinery in *E. cuniculi* (Table 10.2) once again point to a role for the mitosomal organelle in this biosynthetic process.

Microsporidian metabolism is highly reduced. The complete genome of *E. cuniculi* lacks genes for enzymes in the TCA cycle, the mitochondrial electron transport chain and oxidative phosphorylation (Katinka et al. 2001). However, like other amitochondriates, Microsporidia have retained the glycolytic pathway (Katinka et al. 2001). Thus, ATP production may be accomplished by substrate-level phosphorylation, although it has been proposed that these parasites could use host-derived ATP (Katinka et al. 2001). At first glance, the metabolism of Microsporidia resembles the metabolism of other amitochondriates, but there are some important differences. For instance, they lack Fe-hydrogenase (Katinka et al. 2001) and they appear to contain the E1 subunits of PDH rather than the PFO enzyme that is found in other amitochondriate taxa (Katinka et al. 2001; Fast and Keeling 2001). It is unclear how pyruvate decarboxylation is accomplished as only α and β subunits of the E1 subunit of PDH have been characterized in *A. locustae* and *E. cuniculi* (Katinka et al. 2001; Fast and Keeling 2001); the E2 and E3 subunits that compose the full PDH complex have either been completely lost or are so divergent in sequence as to be unrecognizable. Furthermore, it is also unclear how fermentation might be accomplished in Microsporidia, as no genes encoding lactate dehydrogenase or alchohol dehydrogenase have been identified in the genome of *E. cuniculi*. Through analogy with hydrogenosomal PFO, a system based on PDH, ferredoxin and NAD(P)H ferredoxin:oxidoreductase (FOR) is theorized to produce acetate within the mitosomes of microsporidia (Katinka et al. 2001; Keeling and Fast 2002; Vivares et al. 2002). However, it has yet to be demonstrated that that these proteins are located within microsporidian mitosomes. The presence of a transporter that could import pyruvate into the mitosome of *A. locustae* (Williams and Keeling 2005) supports the idea that the last reactions of glucose metabolism are compartmentalized in this organelle.

The protein import system present in the microsporidian mitosomes is homologous to other mitochondrial and hydrogenosomal pathways but the simplicity of the system makes it clearly different. Most of the available data come from the complete genome sequence of *E. cuniculi* (Katinka et al. 2001). Although the mtHsp70 gene, which is an essential component of the TIM system, is present in many Microsporidia (Germot et al. 1997; Hirt et al. 1997; Peyretaillade et al. 1998), *E. cuniculi* lacks its partner mtHsp40. Surprisingly, Cpn60 is not present in any microsporidian (it is the only eukaryote currently known to lack mitochondrial Cpn60). Similarly, although a few components of the translocases of the inner and outer membranes have been identified (e.g. TIM22 and TIM70), no other proteins in these complexes have been identified in microsporidia. Recently, an inner-membrane peptidase 2 (IMP-2) was shown to be expressed in *A. locustae* (Williams and Keeling 2005), whereas IMP proteins appear to be absent in *E. cuniculi*. Further work on mitosomes from diverse species of Microsporidia should reveal their physiological functions and improve our understanding of the diversity of these organelles within this group.

### 10.2.7 *Cryptosporidium* (Apicomplexa)

*Cryptosporidium*, like *Plasmodium falciparum* or *Toxoplasma gondii*, is an apicomplexan parasite that infects both humans and animals. Phylogenetic analyses indicate that the phylum Apicomplexa shares a common ancestor with ciliates and dinoflagellates, which form the clade Alveolata (Ellis et al. 1998; Fig. 10.2).

However, unlike other apicomplexa, *Cryptosporidium* does not appear to contain a standard aerobic mitochondrion. The presence of a mitochondrion-like structure in *Cryptosporidium* was first confirmed in sporozoites of *Cryptosporidium parvum* by both phylogenetic and ultrastructural analyses (Riordan et al. 1999). They contain only a single organelle per cell that is very small, bounded by two membranes, enveloped in rough endoplasmic reticulum, and is closely associated with the nucleus (Riordan et al. 1999; Keithly et al. 2005). mtHsp70 and Cpn60 phylogenies place *Cryptosporidium* in the apicomplexa clade, nested within mitochondrial homologs (Riordan et al. 1999; Slapeta and Keithly 2004; Fig. 10.3a, b), and immunolocalization of these proteins confirms their presence in the organelle (Riordan et al. 2003; Slapeta and Keithly 2004; Putignani et al. 2004). This mitosome lacks canonical aerobic mitochondrial function and a genome (Abrahamsen et al. 2004) and its functions are poorly understood. Some lines of evidence suggest that, like *Giardia* and Microsporidia, the mitosome has been retained in *Cryptosporidium* for the assembly of FeS clusters (LaGier et al. 2003). Indeed, expression of proteins of the ISC system has been reported in *C. parvum* sporozoites (LaGier et al. 2003; Abrahamsen et al. 2004; Table 10.2) and all of them have predicted targeting peptides. The phylogeny of IscS shows that the *Cryptosporidium* sequences cluster with homologues of *G. intestinalis*,

*T. vaginalis* and other apicomplexa within the mitochondrion-bearing eukaryotes, once again suggesting a common origin of all mitochondrion-derived organelles (LaGier et al. 2003; Fig. 10.3).

Although complete genome sequences of both *C. parvum* and *C. hominis* have been determined, their carbohydrate metabolism is not fully understood (Abrahamsen et al. 2004; Xu et al. 2004). Glycolysis appears to be the main energy metabolic pathway, as they lack a functional TCA cycle and oxidative phosphorylation, although some components of the respiratory chain are present (Abrahamsen et al. 2004; Xu et al. 2004). Interestingly the conversion of pyruvate to acetyl-CoA is catalysed by an atypical pyruvate:NADPH oxidoreductase (PNO) that contains an N-terminal PFO domain fused with a C-terminal NADPH–cytochrome P450 reductase (CPR) domain (Rotte et al. 2001; Abrahamsen et al. 2004; Xu et al. 2004). Surprisingly, the same PFO–CPR fusion protein has been reported in *Euglena gracilis* (Rotte et al. 2001), a euglenozoan protist distantly related to apicomplexa. However, *Euglena* PNO appears to be a mitochondrial protein, whereas the function and cellular location of *Cryptosporidium* PNO is unknown. A Narf-like Fe-hydrogenase has been identified in *C. hominis*, and may function in reducing the NADP produced by PNO during pyruvate decarboxylation (Xu et al. 2004).

*Cryptosporidium* is also distinctive in containing an AOX (Abrahamsen et al. 2004; Putignani et al. 2004; Roberts et al. 2004; Xu et al. 2004), an enzyme that has been found in plants, fungi and some protists, and which functions in an alternative respiratory pathway. As with canonical aerobic respiration, the electrons from complex I pass to ubiquinone. At this point, rather than moving to complex II and onward, the electrons on ubiquinone are passed to AOX, which reduces oxygen into water (Vanlerberghe and McIntosh 1997; Roberts et al. 2004). Phylogenetic studies place *Cryptosporidium* AOX in the eukaryotic cluster grouping together with α-proteobacterial AOX, suggesting a possible mitochondrial origin of AOX (Roberts et al. 2004). In addition, *Cryptosporidium* AOX has a predicted targeting peptide, and thus may function in the mitosome as a terminal electron acceptor in this alternative respiratory pathway, although more data will be needed to evaluate this hypothesis (Roberts et al. 2004; Henriquez et al. 2005).

Although it has been suggested that synthesis of ATP occurs predominantly via fermentation (Entrala and Mascaro 1997), α and β subunits of F1 ATP synthase (complex V) have recently been reported in *Cryptosporidium* and are predicted to be targeted to the mitosome (Abrahamsen et al. 2004; Putignani et al. 2004; Henriquez et al. 2005). Thus, proton-gradient-coupled oxidative phosphoryolation may occur in this organism to produce ATP. Indeed, *C. parvum* possesses a PNT that may contribute to generating a membrane potential. Clearly, functional analyses of all of these proteins will be necessary to confirm their cellular locations and physiological roles.

In any case, evidence suggests that *Cryptosporidium* mitosomes have a complete protein import machinery that shares common features with mitochondrial and hydrogenosomal systems. Once again, both Cpn60 and mtHsp70 are targeted to the *C. parvum* organelle, probably via an N-terminal extension (LaGier et al. 2003; Riordan et al. 2003; Slapeta and Keithly 2004). The N-terminal presequences of Cpn60, mtHsp70, IscS and IscU are sufficient to target GFP to yeast mitochondria (LaGier et al. 2003; Riordan et al. 2003; Slapeta and Keithly 2004). The completion of the genome sequences of two *Cryptosporidium* species has allowed the identification of components of the TIM and TOM import systems and both the α and the β subunits of the MPP (Abrahamsen et al. 2004; Xu et al. 2004). However, like in the microsporidian *E. cuniculi*, there are several components of the import machinery that appear to be missing or unrecognizable in the *C. parvum* and *C. hominis* genomes.

### 10.2.8 *Blastocystis hominis* (Heterokonts/Stramenopiles)

*Blastocystis hominis* is a unicellular anaerobic organism that inhabits the human intestinal tract, and whose metabolism and life cycle are not completely understood. Molecular phylogenetic analysis of the SSU rRNA sequences (Silberman et al. 1996) and protein-coding sequences (Arisue et al. 2002c) show *Blastocystis* branching within stramenopiles (heterokonts), a group containing a vast diversity of photosynthetic and heterotrophic protists (Fig. 10.2).

The presence of a mitochondrion-like organelle in the strict anaerobe *B. hominis* has been known for many years. The basic ultrastructure of the organelle has been shown to be very similar to that of archetypical mitochondria with the presence of cristae, some appearing filamentous and branched (Zierdt et al. 1988; Nasirudeen and Tan 2004). Furthermore, biochemical studies indicated that the *Blastocystis* organelle apparently lacked cytochromes, a functional TCA cycle, and a respiratory chain (Zierdt et al. 1988). The *Blastocystis* organelle, however, stains with redox-sensitive dyes, suggesting the presence of a transmembrane potential and metabolic activity (Zierdt et al. 1988; Nasirudeen and Tan 2004). Recently, DNA-specific dyes were used to demonstrate the likely presence of a genome in the *Blastocystis* mitochondrion-derived organelle (Nasirudeen and Tan 2004). If confirmed, this observation, coupled with the earlier biochemical studies, suggests the mitochondrial-derived organelle of *Blastocystis* could turn out to be an excellent model system in which to study biochemical remodelling in adaptation to anaerobiosis.

### 10.2.9 Other Organisms

There remain a number of anaerobic protists that lack classical mitochondrial organelles. Some of these protists, such as the oxymonads and retortamonads, were originally classified as Archezoa (Cavalier-Smith 1983), but consideration of their likely phylogenetic positions (Fig. 10.2) argues against this. Most of the other taxa (e.g. *Carpediemonas*, *Trimastix*, *Postgaardi* and *Psalteriomonas*) were described in the last few decades and their phylogenetic positions also suggest secondary loss of mitochondrial functions (Broers et al. 1990; Simpson and Patterson 1999). All of these protists belong to the new eukaryote "supergroup" the Excavata (Fig. 10.2, Table 10.2).

For oxymonads and retortamonads, there is currently no evidence of any sort for the presence of mitochondrion-derived organelles, however, research on these organisms is currently in progress and their phylogenetic relationships to organelle-containing lineages indicate they too will probably contain relic mitochondria (Fig. 10.2). In most of the other taxa (*Carpediemonas*, *Trimastix*, *Postgaardii*) there is electron microscopy evidence for densely staining double-membrane-bounded bodies reminiscent of hydrogenosomes or mitosomes (Brugerolle and Patterson 1997; Simpson and Patterson 1999; Simpson and Roger 2004). The organelles of *Carpediemonas* are of particular interest as this organism is closely related to diplomonads and retortamonads and could provide insight into the evolutionary trajectory of *Giardia* mitosomes.

The heterolobosean *Psalteriomonas lanterna* is, to our knowledge, the only one of these newly described anaerobic taxa where the mitochondrial-derived organelles have been characterized to some extent. Heteroloboseans are mostly aerobic mitochondriate amoeboflagellates (e.g. *Naegleria*) but some anaerobic species have been identified (Broers et al. 1990; Amaral Zettler et al. 2000). In *Psalteriomonas*, genes encoding a hydrogenosomal ferredoxin and an Fe hydrogenase have been characterized (Broers et al. 1990; Brul et al. 1994), indicating the presence of a hydrogenosome in this organism.

## 10.3 The Origins of Mitochondria, Mitosomes and Hydrogenosomes

From the foregoing section it should be clear that there is vast diversity in the structure, function and evolutionary history of mitochondrion-derived organelles in anaerobic protists. Although much has yet to be learned about these disparate organellar systems, several evolutionary scenarios regarding their evolutionary history seem possible. The two most likely scenarios are shown in Fig. 10.5 and are discussed next.

First, it is possible that enzymes of aerobic and anaerobic energy metabolism found in mitochondrion-related organelles of present-day eukaryotes originated with the α-proteobacterial symbiont that gave rise to mitochondria (Fig. 10.5, top), which could have been a facultative aerobe (e.g. Martin and Müller 1998 and Rotte et al. 2000 provide a biochemical rationale for this view). After the initial symbiotic integration of the mitochondrial ancestor, many endosymbiont genes were transferred to the nucleus, some of which were

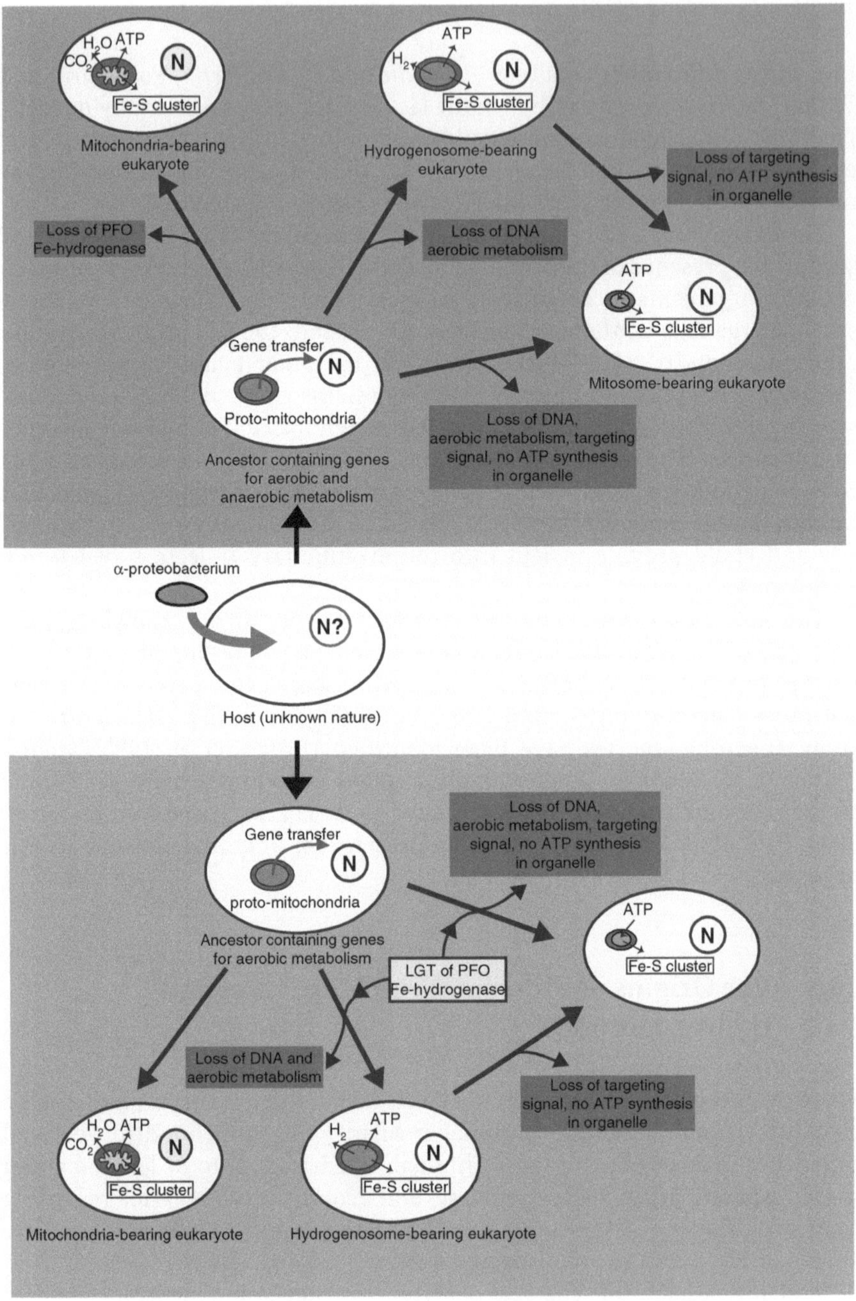


**Fig. 10.5.** Evolution of mitochondria and anaerobic organelles. Two possible scenarios for the evolution of diverse mitochondrial-derived organelles. (**a**) The aerobic and anaerobic metabolic pathways were present in the mitochondrial ancestor, followed by differential loss of functions in different extant eukaryote lineages. (**b**) The mitochondrial ancestor contained enzymes for aerobic metabolism, and the origin of anaerobes occurred via the acquisition of enzymes of anaerobic metabolism by lateral gene transfer (*LGT*). The order of gains and losses is unknown and the order depicted in the diagram is arbitrary

retargeted to the endosymbiont-derived compartment. Subsequently, during the diversification of eukaryotes, predominantly aerobic lineages lost the capacity to perform anaerobic energy metabolism (i.e. they lost enzymes like PFO and hydrogenase), retaining only aerobic respiratory functions and other processes such as FeS cluster biogenesis. Other lineages became specifically adapted to anaerobic niches, and lost aerobic respiratory functions and their mitochondrial genome, but retained the endosymbiont as an ATP-producing hydrogenosome (also retaining the ISC system for FeS cluster biogenesis). Further reductions of hydrogenosomes, or direct evolutionary reduction from the common mitochondrial ancestor, could have led to the origin of mitosomes in disparate anaerobic or parasitic lineages (Fig. 10.5, top).

An alternative scenario would have the common ancestral mitochondrial endosymbiont contribute only the capacity for ATP generation via aerobic respiration as well as the other common mitochondrial processes. In this view, mitochondria in aerobic eukaryotic lineages reflect the ancestral functions that the endosymbiont was originally selected for. Subsequently, some lineages of eukaryotes acquired the capacity to perform anaerobic metabolism, perhaps by acquiring enzymes such as PFO and hydrogenase via LGT. Some anaerobes utilized these enzymes in the cytosol, and, as they become exclusively anaerobic, their mitochondria degenerated to mitosomes, carrying out only essential functions such as FeS cluster biogenesis. In other anaerobes, these proteins acquired mitochondrial-targeting signals yielding hydrogenosomes that, in most cases, lost the capacity for aerobic respiration and their genome. This scenario and other minor variants of this view are shown in Fig. 10.5, bottom.

Both of these hypotheses are consistent with the observed diversity of organelle types. For instance, the existence of transitional hydrogenosome/mitochondrial organelles such as those of *Nyctotherus* would be expected under either scenario. One way to test between them is to consider the phylogenies of the proteins involved in aerobic and anaerobic energy metabolism. A $\alpha$-proteobacterial origin can be inferred for quite a few proteins involved in mitochondrial aerobic energy metabolism, although not for all of them (Esser et al. 2004; J. Leigh and A.J. Roger, unpublished results). The phylogenies of PFO and hydrogenase are somewhat more confusing (Fig. 10.4). The eukaryote PFO/PNO sequences seem to form a weakly supported monophyletic group, but the branching order within this group and the affinities of the clade as a whole within the eubacteria are poorly resolved (Fig. 10.4a). The situation for Fe-hydrogenase is even worse, as eukaryotes are not monophyletic but branch in a similar part of the phylogeny. Furthermore, their relationships to each other and to bacteria are completely unresolved. The only well-resolved relationship of note is the emergence of one of the *Entamoeba* homologs from within the diplomonad group, suggesting its recent origin by LGT. It is possible that much better taxonomic sampling of eukaryote and prokaryote homologs may help resolve these phylogenies. On a more sceptical note, it seems equally possible that deep relationships within these trees have been erased by saturation of sequence changes, and will forever remain a mystery.

## 10.4 Concluding Remarks

Regardless of which of these scenarios is correct, it is clear that there have been multiple origins of organelles with exclusively hydrogenosomal or mitosomal properties within eukaryotes. The bewildering diversity of these organelles and their properties has been revealed only within the last decade, as a number of diverse anaerobic lineages came under intense study. However, there are at least as many independent lineages of anaerobic protists about which virtually nothing is known except their phylogenetic affinities and the existence of putative mitochondrial-derived organelles. Furthermore, there are many additional lineages of anaerobic protists (e.g. in deep-sea environments) (Bernhard et al. 2000) that have only recently been discovered and have yet to be cultivated, or even characterized on an electron microscopy level. Thus, we should expect the diversity in function of the mitochondrial compartment to continue to increase as cell biological and genomic studies of these anaerobes are initiated. The characterization of anaerobic organelles, hydrogenosomes or mitosomes in these organisms should provide a better understanding of the mechanisms of adaptation to anoxic environments in eukaryote evolution and the evolutionary plasticity of the mitochondrial compartment.

*Acknowledgements:* We thank A. Stechmann, J.D. Silberman, M. van der Giezen, J. Tovar, G. Clark, W. Martin, and M. Embley for useful discussions on this topic. M.J.B. and the research described in this chapter were supported by Canadian Institute of Health Research (CIHR) operating grant MOP-62809 awarded to A.J.R. I.R.-T. was supported by a long-term postdoctoral fellowship from EMBO (ALTF230-2003) and CIHR (200502MFE). J.L. was supported by a graduate studentship from the Nova Scotia Health Research Foundation. L.A.H. was supported by a graduate studentship from the Natural Sciences and Engineering Research Council and the Killam Foundation. A.J.R. was supported by a fellowship from the Alfred P. Sloan Foundation and the Peter Lougheed/CIHR New Investigator Award. We also thank Mariona.

## References

Abrahamsen MS, Templeton TJ, Enomoto S, Abrahante JE, Zhu G, Lancto CA, Deng M, Liu C, Widmer G, Tzipori S, Buck GA, Xu P, Bankier AT, Dear PH, Konfortov BA, Spriggs HF, Iyer L, Anantharaman V, Aravind L, Kapur V (2004) Complete genome sequence of the apicomplexan, *Cryptosporidium parvum*. Science 304:441–445

Akhmanova A, Voncken FG, Harhangi H, Hosea KM, Vogels GD, Hackstein JH (1998a) Cytosolic enzymes with a mitochondrial ancestry from the anaerobic chytrid *Piromyces* sp. E2. Mol Microbiol 30:1017–1027

Akhmanova A, Voncken F, van Alen T, van Hoek A, Boxma B, Vogels G, Veenhuis M, Hackstein JH (1998b) A hydrogenosome with a genome. Nature 396:527–528

Akhmanova A, Voncken FG, Hosea KM, Harhangi H, Keltjens JT, op den Camp HJ, Vogels GD, Hackstein JH (1999) A hydrogenosome with pyruvate formate-lyase: anaerobic chytrid fungi use an alternative route for pyruvate catabolism. Mol Microbiol 32:1103–1114

Ali V, Shigeta Y, Tokumoto U, Takahashi Y, Nozaki T (2004) An intestinal parasitic protist, *Entamoeba histolytica*, possesses a non-redundant nitrogen fixation-like system for iron-sulfur cluster assembly under anaerobic conditions. J Biol Chem 279:16863–16874

Amaral Zettler LA, Nerad TA, O'Kelly CJ, Peglar MT, Gillevet PM, Silberman JD, Sogin ML (2000) A molecular reassessment of the *Leptomyxid* amoebae. Protist 151:275–282

Anderson IJ, Loftus BJ (2005) *Entamoeba histolytica*: observations on metabolism based on the genome sequence. Exp Parasitol 110:173–177

Andersson JO, Sjogren AM, Davis LA, Embley TM, Roger AJ (2003) Phylogenetic analyses of diplomonad genes reveal frequent lateral gene transfers affecting eukaryotes. Curr Biol 13:94–104

Arisue N, Sanchez LB, Weiss LM, Muller M, Hashimoto T (2002a) Mitochondrial-type hsp70 genes of the amitochondriate protists, *Giardia intestinalis*, *Entamoeba histolytica* and two microsporidians. Parasitol Int 51:9–16

Arisue N, Hashimoto T, Lee JA, Moore DV, Gordon P, Sensen CW, Gaasterland T, Hasegawa M, Müller M (2002b) The phylogenetic position of the pelobiont *Mastigamoeba balamuthi* based on sequences of rDNA and translation elongation factors EF-1alpha and EF-2. J Eukaryot Microbiol 49:1–10

Arisue N, Hashimoto T, Yoshikawa H, Nakamura Y, Nakamura G, Nakamura F, Yano TA, Hasegawa M (2002c) Phylogenetic position of *Blastocystis hominis* and of stramenopiles inferred from multiple molecular sequence data. J Eukaryot Microbiol 49:42–53

Bakatselou C, Clark CG (2000) A mitochondrial-type hsp70 gene of *Entamoeba histolytica*. Arch Med Res 31:S176–S177

Bakker-Grunwald T, Wostmann C (1993) *Entamoeba histolytica* as a model for the primitive eukaryotic cell. Parasitol Today 9:27–31

Bapteste E, Brinkmann H, Lee JA, Moore DV, Sensen CW, Gordon P, Durufle L, Gaasterland T, Lopez P, Muller M, Philippe H (2002) The analysis of 100 genes supports the grouping of three highly divergent amoebae: *Dictyostelium*, *Entamoeba*, and *Mastigamoeba*. Proc Natl Acad Sci USA 99:1414–1419

Benchimol M, Durand R, Almeida JC (1997) A double membrane surrounds the hydrogenosomes of the anaerobic fungus *Neocallimastix frontalis*. FEMS Microbiol Lett 154:277–282

Bernhard JM, Buck KR, Farmer MA, Bowser SS (2000) The Santa Barbara Basin is a symbiosis oasis. Nature 403:77–80

Boxma B, Voncken F, Jannink S, van Alen T, Akhmanova A, van Weelden SW, van Hellemond JJ, Ricard G, Huynen M, Tielens AG, Hackstein JH (2004) The anaerobic chytridiomycete fungus *Piromyces* sp. E2 produces ethanol via pyruvate:formate lyase and an alcohol dehydrogenase E. Mol Microbiol 51:1389–1399

Boxma B, de Graaf RM, van der Staay GW, van Alen TA, Ricard G, Gabaldon T, van Hoek AH, Moon-van der Staay SY, Koopman WJ, van Hellemond JJ, Tielens AG, Friedrich T, Veenhuis M, Huynen MA, Hackstein JH (2005) An anaerobic mitochondrion that produces hydrogen. Nature 434:74–79

Bozner P (1997) Immunological detection and subcellular localization of Hsp70 and Hsp60 homologs in *Trichomonas vaginalis*. J Parasitol 83:224–229

Brinkmann H, van der Giezen M, Zhou Y, Poncelin de Raucourt G, Philippe H (2005) An empirical assessment of long-branch attraction artefacts in deep eukaryotic phylogenomics. Syst Biol 54:743–757

Broers CAM, Stumm CK, Vogels GD, Brugerolle G (1990) *Psalteriomonas lanterna* gen. nov., sp. nov., a free-living amoeboflagellate isolated from freshwater anaerobic sediments. Eur J Protistol 25:369–380

Brown JR, Doolittle WF (1997) Archaea and the prokaryote-to-eukaryote transition. Microbiol Mol Biol Rev 61:456–502

Brugerolle G, Patterson DJ (1997) Ultrastructure of *Trimastix convexa* Hollande, an amitochondriate anaerobic flagellate with a previously undescribed organization. Eur J Protistol 33:121–130

Brul S, Veltman RH, Lombardo MC, Vogels GD (1994) Molecular cloning of hydrogenosomal ferredoxin cDNA from the anaerobic amoeboflagellate *Psalteriomonas lanterna*. Biochim Biophys Acta 1183:544–546

Bui ET, Johnson PJ (1996) Identification and characterization of [Fe]-hydrogenases in the hydrogenosome of *Trichomonas vaginalis*. Mol Biochem Parasitol 76:305–310

Bui ET, Bradley PJ, Johnson PJ (1996) A common evolutionary origin for mitochondria and hydrogenosomes. Proc Natl Acad Sci USA 93:9651–9656

Cavalier-Smith T (1983) A 6-kingdom classification and a unified phylogeny. In: Schwemmler W, Schenk HEA (eds) Endocytobiology. de Gruyter, Berlin, pp 1027–1034

Cavalier-Smith T (1987) Eukaryotes with no mitochondria. Nature 326:332–333

Cavalier-Smith T (1991) Archamoebae: the ancestral eukaryotes? BioSystems 25:25–38

Cavalier-Smith T, Chao EE (1996) Molecular phylogeny of the free-living archezoan *Trepomonas agilis* and the nature of the first eukaryote. J Mol Evol 43:551–562

Chan KW, Slotboom DJ, Cox S, Embley TM, Fabre O, van der Giezen M, Harding M, Horner DS, Kunji ER, Leon-Avila G, Tovar J (2005) A novel ADP/ATP transporter in the mitosome of the microaerophilic human parasite *Entamoeba histolytica*. Curr Biol 15:737–742

Chávez LA, Balamuthi W, Gong T (1986) A light and electron microscopical study of a new, polymorphic free-living amoeba, *Phreatamoeba balamuthi* n. g., n. sp. J Protozool. 33: 397– 404

Clark CG, Roger AJ (1995) Direct evidence for secondary loss of mitochondria in *Entamoeba histolytica*. Proc Natl Acad Sci USA 92:6518–6521

Davidson EA, van der Giezen M, Horner DS, Embley TM, Howe CJ (2002) An [Fe] hydrogenase from the anaerobic hydrogenosome-containing fungus *Neocallimastix frontalis* L2. Gene 296:45–52

de Duve C (1969) Evolution of the peroxisome. Ann N Y Acad Sci 168:369–381

Dolezal P, Smid O, Rada P, Zubacova Z, Bursac D, Sutak R, Nebesarova J, Lithgow T, Tachezy J (2005) *Giardia* mitosomes and trichomonad hydrogenosomes share a common mode of protein targeting. Proc Natl Acad Sci USA 102:10924–10929

Dyall SD, Johnson PJ (2000) Origins of hydrogenosomes and mitochondria: evolution and organelle biogenesis. Curr Opin Microbiol 3:404–411

Dyall SD, Koehler CM, Delgadillo-Correa MG, Bradley PJ, Plumper E, Leuenberger D, Turck CW, Johnson PJ (2000) Presence of a member of the mitochondrial carrier family in hydrogenosomes: conservation of membrane-targeting pathways between hydrogenosomes and mitochondria. Mol Cell Biol 20:2488–2497

Dyall SD, Yan W, Delgadillo-Correa MG, Lunceford A, Loo JA, Clarke CF, Johnson PJ (2004) Non-mitochondrial complex I proteins in a hydrogenosomal oxidoreductase complex. Nature 431:1103–1107

Edgar RC (2004) MUSCLE: a multiple sequence alignment method with reduced time and space complexity. BMC Bioinform 5:113

Ellis TJ, Morrison DA, Jeffries AC (1998) The phylum Apicomplexa: an update on the molecular phylogeny. In: Coombs GH, Vickerman K, SleighMA, Warren A (eds) Evolutionary relationships among the Protozoa. Kluwer, Boston, pp 255–274

Embley TM, Hirt RP (1998) Early branching eukaryotes? Curr Opin Genet Dev 8:624–629

Embley TM, Finlay BJ, Dyal PL, Hirt RP, Wilkinson M, Williams AG (1995) Multiple origins of anaerobic ciliates with hydrogenosomes within the radiation of aerobic ciliates. Proc R Soc Lond B Biol Sci 262:87–93

Embley TM, van der Giezen M, Horner DS, Dyal PL, Bell S, Foster PG (2003) Hydrogenosomes, mitochondria and early eukaryotic evolution. IUBMB Life 55:387–395

Emelyanov VV (2001) Evolutionary relationship of Rickettsiae and mitochondria. FEBS Lett 501:11–18

Entrala E, Mascaro C (1997) Glycolytic enzyme activities in *Cryptosporidium parvum* oocysts. FEMS Microbiol Lett 151:51–57

Esser C, Ahmadinejad N, Wiegand C, Rotte C, Sebastiani F, Gelius-Dietrich G, Henze K, Kretschmann E, Richly E, Leister D, Bryant D, Steel MA, Lockhart PJ, Penny D, Martin W (2004) A genome phylogeny for mitochondria among alpha-proteobacteria and a predominantly eubacterial ancestry of yeast nuclear genes. Mol Biol Evol 21:1643–1660

Fast NM, Keeling PJ (2001) Alpha and beta subunits of pyruvate dehydrogenase E1 from the microsporidian *Nosema locustae*: mitochondrion-derived carbon metabolism in microsporidia. Mol Biochem Parasitol 117:201–209

Fast NM, Logsdon JM Jr, Doolittle WF (1999) Phylogenetic analysis of the TATA box binding protein (TBP) gene from *Nosema locustae*: evidence for a microsporidia-fungi relationship and spliceosomal intron loss. Mol Biol Evol 16:1415–1419

Fenchel T, Finlay BJ (1995) Ecology and evolution in anoxic worlds. Oxford University Press, Oxford

Field J, Rosenthal B, Samuelson J (2000) Early lateral transfer of genes encoding malic enzyme, acetyl-CoA synthetase and alcohol dehydrogenases from anaerobic prokaryotes to *Entamoeba histolytica*. Mol Microbiol 38:446–455

Fitzpatrick DA, Creevey CJ, McInerney JO (2005) Genome phylogenies indicate a meaningful α-proteobacterial phylogeny and support a grouping of the mitochondria with the Rickettsiales. Mol Biol Evol 23:74–85

Foster PG, Hickey DA (1999) Compositional bias may affect both DNA-based and protein-based phylogenetic reconstructions. J Mol Evol 48:284–290

Foster PG, Jermiin LS, Hickey DA (1997) Nucleotide composition bias affects amino acid content in proteins coded by animal mitochondria. J Mol Evol 44:282–288

Gelius-Dietrich G, Henze K (2004) Pyruvate formate lyase (PFL) and PFL activating enzyme in the chytrid fungus *Neocallimastix frontalis*: a free-radical enzyme system conserved across divergent eukaryotic lineages. J Eukaryot Microbiol 51:456–463

Germot A, Philippe H, Le Guyader H (1996) Presence of a mitochondrial-type 70-kDa heat shock protein in *Trichomonas vaginalis* suggests a very early mitochondrial endosymbiosis in eukaryotes. Proc Natl Acad Sci USA 93:14614–14617

Germot A, Philippe H, Le Guyader H (1997) Evidence for loss of mitochondria in Microsporidia from a mitochondrial-type HSP70 in *Nosema locustae*. Mol Biochem Parasitol 87:159–168

Ghosh S, Field J, Rogers R, Hickman M, Samuelson J (2000) The *Entamoeba histolytica* mitochondrion-derived organelle (crypton) contains double-stranded DNA and appears to be bound by a double membrane. Infect Immun 68:4319–4322

Gray MW, Burger G, Lang BF (1999) Mitochondrial evolution. Science 283:1476–1481

Guindon S, Gascuel O (2003) A simple, fast, and accurate algorithm to estimate large phylogenies by maximum likelihood. Syst Biol 52:696–704

Hashimoto T, Nakamura Y, Kamaishi T, Hasegawa M (1997) Early evolution of eukaryotes inferred from protein phylogenies of translation elongation factors 1-alpha and 2. Arch Protistenk 148:287–295

Hashimoto T, Sanchez LB, Shirakura T, Muller M, Hasegawa M (1998) Secondary absence of mitochondria in Giardia lamblia and Trichomonas vaginalis revealed by valyl-tRNA synthetase phylogeny. Proc Natl Acad Sci USA 95:6860–6865

Henriquez FL, Richards TA, Roberts F, McLeod R, Roberts CW (2005) The unusual mitochondrial compartment of *Cryptosporidium parvum*. Trends Parasitol 21:68–74

Hinkle G, Leipe DD, Nerad TA, Sogin ML (1994) The unusually long small subunit ribosomal RNA of *Phreatamoeba balamuthi*. Nucleic Acids Res 22:465–469

Hirt RP, Healy B, Vossbrinck CR, Canning EU, Embley TM (1997) A mitochondrial Hsp70 orthologue in *Vairimorpha necatrix*: molecular evidence that microsporidia once contained mitochondria. Curr Biol 7:995–998

Hirt RP, Logsdon JM Jr, Healy B, Dorey MW, Doolittle WF, Embley TM (1999) Microsporidia are related to Fungi: evidence from the largest subunit of RNA polymerase II and other proteins. Proc Natl Acad Sci USA 96:580–585
Horner DS, Hirt RP, Kilvington S, Lloyd D, Embley TM (1996) Molecular data suggest an early acquisition of the mitochondrion endosymbiont. Proc R Soc Lond B Biol Sci 263:1053–1059
Horner DS, Foster PG, Embley TM (2000) Iron hydrogenases and the evolution of anaerobic eukaryotes. Mol Biol Evol 17:1695–1709
Hrdy I, Muller M (1995) Primary structure and eubacterial relationships of the pyruvate:ferredoxin oxidoreductase of the amitochondriate eukaryote *Trichomonas vaginalis*. J Mol Evol 41:388–396
Hrdy I, Hirt RP, Dolezal P, Bardonova L, Foster PG, Tachezy J, Embley TM (2004) *Trichomonas* hydrogenosomes contain the NADH dehydrogenase module of mitochondrial complex I. Nature 432:618–622
Johnson DC, Dean DR, Smith AD, Johnson MK (2005) Structure, function, and formation of biological iron-sulfur clusters. Annu Rev Biochem 74:247–281
Katinka MD, Duprat S, Cornillot E, Metenier G, Thomarat F, Prensier G, Barbe V, Peyretaillade E, Brottier P, Wincker P, Delbac F, El Alaoui H, Peyret P, Saurin W, Gouy M, Weissenbach J, Vivares CP (2001) Genome sequence and gene compaction of the eukaryote parasite *Encephalitozoon cuniculi*. Nature 414:450–453
Keeling PJ (2003) Congruent evidence from alpha-tubulin and beta-tubulin gene phylogenies for a zygomycete origin of microsporidia. Fungal Genet Biol 38:298–309
Keeling PJ, Doolittle WF (1996) Alpha-tubulin from early-diverging eukaryotic lineages and the evolution of the tubulin family. Mol Biol Evol 13:1297–1305
Keeling PJ, Doolittle WF (1997) Evidence that eukaryotic triosephosphate isomerase is of alpha-proteobacterial origin. Proc Natl Acad Sci USA 94:1270–1275
Keeling PJ, Fast NM (2002) Microsporidia: biology and evolution of highly reduced intracellular parasites. Annu Rev Microbiol 56:93–116
Keithly JS, Langreth SG, Buttle KF, Mannella CA (2005) Electron tomographic and ultrastructural analysis of the *Cryptosporidium parvum* relict mitochondrion, its associated membranes, and organelles. J Eukaryot Microbiol 52:132–140
LaGier MJ, Tachezy J, Stejskal F, Kutisova K, Keithly JS (2003) Mitochondrial-type iron-sulfur cluster biosynthesis genes (IscS and IscU) in the apicomplexan *Cryptosporidium parvum*. Microbiology 149:3519–3530
Leon-Avila G, Tovar J (2004) Mitosomes of *Entamoeba histolytica* are abundant mitochondrion-related remnant organelles that lack a detectable organellar genome. Microbiology 150:1245–1250
Lill R, Muhlenhoff U (2005) Iron-sulfur-protein biogenesis in eukaryotes. Trends Biochem Sci 30:133–141
Lindmark DG, Muller M (1973) Hydrogenosome, a cytoplasmic organelle of the anaerobic flagellate *Tritrichomonas foetus*, and its role in pyruvate metabolism. J Biol Chem 248:7724–7728
Lloyd D, Ralphs JR, Harris JC (2002) *Giardia intestinalis*, a eukaryote without hydrogenosomes, produces hydrogen. Microbiology 148:727–733
Loftus B, Anderson I, Davies R, Alsmark UC, Samuelson J, Amedeo P, Roncaglia P, Berriman M, Hirt RP, Mann BJ, Nozaki T, Suh B, Pop M, Duchene M, Ackers J, Tannich E, Leippe M, Hofer M, Bruchhaus I, Willhoeft U, Bhattacharya A, Chillingworth T, Churcher C, Hance Z, Harris B, Harris D, Jagels K, Moule S, Mungall K, Ormond D, Squares R, Whitehead S, Quail MA, Rabbinowitsch E, Norbertczak H, Price C, Wang Z, Guillen N, Gilchrist C, Stroup SE, Bhattacharya S, Lohia A, Foster PG, Sicheritz-Ponten T, Weber C, Singh U, Mukherjee C, El-Sayed NM, Petri WA Jr, Clark CG, Embley TM, Barrell B, Fraser CM, Hall N (2005) The genome of the protist parasite *Entamoeba histolytica*. Nature 433:865–868
Mai Z, Ghosh S, Frisardi M, Rosenthal B, Rogers R, Samuelson J (1999) Hsp60 is targeted to a cryptic mitochondrion-derived organelle ("crypton") in the microaerophilic protozoan parasite *Entamoeba histolytica*. Mol Cell Biol 19:2198–2205

Markos A, Miretsky A, Muller M (1993) A glyceraldehyde-3-phosphate dehydrogenase with eubacterial features in the amitochondriate eukaryote, *Trichomonas vaginalis.* J Mol Evol 37:631–643
Martin W, Müller M (1998) The hydrogen hypothesis for the first eukaryote. Nature 392:37–41
Martin W, Hoffmeister M, Rotte C, Henze K (2001) An overview of endosymbiotic models for the origins of eukaryotes, their ATP-producing organelles (mitochondria and hydrogenosomes), and their heterotrophic lifestyle. Biol Chem 382:1521–1539
Müller M (1998) Enzymes and compartmentation of core energy metabolism of anaerobic protists: a special case in eukaryotic evolution. In: Coombs GH, Vickerman K, Sleigh MA, Warren A (eds) Evolutionary relationships among protozoa. Chapman and Hall, London, pp 109–127
Nasirudeen AM, Tan KS (2004) Isolation and characterization of the mitochondrion-like organelle from *Blastocystis hominis.* J Microbiol Methods 58:101–109
Nixon JE, Wang A, Field J, Morrison HG, McArthur AG, Sogin ML, Loftus BJ, Samuelson J (2002) Evidence for lateral transfer of genes encoding ferredoxins, nitroreductases, NADH oxidase, and alcohol dehydrogenase 3 from anaerobic prokaryotes to *Giardia lamblia* and *Entamoeba histolytica.* Eukaryot Cell 1:181–190
Nixon JE, Field J, McArthur AG, Sogin ML, Yarlett N, Loftus BJ, Samuelson J (2003) Iron-dependent hydrogenases of *Entamoeba histolytica* and *Giardia lamblia*: activity of the recombinant entamoebic enzyme and evidence for lateral gene transfer. Biol Bull 204:1–9
Peyretaillade E, Broussolle V, Peyret P, Metenier G, Gouy M, Vivares CP (1998) Microsporidia, amitochondrial protists, possess a 70-kDa heat shock protein gene of mitochondrial evolutionary origin. Mol Biol Evol 15:683–689
Philippe H, Laurent J (1998) How good are deep phylogenetic trees? Curr Opin Genet Dev 8:616–623
Putignani L, Tait A, Smith HV, Horner D, Tovar J, Tetley L, Wastling JM (2004) Characterization of a mitochondrion-like organelle in *Cryptosporidium parvum.* Parasitology 129:1–18
Reeves RE (1984) Metabolism of *Entamoeba histolytica* Schaudinn, 1903. Adv Parasitol 23:105–142
Regoes A, Zourmpanou D, Leon-Avila G, van der Giezen M, Tovar J, Hehl AB (2005) Protein import, replication, and inheritance of a vestigial mitochondrion. J Biol Chem 280:30557–30563
Rensing SA, Obrdlik P, Rober-Kelber N, Müller SB, Hofmann CJB, van de Peer Y, Maier U- (1997) Molecular phylogeny of the stress-70 protein family with reference to algal relationships. Eur J Phycol 32:279–285
Riordan CE, Langreth SG, Sanchez LB, Kayser O, Keithly JS (1999) Preliminary evidence for a mitochondrion in *Cryptosporidium parvum*: phylogenetic and therapeutic implications. J Eukaryot Microbiol 46:52S–55S
Riordan CE, Ault JG, Langreth SG, Keithly JS (2003) *Cryptosporidium parvum* Cpn60 targets a relict organelle. Curr Genet 44:138-147
Rivera MC, Jain R, Moore JE, Lake JA (1998) Genomic evidence for two functionally distinct gene classes. Proc Natl Acad Sci USA 95:6239–6244
Roberts CW, Roberts F, Henriquez FL, Akiyoshi D, Samuel BU, Richards TA, Milhous W, Kyle D, McIntosh L, Hill GC, Chaudhuri M, Tzipori S, McLeod R (2004) Evidence for mitochondrial-derived alternative oxidase in the apicomplexan parasite *Cryptosporidium parvum*: a potential anti-microbial agent target. Int J Parasitol 34:297–308
Rodriguez MA, Garcia-Perez RM, Mendoza L, Sanchez T, Guillen N, Orozco E (1998) The pyruvate:ferredoxin oxidoreductase enzyme is located in the plasma membrane and in a cytoplasmic structure in *Entamoeba.* Microb Pathog 25:1–10
Roger AJ (1999) Reconstructing early events in eukaryotic evolution. Am Nat 154:S146–S163
Roger AJ, Clark CG, Doolittle WF (1996) A possible mitochondrial gene in the early-branching amitochondriate protist Trichomonas vaginalis. Proc Natl Acad Sci USA 93:14618–14622
Roger AJ, Svard SG, Tovar J, Clark CG, Smith MW, Gillin FD, Sogin ML (1998) A mitochondrial-like chaperonin 60 gene in *Giardia lamblia*: evidence that diplomonads once harbored an

endosymbiont related to the progenitor of mitochondria. Proc Natl Acad Sci USA 95:229–234
Roger AJ, Morrison HG, Sogin ML (1999) Primary structure and phylogenetic relationships of a malate dehydrogenase gene from *Giardia lamblia*. J Mol Evol 48:750–755
Ronquist F, Huelsenbeck JP (2003) MrBayes 3: Bayesian phylogenetic inference under mixed models. Bioinformatics 19:1572–1574
Rosenthal B, Mai Z, Caplivski D, Ghosh S, de la Vega H, Graf T, Samuelson J (1997) Evidence for the bacterial origin of genes encoding fermentation enzymes of the amitochondriate protozoan parasite *Entamoeba histolytica*. J Bacteriol 179:3736–3745
Rotte C, Henze K, Muller M, Martin W (2000) Origins of hydrogenosomes and mitochondria. Curr Opin Microbiol 3:481–486
Rotte C, Stejskal F, Zhu G, Keithly JS, Martin W (2001) Pyruvate:$NADP^+$ oxidoreductase from the mitochondrion of *Euglena gracilis* and from the apicomplexan *Cryptosporidium parvum*: a biochemical relic linking pyruvate metabolism in mitochondriate and amitochondriate protists. Mol Biol Evol 18:710–720
Sagan L (1967) On the origin of mitosing cells. J Theor Biol 14:255–274
Seravin LN, Goodkov AV (1987) Cytoplastmic microbody-like granules of the amoeba Pelomyxa palustris. Tsitologiia 29:600– 603 (in Russian)
Silberman JD, Sogin ML, Leipe DD, Clark CG (1996) Human parasite finds taxonomic home. Nature 380:398
Simpson AGB, Patterson DJ (1999) The ultrastructure of *Carpediemonas membranifera* (Eukaryota) with reference to the excavate hypothesis. Eur J Protistol 35:353–370
Simpson AGB, Roger AJ (2004) Excavata and the origin of amitochondriate eukaryotes. In: Hirt RP, Horner DS (eds) Organelles, genomes and eukaryote phylogeny: an evolutionary synthesis in the age of genomics. CRC, Boca Raton, pp 27–53
Simpson AG, Roger AJ, Silberman JD, Leipe DD, Edgcomb VP, Jermiin LS, Patterson DJ, Sogin ML (2002) Evolutionary history of "early-diverging" eukaryotes: the excavate taxon *Carpediemonas* is a close relative of *Giardia*. Mol Biol Evol 19:1782–1791
Slapeta J, Keithly JS (2004) *Cryptosporidium parvum* mitochondrial-type HSP70 targets homologous and heterologous mitochondria. Eukaryot Cell 3:483–494
Sogin ML (1989) Evolution of eukaryotic microorganisms and their small subunit ribosomal RNAs. Am Zool 29:487–499
Sogin ML (1991) Early evolution and the origin of eukaryotes. Curr Opin Genet Dev 1:457–463
Sogin ML (1997) History assignment: when was the mitochondrion founded? Curr Opin Genet Dev 7AS0318:792–799
Stanier RY (1970) Some aspects of the biology of cells and their possible evolutionary significance. In: Charles HP, Knight BD (eds) Organization and control in prokaryotic and eukaryotic cells: 20th symposium of the Society for General Microbiology. Cambridge University Press, London, pp 1–38
Stiller JW, Duffield EC, Hall BD (1998) Amitochondriate amoebae and the evolution of DNA-dependent RNA polymerase II. Proc Natl Acad Sci USA 95:11769–11774
Sutak R, Dolezal P, Fiumera HL, Hrdy I, Dancis A, Delgadillo-Correa M, Johnson PJ, Muller M, Tachezy J (2004) Mitochondrial-type assembly of FeS centers in the hydrogenosomes of the amitochondriate eukaryote *Trichomonas vaginalis*. Proc Natl Acad Sci USA 101:10368–10373
Tachezy J, Sanchez LB, Muller M (2001) Mitochondrial type iron-sulfur cluster assembly in the amitochondriate eukaryotes *Trichomonas vaginalis* and *Giardia intestinalis*, as indicated by the phylogeny of IscS. Mol Biol Evol 18:1919–1928
Thomarat F, Vivares CP, Gouy M (2004) Phylogenetic analysis of the complete genome sequence of *Encephalitozoon cuniculi* supports the fungal origin of microsporidia and reveals a high frequency of fast-evolving genes. J Mol Evol 59:780–791
Tielens AG, Rotte C, van Hellemond JJ, Martin W (2002) Mitochondria as we don't know them. Trends Biochem Sci 27:564– 572
Tjaden J, Haferkamp I, Boxma B, Tielens AG, Huynen M, Hackstein JH (2004) A divergent ADP/ATP carrier in the hydrogenosomes of *Trichomonas gallinae* argues for an independent origin of these organelles. Mol Microbiol 51:1439–1446

Tovar J, Fischer A, Clark CG (1999) The mitosome, a novel organelle related to mitochondria in the amitochondrial parasite *Entamoeba histolytica*. Mol Microbiol 32:1013–1021
Tovar J, Leon-Avila G, Sanchez LB, Sutak R, Tachezy J, van der Giezen M, Hernandez M, Muller M, Lucocq JM (2003) Mitochondrial remnant organelles of *Giardia function* in iron-sulphur protein maturation. Nature 426:172–176
van der Giezen M, Sjollema KA, Artz RR, Alkema W, Prins RA (1997a) Hydrogenosomes in the anaerobic fungus *Neocallimastix frontalis* have a double membrane but lack an associated organelle genome. FEBS Lett 408:147–150
van der Giezen M, Rechinger KB, Svendsen I, Durand R, Hirt RP, Fevre M, Embley TM, Prins RA (1997b) A mitochondrial-like targeting signal on the hydrogenosomal malic enzyme from the anaerobic fungus *Neocallimastix frontalis*: support for the hypothesis that hydrogenosomes are modified mitochondria. Mol Microbiol 23:11–21
van der Giezen M, Kiel JA, Sjollema KA, Prins RA (1998) The hydrogenosomal malic enzyme from the anaerobic fungus *Neocallimastix frontalis* is targeted to mitochondria of the methylotrophic yeast *Hansenula polymorpha*. Curr Genet 33:131–135
van der Giezen M, Slotboom DJ, Horner DS, Dyal PL, Harding M, Xue GP, Embley TM, Kunji ER (2002) Conserved properties of hydrogenosomal and mitochondrial ADP/ATP carriers: a common origin for both organelles. EMBO J 21:572–579
van der Giezen M, Birdsey GM, Horner DS, Lucocq J, Dyal PL, Benchimol M, Danpure CJ, Embley TM (2003) Fungal hydrogenosomes contain mitochondrial heat-shock proteins. Mol Biol Evol 20:1051–1061
van der Giezen M, Cox S, Tovar J (2004) The iron-sulfur cluster assembly genes IscS and IscU of *Entamoeba histolytica* were acquired by horizontal gene transfer. BMC Evol Biol 4:7
van der Giezen M, Tovar J, Clark CG (2005) Mitochondrion-derived organelles in protists and fungi. Int Rev Cytol 244:175–225
van Hoek AH, Akhmanova AS, Huynen MA, Hackstein JH (2000) A mitochondrial ancestry of the hydrogenosomes of *Nyctotherus ovalis*. Mol Biol Evol 17:202–206
Vanlerberghe GC, McIntosh L (1997) Alternative oxidase: From gene to function. Annu Rev Plant Physiol Plant Mol Biol 48:703–734
Vinh le S, von Haeseler A (2004) IQPNNI: moving fast through tree space and stopping in time. Mol Biol Evol 21:1565–1571
Vivares CP, Gouy M, Thomarat F, Metenier G (2002) Functional and evolutionary analysis of a eukaryotic parasitic genome. Curr Opin Microbiol 5:499–505
von Dohlen CD, Kohler S, Alsop ST, McManus WR (2001) Mealybug beta-proteobacterial endosymbionts contain gamma-proteobacterial symbionts. Nature 412:433–436
Voncken F, Boxma B, Tjaden J, Akhmanova A, Huynen M, Verbeek F, Tielens AG, Haferkamp I, Neuhaus HE, Vogels G, Veenhuis M, Hackstein JH (2002a) Multiple origins of hydrogenosomes: functional and phylogenetic evidence from the ADP/ATP carrier of the anaerobic chytrid *Neocallimastix* sp. Mol Microbiol 44:1441–1454
Voncken FG, Boxma B, van Hoek AH, Akhmanova AS, Vogels GD, Huynen M, Veenhuis M, Hackstein JH (2002b) A hydrogenosomal [Fe]-hydrogenase from the anaerobic chytrid *Neocallimastix* sp. L2. Gene 284:103–112
Williams BA, Keeling PJ (2005) Microsporidian mitochondrial proteins: expression in *Antonospora locustae* spores and identification of genes coding for two further proteins. J Eukaryot Microbiol 52:271–276
Williams BA, Hirt RP, Lucocq JM, Embley TM (2002) A mitochondrial remnant in the microsporidian *Trachipleistophora hominis*. Nature 418:865–869
Xu P, Widmer G, Wang Y, Ozaki LS, Alves JM, Serrano MG, Puiu D, Manque P, Akiyoshi D, Mackey AJ, Pearson WR, Dear PH, Bankier AT, Peterson DL, Abrahamsen MS, Kapur V, Tzipori S, Buck GA (2004) The genome of *Cryptosporidium hominis*. Nature 431:1107–1112
Yarlett N, Orpin CG, Munn EA, Yarlett NC, Greenwood CA (1986) Hydrogenosomes in the rumen fungus *Neocallimastix patriciarum*. Biochem J 236:729–739
Zierdt CH, Donnolley CT, Muller J, Constantopoulos G (1988) Biochemical and ultrastructural study of *Blastocystis hominis*. J Clin Microbiol 26:965–970

# 11 Mitosomes of Parasitic Protozoa: Biology and Evolutionary Significance

JORGE TOVAR


## Summary

Mitosomes are mitochondrion-related organelles that exist in a range of anaerobic/microaerophilic parasitic protozoa which lack typical mitochondria. While hydrogenosomes (their sister organelles) generate ATP and molecular hydrogen as end products of their metabolism, mitosomes appear to have lost their capacity for biological energy generation and do not produce hydrogen. Despite their highly derived nature, mitosomes share several morphological, biochemical and physiological characteristics with mitochondria and hydrogenosomes. They are surrounded by two limiting membranes, contain mitochondrial marker proteins, possess protein-import mechanisms functionally conserved with those of mitochondria and hydrogenosomes and appear to fulfil the only essential biosynthetic function of mitochondria, i.e. the assembly of iron–sulphur (FeS) clusters critical for the maturation of cellular FeS proteins. Much remains to be investigated about the biology of mitosomes but their widespread presence in anaerobic/microaerophilic eukaryotes, their remnant metabolism and their mutually exclusive distribution with mitochondria and hydrogenosomes suggest that mitosomes of parasitic protozoa represent vestigial anaerobic mitochondria.


## 11.1 Introduction

Mitochondria are energy-generating organelles of eukaryotic cells. Biochemical studies and phylogenetic analyses of their remnant genome have demonstrated that these organelles arose from a unique endosymbiotic interaction between an α-proteobacterium and a host cell (Andersson et al. 2003b; Gray et al. 1999). The evolution of mitochondria is thus intrinsically linked with that of the eukaryotic cell. In all current theories of eukaryotic evolution, the establishment of the original mitochondrial endosymnbiont is placed either at the origin of the first eukaryotic cell or soon after its early diversification over 1,500 million years ago (Martin et al. 2001). Aerobic mitochondria are by far the best characterised type of energy-generating organelle of eukaryotes. During aerobic respiration, oxygen is used as the final acceptor

Origin of Mitochondria and Hydrogenosomes (ed. by William F. Martin and Miklós Müller)

of electrons and ATP, $CO_2$ and $H_2O$ are generated in the mitochondrion as metabolic end products. But there are many eukaryotic organisms across the taxonomic range – including unicellular fungi and euglenoids as well as multicellular molluscs, flatworms, snails and crustaceans – that spend at least part of their lives in oxygen-deprived environments and have adapted their mitochondria to function under both aerobic and anaerobic conditions to suit their lifestyles. So-called facultative anaerobes, these organisms contain modified mitochondria able to generate ATP through proton-pumping transport of electrons from organic or inorganic electron donors in the presence or absence of oxygen (Tielens et al. 2002). Such chimaeric energy metabolism and the widely accepted view that mitochondria are monophyletic in origin suggest that facultative anaerobic mitochondria arose from the original mitochondrial endosymbiont through reductive evolution, likely aided by the generation of novel cellular functions through the adaptive reshuffling of preexisting protein domains and/or lateral acquisition of prokaryotic and eukaryotic gene products, either directly from the host or from external sources (Doolittle et al. 2003; Gray et al. 1999; Tielens et al. 2002).

That such reductive and adaptive tendencies may lead to the evolution of degenerate mitochondria no longer able to generate ATP through electron transport and oxidative phosphorylation – but instead exclusively by substrate-level phosphorylation – was recently demonstrated in the hydrogenosome-containing anaerobic ciliate *Nyctotherus ovalis*. Phylogenetic analyses of gene products encoded by its remnant hydrogenosomal genome provided unequivocal evidence that *Nyctotherus*'s hydrogenosomes are modified mitochondria (Boxma et al. 2005). But how far can the reductive evolutionary tendency of mitochondria be taken? A clue to such an intriguing question may be found in the recently discovered mitosomes of parasitic protozoa, another type of mitochondrion-related organelle. Like mitochondria and hydrogenosomes, mitosomes appear to be morphologically and functionally heterogeneous but seem to have lost their capacity for biological energy generation along with most other mitochondrial functions. This chapter reviews the history and current status of mitosome research, highlights the diverse biology of mitosomes in various eukaryotic lineages and explores the evolutionary significance of these unique organelles.

## 11.2 Discovery of Mitosomes: a Brief History

To place the research leading to the discovery of mitosomes in a historical context it is important to remember that all eukaryotic organisms in which typical mitochondria could not be recognised by microscopic or biochemical methods were once grouped within the kingdom Archezoa (Cavalier-Smith 1983). In its original formulation, the Archezoa hypothesis proposed that members of four heterogeneous amitochondriate protist groups – namely Metamonads, Microsporidia, Parabasalia and Archamoebae – were the direct descendants of the proto-eukaryotic cell, i.e. a primitive nucleated cell proposed to have existed prior to the endosymbiotic acquisition of the mitochondrion. Although this hypothesis has now been thoroughly disproved and since abandoned by its original proposer (Cavalier-Smith 2002), the Archezoa hypothesis triggered much of the research that paved the way for the discovery of mitosomes in parasitic protozoa as diverse as *Entamoeba histolytica*, *Trachipleistophora hominis* and *Giardia intestinalis.* Although various names have been used to describe these organelles (e.g. mitosomes, cryptons, relict mitochondria, mitochondrial remnants, mitochondrion-like organelles), the generic term mitosome is used here – as in other reviews – to describe all mitochondrion-related organelles that have lost their capacity for aerobic or anaerobic ATP biosynthesis (van der Giezen et al. 2005; van der Giezen and Tovar 2004, 2005). An attempt is made to present the research leading to the discovery of mitosomes in chronological order, with the caveat that many of these studies were conducted in parallel and reported simultaneously by various laboratories worldwide.

### 11.2.1 *Entamoeba histolytica*

Mitosomes were first discovered in the microaerophilic human pathogen *E. histolytica*, the causative agent of diarrhoeal disease known as amoebiasis and a member of the Archamoebae, sensus Cavalier-Smith (1983). The absence of typical mitochondria and a very simple cellular architecture led to the suggestion that *E. histolytica* was a very primitive eukaryote (Cavalier-Smith 1991; Meza 1992). The first clue that *Entamoeba* might not be primitively amitochondrial came from the phylogenetic analysis of its small-subunit ribosomal RNA which placed *E. histolytica* as a late eukaryotic branch, well after the evolutionary divergence of mitochondrion-containing groups such as the euglenozoa and heterolobosea (Sogin et al. 1989; Sogin 1991). The subsequent demonstration that genes encoding putative mitochondrial proteins such as chaperonin 60 (Cpn60) and pyridine nucleotide transhydrogenase (PNT) are present in the nuclear genome of *Entamoeba* strengthens the view that this organism might not be primitively amitochondrial. Seminal work by Clark and Roger (1995) showed that in phylogenetic analyses Cpn60 and its encoding gene clustered together with mitochondrial homologues, to the exclusion of bacterial sequences. Analysis of their conceptually translated products documented the presence of amino-terminal sequences rich in hydroxylated and basic amino acids reminiscent of mitochondrial targeting presequences in both Cpn60 and PNT, further suggesting their mitochondrial nature. In simultaneous but independent studies, purified recombinant EhCpn60 or a synthetic carboxyl-terminal peptide specific for EhCpn60 was used to generate specific antibodies against the amoebal Cpn60. Used in cell fractionation and immunolocalisation experiments, these antibodies were critical in the identification and subsequent characterisation of *Entamoeba* mitosomes (Mai et al. 1999; Tovar et al. 1999).

### 11.2.2 *Trachipleistophora hominis*

A member of the Microsporidia, *Trachipleistophora* is an obligate intracellular parasite reported to cause myositis in HIV-infected individuals (Field et al. 1996). Given their minute genome sizes and the apparently chimaeric nature of microsporidian ribosomal RNA genes, members of this group were thought to represent the descendants of primitive premitochondrial eukaryotes (Biderre et al. 1995; Vossbrinck et al. 1987; Vossbrinck and Woese 1986). Evidence to the contrary came with the identification and phylogenetic analyses of mitochondrial-type Hsp70 (mtHsp70) genes in a number of microsporidian groups and from the combined phylogenetic reconstructions of other microsporidian proteins (Baldauf et al. 2000; Hirt et al. 1999; Keeling and Fast 2002; Van De Peer et al. 2000). Together, these data provided strong support for the postmitochondrial evolution of this group and their affinity to fungi. Identification of a significant number of genes encoding putative mitochondrial proteins in the fully sequenced *Encephalitozoọn cuniculi* genome led to the hypothesis that mitosomes might exist in this parasite (Katinka et al. 2001; Keeling 2001). Direct physical evidence for the existence of microsporidian mitosomes came from cell fractionation and immunomicroscopy experiments where specific antibodies against *Trachipleistophora* mtHsp70 were used to identify intracellular compartments housing this mitochondrial protein (Williams et al. 2002).

### 11.2.3 *Giardia intestinalis*

*Giardia* is perhaps the best known member of the Diplomonads, frequently described as an ancient protist group whose primitive nature is suggested by the apparent lack of typical mitochondria and peroxisomes, the presence of a poorly developed endomembrane system and by their early branching in a number of gene phylogenies (Adam 2001; Hehl and Marti 2004; Sogin 1991). Perhaps the first suggestive evidence that *Giardia* might not be primitively amitochondrial came from immunomicroscopy experiments that employed a polyclonal antibody specific for the human mitochondrial Cpn60 (Soltys and Gupta 1994). Despite the punctuate distribution of fluorescent antigen observed in confocal microscopy images, no evidence of membrane association could be detected in electron microscopy micrographs, suggesting that, although aggregated, the cross-reactive antigen of *Giardia* was not compartmentalised. The subsequent identification and phylogenetic analyses of the *Giardia* gene encoding the mitochondrial Cpn60 as well as the mitochondrial proteins mtHsp70 and the iron–sulphur cluster assembly proteins IscS (cysteine desulphurase) and IscU provided strong evidence for the postmitochondrial evolutionary origin of *Giardia* (Arisue et al. 2002b; Morrison et al. 2001; Roger et al. 1998; Tachezy et al. 2001). The use of specific antibodies against *Giardia* IscS and IscU in cell fractionation and immunomicroscopy experiments led to the identification of mitosomes, thus providing a direct demonstration that *Giardia* too evolved from mitochondrial ancestors (Tovar et al. 2003).

### 11.2.4 *Cryptosporidium parvum*

*Cryptosporidium* is a member of the Apicomplexa, a group of parasitic protozoa with mitochondrion-bearing members (e.g. *Toxoplasma gondii*, *Plasmodium falciparum*) that cause devastating diseases in human populations around the world. Although lacking typical mitochondria (Coombs 1999; Tetley et al. 1998), its taxonomical association with *Plasmodium* and *Toxoplasma* suggested that the lack of the organelle was likely due to secondary loss rather than primitive absence. Initial microscopic evidence for the existence of a mitochondrial relict organelle (Riordan et al. 1999) was confirmed by the identification and phylogenetic analysis of the *C. parvum* *cpn60* gene and by specific labelling of the suspect intracellular compartment with specific antibodies against the mitochondrial marker protein Cpn60 (Putignani et al. 2004; Riordan et al. 2003). Because there is no evidence that this organelle could participate in ATP biosynthesis, it is considered to be a mitosome under the definition used in this chapter.

### 11.2.5 *Blastocystis hominis*

*Blastocystis hominis* is an intestinal parasite of humans associated with diarrhoeal disease and irritable bowel syndrome. A member of the Stramenopiles – presumed close relatives of the alveolates (e.g. *Cryptosporidium*) – its taxonomic classification has been a matter of debate and speculation (Arisue et al. 2002a; Nakamura et al. 1996; Silberman et al. 1996; Zierdt 1991, 1993). Despite being a strictly anaerobic organism, *B. hominis* contains recognisable mitochondria, but it is tentatively included here as a mitosome-containing organism on the basis of the lack of evidence for the involvement of its anaerobic mitochondria in energy metabolism (Nasirudeen and Tan 2004; Zierdt 1991).

## 11.3 Mitosome Biology

Mitosome research has benefited enormously from the information generated by genome-wide sequencing projects of parasitic protozoa, including those of *Giardia*, *Entamoeba*, *Cryptosporidium* and *Encephalitozoon*. Such information has been instrumental in the design of testable hypotheses and in the implementation of informed research strategies. Much of our current understanding of the biology of mitosomes stems from classical biochemical studies and from the use of specific antibodies against mitosomal proteins in cell fractionation and immunomicroscopy experiments.

### 11.3.1 Morphology

Like mitochondria and hydrogenosomes, mitosomes are morphologically heterogeneous but share the one defining characteristic of all endosymbiosis-derived organelles: they are surrounded by two closely apposed limiting membranes. *Trachipleistophora* and *Giardia* mitosomes are abundant organelles (average of 28 and 55 organelles per cell, respectively), and are the smallest mitochondrion-related organelles thus far reported. In immunoelectron microscopy micrographs they appear elongated in profile and measure an average of 50 mm × 90 nm and 60 mm × 140 nm, respectively (Tovar et al. 2003; Williams et al. 2002). No invaginations of the inner membrane (cristae) are present in these organelles. *Entamoeba* mitosomes appear in electron microscopy micrographs of enriched preparations as ovoid double-membrane-bounded organelles lacking cristae and measuring an estimated 0.5–1.0 μm in diameter (Ghosh et al. 2000). Size-restricted confocal imaging of fixed trophozoites has revealed the presence of over 150 mitosomes per cell and suggests that most of these organelles are smaller than half a micron in size (León-Avila and Tovar 2004). In contrast to the mitosomes of other organisms, the *Entamoeba* organelles have not been unequivocally identified by immunoelectron microscopy as a good mitosome-specific antibody suitable for this purpose has yet to be identified. In *Cryptosporidum*, a single oval mitosome estimated in different studies at around 200 and 500 nm in diameter has been identified by electron microscopy (Putignani et al. 2004; Riordan et al. 2003). This mitosome is invariably located between the crystalloid body and the nucleus and appears surrounded by the endoplasmic reticulum. Electron tomographic reconstruction has documented the existence of atrophic cristae-like membrane networks within the organelle but their functional significance is unclear as no oxidative phosphorylation occurs in this parasite (Keithly et al. 2005). The mitosomes of *Blastocystis* are variable in number, from just a few in rapidly dividing cells to well over 100 in older giant cells, and display unusually short goblet-shaped cristae that can also be filamentous and branched (Nasirudeen and Tan 2004; Zierdt 1991). Such patent diversity in mitosome morphology, akin to the morphological variability observed amongst mitochondria and amongst hydrogensomes of unrelated taxa, may result from niche-specific environmental pressures driving organelle evolution in diverse taxonomic groups.

### 11.3.2 Organelle Biochemistry and Protein Complement

The physiology of mitosomes appears to be diverse too. Sequence information from the genome projects of parasitic protozoa has revealed the extent of reductive evolution in the mitosomes of different taxa, an indication of how these parasites have streamlined their biology in their adaptation to a parasitic lifestyle. Their metabolic range appears to be severely restricted and, although the identity and function of a fair proportion of putative open reading frames found in their genomes remain unassigned, the general perception is that most of the biochemistry present in the original endosymbiont has been lost from mitosomes through reductive evolution. This characteristic makes them ideal models for the study of organelle evolution and for determining the minimal protein complement required for the viability of a self-replicating, endosymbiosis-derived organelle.

Most of the main physiological functions of mitochondria seem to be absent in mitosomes, including haem biosynthesis, the urea cycle, the oxidative degradation of fatty acids and the decarboxylation of pyruvate to acetyl-coenzyme A (acetyl-CoA). Also absent are the citric acid cycle-dependent generation of reducing equivalents and concomitant production of amino acid and nucleotide precursors and the generation of an electrochemical gradient by proton-pumping electron transport necessary for ATP biosynthesis. Current evidence indicates that in most mitosome-containing organisms the pyruvate:ferredoxin oxidoreductase (PFO) mediated decarboxylation of pyruvate to acetyl-CoA has been transferred to the cytosol, where ATP biosynthesis by substrate-level phosphorylation takes place. In addition, other mitochondrial functions such as calcium and metal homeostasis may have also been lost from mitosomes. So what then is left in these highly degenerate mitochondrion-related organelles?

### 11.3.3 Iron–Sulphur Cluster Biosynthesis

FeS proteins are important components of prokaryotic and eukaryotic cells that participate in electron transport, metabolic regulation and enzyme catalysis. The assembly of their FeS moieties has been reported as the only essential biosynthetic function of the mitochondrion (Lill and Kispal 2000). Recently, this biosynthetic function has also been demonstrated in mitosomes and hydrogensomes (Sutak et al. 2004; Tovar et al. 2003). Cysteine desulphurase (IscS/NifS) and the iron-binding scaffold protein IscU/NifU play central roles in the biosynthesis of FeS clusters. Through their concerted action and that of other proteins such as ferredoxin, organellar Hsp70 and frataxin, molecular iron and the sulphur atom of cysteine are brought together into assembled clusters that are either incorporated into organellar FeS proteins or exported from the mitochondrion via an ABC-type transporter for the maturation of extra-organellar FeS proteins, presumably in the cell cytosol. Nuclear genes encoding homologues of IscS/NifS and IscU/NifU have been identified in the genomes of *E. histolytica*, *G. intestinalis*, *E. cuniculi* and *C. parvum* (Ali et al. 2004; Katinka et al. 2001; LaGier et al. 2003; Tachezy et al. 2001; Tovar et al. 2003; van der Giezen et al. 2004). With one exception (see later), all mitosome-containing organisms harbour mitochondrial-type IscS/NifS and IscU/NifU proteins (Katinka et al. 2001; LaGier et al. 2003; Tachezy et al. 2001; Tovar et al. 2003). Other genes whose encoded products are known to participate in this process have also been identified, including ferredoxin and frataxin in *C. parvum* and ferredoxin, an SSQ-type Hsp70 chaperone and a homologue of the ABC-type ATM-1 transporter (FeS cluster export) in *E. cuniculi* (Abrahamsen et al. 2004; Katinka et al. 2001).

Interestingly, *E. histolytica* contains bacterial-type NifS and NifU proteins closely related to ε-proteobacterial homologues, suggesting that *Entamoeba* acquired these genes by horizontal gene transfer (Ali et al. 2004; van der Giezen et al. 2004). This event must have occurred prior to the diversification of the *Entamoeba* genus because other *Entamoeba* species (e.g., *E. invadens*, *E. dispar*, *E. moshkovski*) appear to have bacterial-type Nif genes too (van der Giezen and Tovar, unpublished observations). The absence of any additional mitochondrial-type FeS cluster (Isc) assembly genes in the fully sequenced *E. histolytica* genome suggests that the bacterial genes acquired by *Entamoeba* served as a direct replacement for those that must have been present in the original mitochondrial endosymbiont (Loftus et al. 2005). Although the cellular localisation of these *Entamoeba* proteins remains to be unequivocally established, it is reasonable to hypothesise that they too might function inside mitosomes. If this is proved not the case however, *Entamoeba* would represent the only characterised eukaryotic group without compartmentalised FeS cluster biosynthesis.

### 11.3.4 Molecular Chaperones

Most proteins that function inside the mitochondrion undergo posttranslational import into the organelle. This process employs mitochondrial-type Hsp70 to "pull" precursor proteins from the inner-membrane-bound translocation complexes into the organelle and Cpn60 to assist in their proper refolding (Matouschek et al. 2000; Sigler et al. 1998). As such, these ATP-dependent molecular chaperones are considered the quintessential molecular markers of mitochondria. Genes encoding the mtHsp70 chaperone have been identified in the genomes of *E. histolytica*, *G. intestinalis*, *T. hominis*, *E. cuniculi* and *C. parvum* (Abrahamsen et al. 2004; Arisue et al. 2002b; Bakatselou et al. 2000; Katinka et al. 2001; Morrison et al. 2001; Williams et al. 2002) and to date the mitosomal localisation of either native mtHsp70 or epitope-tagged versions of the protein has been demonstrated through the use of suitable specific antibodies in *T. hominis*, *G. intestinalis* and *C. parvum* (Regoes et al. 2005; Slapeta and Keithly 2004; Williams et al. 2002). Interestingly, while mitochondrial-type *cpn60* genes have been detected in *E. histolytica*, *G. intestinalis* and *C. parvum*, this gene appears to have been lost from the genome of the microsporidian *E. cuniculi* and possibly from that of *T. hominis* (Katinka et al. 2001; Williams et al. 2002). This observation suggests the existence of Cpn60-independent organelle protein folding in microsporidian mitosomes similar to that reported in Cpn60 knockout yeast mutants (Rospert et al. 1996). As with mtHsp70, the mitosomal localisation of either native Cpn60 or epitope-tagged versions of the protein has been recently demonstrated in *C. parvum*, *E. histolytica* and *G. intestinalis* through the use of specific homologous antibodies against the corresponding native proteins (León-Avila and Tovar 2004; Regoes et al. 2005; Riordan et al. 2003). The recent demonstration that *Giardia* Cpn60 does indeed reside in mitosomes (Regoes et al. 2005) resolved the controversy arising from the previous mislocalisation of *Giardia* Cpn60 in studies where heterologous antibodies against human or *Entamoeba* Cpn60 were used (Knight 2004; Soltys and Gupta 1994; Tovar et al. 2003). Further, it highlights the importance of using homologous antibodies when investigating the cellular localisation of proteins in highly derived parasites such as *Giardia*.

### 11.3.5 Adenine Nucleotide Transporters

The physiological functions of both mtHsp70 and Cpn60 require ATP hydrolysis. Since mitosomes appear to lack the capacity for compartmentalised energy generation, a suitable mechanism of ATP exchange with the cytosol must exist in mitosome-containing organisms to satisfy this need. Putative ADP/ATP transporter genes have been identified in the genomes of *E. histolytica*, *C. parvum* and *E. cuniculi* (Abrahamsen et al. 2004; Katinka et al. 2001; Loftus et al. 2005) but only the *Entamoeba* carrier has been characterised in some detail (Chan et al. 2005). Phylogenetic analysis and biochemical characterisation of this protein suggest a different origin from that of the classical mitochondrial-type nucleotide transporters, as reflected in its functional independence from a membrane potential, its different substrate specificity and its lack of susceptibility to carboxyatractyloside and bongkrekic acid – specific inhibitors of mitochondrial-type transporters. Collectively these carrier proteins are known to be a eukaryotic evolutionary invention not derived from the original mitochondrial endosymbiont.

### 11.3.6 Electron Transport

As indicated already, membrane-associated electron transport has not been demonstrated in mitosomes. Neither cytochromes nor haem proteins have been detected by chemical or spectroscopic analyses in anaerobic protozoa and no components of respiratory complexes I–IV have been identified in the genomes of *C. parvum*, *E. cuniculi*, *E. histolytica* and *G. intestinalis* (Abrahamsen et al. 2004; Katinka et al. 2001; Loftus et al. 2005; McArthur et al. 2000; Müller 2003). However, low levels of ubiquinone have been detected in *Giardia* and *Entamoeba* (Ellis et al. 1994). A limited level of membrane-associated electron transport activity in *C. parvum* is suggested by the presence of a gene encoding alternative oxidase (AO) in its genome. This enzyme transfers electrons to molecular oxygen and its catalytic activity is associated with thioredoxin reductase and glutathione peroxidase in plants and fungi (Abrahamsen et al. 2004; Putignani et al. 2004). Because these two enzymes are also present in the *C. parvum* genome, a previously unidentified oxygen-scavenging mechanism might operate in this organism; its functionality and the precise cellular localisation of AO remain to be determined experimentally.

The observed reduction of 5-cyano-2,3,-ditolyl tetrazolium chloride (CTC) and concomitant formazan precipitation in *Giardia* was recently taken as evidence for the presence of mitochondrial electron transport in this organism (Lloyd et al. 2002a). However, the specificity of tetrazolium salt reduction in *Giardia* is questionable as tetrazolium salt reducing systems have been demonstrated in the internal membranes and in the plasma membrane of mammalian cells (Bernas and Dobrucki 2000). Co-localisation experiments with organelle-specific molecular markers are required to ascertain the nature of CTC reduction in *Giardia*. Uptake of rhodamine 123, another fluorescent cation, has also been used in *Giardia* and *Blastocystis* as evidence for the presence of membrane potential in mitosomes (Lloyd et al. 2002a; Nasirudeen and Tan 2004; Zierdt et al. 1988). Staining of *Blastocystis* mitosomes is inhibited by the addition of sodium azide, which affects the membrane potential through inhibition of cytochrome *c* oxidase, an activity previously undetected in this organism (Nasirudeen and Tan 2004; Zierdt et al. 1988). As with CTC reduction, the specificity of rhodamine 123 staining in *Giardia* remains to be validated by co-localisation experiments using organelle-specific molecular markers.

### 11.3.7 Other Putative Organellar Functions

There is some evidence for the presence of other physiological functions traditionally associated with mitochondria in some mitosome-containing parasitic protozoa. Amongst others, genes encoding PNT – an inner-membrane mitochondrial protein – have been identified in *E. histolytica* and in *C. parvum* (Abrahamsen et al. 2004; Clark and Roger 1995; Loftus et al. 2005). While PNT catalytic activity has been demonstrated in the former, the cellular localisation of the protein and its putative involvement in proton translocation across the organellar membrane remains to be determined in both organisms (Weston et al. 2002). Subunits of pyruvate hydrogenase (PDH) and of F1 ATP synthase have also been identified in the genomes of *E. cuniculi* and *C. parvum* respectively (Abrahamsen et al. 2004; Katinka et al. 2001). Although it is possible that these proteins may contribute to the overall scheme of cellular energy metabolism in these organisms, the question of whether or not they localise to mitosomes remains a matter of speculation. Recently, metronidazole was shown to induce morphological and biochemical changes in *Blastocystis* similar to those observed during apoptotic cell death in mitochondrial eukaryotes, including nuclear condensation, DNA fragmentation and reduced cytosolic volume (Desagher and Martinou 2000). Because of the involvement of mitochondria in apoptosis and because metronidazole is activated through PFO in anaerobic protozoa, it is possible that PFO might be compartmentalised in the mitosomes of *B. hominis*. The putative role of these organelles in metronidazole-induced apoptotic-like cell death warrants further investigation.

### 11.3.8 Mitosome Biogenesis

As mentioned earlier, luminal mitochondrial and hydrogensomal proteins are imported into their respective organelle following their biosynthesis in cytosolic ribosomes. The demonstration that mitosomes harbour a collection of mitochondrial marker proteins would suggest the presence of equivalent organelle protein-import mechanisms in mitosome-containing parasitic protozoa. Investigating how mitosomal proteins are imported into these highly degenerate cellular organelles and identifying the molecular components involved in the process is a matter of considerable research activity in the field.

Of great interest too is the issue of how mitosomes divide and segregate in replicating cells. Although in its infancy, early research on this topic is providing intriguing clues as to the nature of organelle replication and the likely molecular components that aid the faithful segregation of these minimal mitochondrial organelles in parasitic protozoa. These topics are discussed at length in the next section.

## 11.4 Protein Import

Current evidence suggests that luminal mitosomal proteins can be imported into the organelle via two clearly distinct routes: one that requires the presence of amino-terminal targeting presequences equivalent to those present in mitochondrial and hydrogenosomal proteins and an alternative presequence-independent pathway that seems to rely on the recognition of internal "targeting" sequences (Bradley et al. 1997; Wiedemann et al. 2004). The experimental evidence for the existence of each of these pathways is discussed next.

### 11.4.1 Presequence-Dependent Import

The early observation that amino-terminal sequences rich in hydroxylated and basic amino acids reminiscent of mitochondrial targeting signals were present in the conceptually translated products of the *E. histolytica* genes encoding mitochondrial-marker proteins Cpn60 and PNT provided the first indication that a presequence-dependent protein import pathway might operate in the then-hypothetical mitosomes (Clark and Roger 1995). Since then, putative amino-terminal mitosome targeting presequences have been identified in a range of mitosomal proteins from *E. histolytica*, *E. cuniculi*, *C. parvum* and *G. intestinalis* (Bakatselou et al. 2000; Katinka et al. 2001; Nixon et al. 2002; Riordan et al. 2003; Slapeta and Keithly 2004; Tachezy et al. 2001; Tovar et al. 2003). Detection of putative mitosome targeting presequences has relied mostly on bioinformatic algorithms trained to recognise mitochondrial targeting signals from model eukaryotes such as yeasts, plants and mammals. Their identification is therefore tentative and much dependent on the stringency used for each search. Given the extent of reductive evolution observed in mitosome-containing eukaryotes, it is much a matter of luck as of individual judgement whether or not a particular amino acid sequence turns out to be functional. As such, the functionality of each putative targeting presequence must be tested individually.

In the absence of an in vitro mitosome protein-import system, all efforts to test mitosome targeting presequence functionality have centred on suitable deletion experiments and the engineering of chimaeric fusion proteins to target reporter proteins into mitosomes in vivo. Further, the cross-functionality of mitosomal, hydrogenosomal and mitochondrial targeting signals has also been tested as a means of obtaining functional evidence in support of the evolutionary relatedness of these organelles. The first evidence for the functional requirement of a mitosome targeting presequence and the functional conservation of mitosomal and mitochondrial protein import came from experiments in *E. histolytica* where deletion of amino acids 2–15 from Cpn60 led to the accumulation of recombinant protein in the cytosol, a mutant phenotype that could be reversed by supplying the mutated protein with a mitochondrial targeting signal from *Trypansoma cruzi* mtHsp70 (Tovar et al. 1999). Since then, the putative organelle targeting signals from *Cryptosporidium* Cpn60, IscS, IscU and mtHsp70 as well as those from *Giardia* IscU and ferredoxin fused to green fluorescent protein (GFP) have all been shown to target the reporter protein into mitosomes and/or into yeast or mammalian mitochondria (LaGier et al. 2003; Regoes et al. 2005; Riordan et al. 2003; Slapeta and Keithly 2004). Moreover, recombinant *Giardia* ferredoxin and IscU proteins tagged with the viral haemaglutinin epitope or with GFP have also been shown to be imported in a presequence-dependent fashion into mitosomes, trichomonad hydrogenosomes and mammalian mitochondria using immunofluorescence microscopy and cell fractionation experiments (Dolezal et al. 2005; Regoes et al. 2005). That organelle targeting is abrogated by deletion of these amino-terminal targeting signals has demonstrated their absolute requirement for protein import into mitosomes.

A defining characteristic of mitochondrial and hydrogensomal protein import is the proteolytic removal of amino-terminal targeting presequences upon organelle import. In *Giardia*, two IscU bands have been observed by western blotting in the organelle fraction that differ in size by about 2 kDa, a minor band of 17 kDa (precursor protein) and a major band of around 15 kDa (mature protein), consistent with the removal of the IscU amino-terminal presequence. The cleavage of IscU and ferredoxin targeting signals has also been observed in recombinant parasites that overexpress GFP fusion proteins containing the corresponding targeting presequences at the amino terminus. In these cells, precursor GFP proteins accumulate in the cytosol, whereas mature proteins are found exclusively in the organellar fraction (Regoes et al. 2005). Such accumulation of precursor protein in the cell cytosol demonstrated that mitosome import, like mitochondrial and hydrogensomal import, is an active and saturable process. Further evidence for the proteolytic processing of giardial IscU presequence comes from in vitro experiments in which radiolabelled IscU was incubated with mitochondrial and hydrogensomal lysates (Dolezal et al. 2005). In these experiments, the targeting presequence is cleaved off in a time-dependent manner and is blocked by addition of EDTA, an inhibitor of metaloproteases. Taken together, these data provide substantial evidence for the existence of a presequence-dependent protein-import pathway in mitosomes which is functionally conserved with those of mitochondria and hydrogensomes.

### 11.4.2 Presequence-Independent Import

A number of luminal mitochondrial proteins are imported into the organelle in the absence of recognisable amino-terminal targeting peptides, including Cpn10, rhodanese and 3-oxoacyl-CoA thiolase (Arakawa et al. 1990; Jarvis et al. 1995; Miller et al. 1991). These proteins contain poorly characterised internal targeting signals that are not removed upon translocation across the organelle's membranes. Initial evidence that a presequence-independent protein-import pathway operates in mitosomes came from the demonstration that *Trachipleistophora* mtHsp70 is imported into mitosomes in the absence of a recognisable amino-terminal peptide (Williams et al. 2002). *Giardia* Cpn60, mtHsp70 and IscS all lack a detectable targeting presequence yet they are all efficiently imported into mitosomes (Arisue et al. 2002b; Morrison et al. 2001; Regoes et al. 2005; Roger et al. 1998; Tachezy et al. 2001; Tovar et al. 2003). Deletion experiments in which the initial five amino acids of Cpn60 were removed from the protein did not affect its mitosomal localisation, confirming that, unlike its mitochondrial and hydrogenosomal homologues, the amino-terminal portion of *Giardia* Cpn60 plays no role in its cellular distribution (Regoes et al. 2005). Further, deletion of either the amino-terminal or the carboxy-terminal halves of *Giardia* IscS had no effect on the in vivo mitosome import competence of the protein, suggesting that two or more internal sequences are important for organelle import and that these are likely distributed throughout the length of the protein (Dolezal et al. 2005). Together, these data demonstrate that a presequence-independent protein-import pathway is fully operational in mitosomes.

The observed natural predisposition of some bacterial proteins for mitochondrial targeting suggests an ancestral presequence-independent protein-import pathway present in early mitochondria, less sophisticated than the presequence-dependent pathway predominantly operational in present-day mitochondria (Lucattini et al. 2004; Wiedemann et al. 2004). The observation that while mitochondrial and hydrogenosomal Cpn60, mtHsp70 and IscS are all imported into their respective organelles in a presequence-dependent manner their corresponding mitosomal homologues from *Giardia* (Cpn60, mtHsp70 and IscS) and from *Trachipleistiphora* (mtHsp70) are all imported into mitosomes in a presequence-independent manner suggests that reductive evolution is favouring a regression to the ancestral protein-import pathway. This hypothesis is also supported by the observation that only two (IscU and ferredoxin) of the five known luminal mitosomal proteins of *Giardia* have retained functional mitosome targeting presequences.

### 11.4.3 Protein Translocases

Given the active and saturable nature of protein import into mitosomes and the observed functional conservation with mitochondrial protein import, the presence in mitosomes of at least degenerate versions of the highly sophisticated protein translocation complexes that operate in mitochondria should almost be a foregone conclusion. However, extensive database searches of all sequenced parasitic protozoal genomes using as query sequences subunits of the translocation complexes of the outer and inner membranes, TOM and TIM respectively, from a wide range of mitochondrial eukaryotes has produced little precious information. For example, of the major subunits of the TOM complex only putative homologues of TOM70 and TOM40 have been identified in some mitosome-containing organisms; TOM70 in *E. cuniculi* and possibly in *G. intestinalis* (Katinka et al. 2001; Regoes et al. 2005) and TOM40 in *C. parvum* and possibly in *E. histolytica* (Abrahamsen et al. 2004; Putignani et al. 2004; van der Giezen et al. 2005). A similar situation exists for the translocators of the inner membrane. While putative TIM44 and TIM17 homologues have been identified in *C. parvum* (Abrahamsen et al. 2004; Putignani et al. 2004), only TIM22 appears to be present in *E. cuniculi* (Katinka et al. 2001). These data paint the picture of a highly degenerate but functional mitosome protein-import system based on remnant proteins – some just recognisable, others beyond recognition by current algorithms. Alternatively, other structurally simpler membrane-bound translocators similar to those of trichomonad hydrogenosomes may operate in mitosomes (Dyall et al. 2003). The use of alternative protein identification strategies including organelle proteomics and mass spectrometry will enhance our understanding of the nature of mitosomal translocases.

### 11.4.4 Organelle Division and Inheritance

Almost nothing is known about mitosome division in parasitic protozoa. Gene homologues encoding putative dynamin-related mechanoenzymes similar to those that participate in mitochondrial fission and fusion have been identified in the genomes of *E. histolytica*, *G. intestinalis*, *C. parvum* and *E. cuniculi*, but their putative involvement in organelle division has not yet been tested (van der Giezen et al. 2005). The initial clues as to how *Giardia* mitosomes divide and segregate into daughter cells are just beginning to emerge. In confocal microscopy images, *Giardia* mitosomes appear as two morphologically distinct structures, a characteristic rod-shaped organelle localised between the nuclei and closely associated with axonemal basal bodies and a collection of spherical organelles distributed throughout the cytosol (Tovar et al. 2003). Time-lapsed confocal microscopy investigations have recently shown that the single central mitosome divides during mitosis and segregates into daughter cells in a cell-cycle-dependent manner, while peripheral mitosomes are maintained at relatively constant numbers and segregate into daughter cells stochastically (Regoes et al. 2005). Although the possibility that peripheral mitosomes may replicate independently cannot be excluded at present, it is possible that peripheral mitosomes might emanate from the central mitosome either by fission or septation. In the same study, specific inhibitors of the cell cytoskeleton were used to demonstrate that positioning and division of the central mitosome is dependent on the integrity of microtubules. The association of the unique central mitosome with the basal bodies of flagellar axonemes during *Giardia* division is reminiscent of the association of a dividing single mitochondrion of *Trypanosoma brucei* with the basal bodies (Ogbadoyi et al. 2003).

## 11.5 Evolutionary Considerations

Mitosomes of parasitic protozoa were discovered at a time when molecular studies had just provided clear supporting evidence for the common evolutionary origins of mitochondria and hydrogenosomes (Bui et al. 1996; Germot et al. 1996; Horner et al. 1996; Roger et al. 1996). The idea that hydrogensomes could have originated from mitochondria was a simple and appealing hypothesis to explain not only the mutually exclusive distribution of mitochondria and hydrogensomes but also the significant observation that – in contrast to the monophyletic origin of mitochondria – hydrogenosomes have arisen several times independently in diverse eukaryotic groups, including some with close mitochondrial relatives (Cavalier-Smith 1987; Embley et al. 1995, 1997; Finlay and Fenchel 1989). Despite its obvious simplicity and appeal, the "vertical descent" hypothesis could not explain at the time how the anaerobic biochemistry of hydrogenosomes might be derived from the aerobic biochemistry of typical mitochondria. The realisation that diverse anaerobic versions of mitochondria with chimaeric aerobic/anaerobic biochemistry exist in eukaryotic lineages across the evolutionary spectrum has helped bridge this gap in understanding (Tielens et al. 2002; see also Chap. 5 by Tielens and van Hellemond in this volume). Although the evolutionary routes that lead from mitochondria to hydrogenosomes remain to be fully mapped, the recent sequencing and phylogenetic analysis of the remanant hydrogenosomal genome of the anaerobic ciliate *N. ovalis* has provided unequivocal genetic evidence that *Nyctotherus*'s hydrogensomes are indeed modified mitochondria (Boxma et al. 2005; Gray 2005). It is therefore no longer a matter of whether or not such evolutionary conversion is possible but a matter of mapping out the precise routes by which a mitochondrion can evolve into a hydrogenosome. In this context, what is the evolutionary significance of mitosomes?

Anaerobic energy metabolism in mitosome-containing protozoa displays remarkable similarity with the energy metabolism of *Trichomonas vaginalis*, a hydrogenosome-containing parasitic protist. However, while energy metabolism is compartmentalised in trichomonad hydrogenosomes, current evidence suggests that key enzymes that participate in this process are not compartmentalised in mitosomes. Nevertheless, given the similar biogenesis, morphology and biochemistry of mitosomes and hydrogenosomes as well as their mutually exclusive distribution, the idea that mitosomes could represent modified hydrogenosomes and, by implication, degenerate mitochondria is appealing (Martin and Müller 1998; Tielens et al. 2002; van der Giezen et al. 2005; van der Giezen and Tovar 2004, 2005). To explore this idea, the nature and cellular distribution of enzymes that participate in key areas of energy metabolism (i.e. pyruvate decarboxylation, ATP biosynthesis and hydrogen production) in mitosome-containing and hydrogenosome-containing parasitic protozoa are considered.

1. Pyruvate decarboxylation to acetyl-CoA in *Trichomonas*, *Giardia* and *Entamoeba* is carried out by PFO. While this enzyme is compartmentalised in trichomonad hydrogenosomes, purification of PFO from the soluble fraction of *E. histolytica* extracts suggests its cytosolic nature in this organism – although a recent immunomicroscopy study suggested association with the plasma membrane (Reeves et al. 1977; Rodríguez et al. 1998). In *Giardia*, the majority of PFO activity was detected and purified from the sedimentable fraction and, on this basis, was hypothesised to be associated with the plasma membrane (Brown et al. 1998; Townson et al. 1996); however, in the light of new knowledge, the possibility that *Giardia* PFO may actually be associated with mitosomes cannot be excluded (Tovar et al. 2003). In *Cryptosporidium*, pyruvate decarboxylation is catalysed by the enzyme pyruvate:NADPH$^+$ oxidoreductase (PNO), a chimaeric protein containing fused PFO and NADPH$^+$ cytochrome P450 reductase domains similar to that present in the anaerobic mitochondria of *Euglena*

(Rotte et al. 2001). The cellular localisation of PNO in *Cryptosporidium* has not been reported. Phylogenetic analyses of PFO and PNO suggest they all share a common eubacterial origin, although at present the possibility that these enzymes may have derived from the original mitochondrial endosymbiont is not supported (Embley et al. 2003; Horner et al. 1999; Rotte et al. 2001).

2. Hydrogenosomes convert acetyl-CoA into acetate with the concomitant generation of ATP by substrate-level phosphorylation. This is achieved through the concerted action of succinyl-CoA synthase and acetate:succinate CoA transferase (ASCT), a catalytic system also operational in anaerobic and in some aerobic mitochondria (Müller 2003; Tielens et al. 2002). Neither of these enzymes is present in *Giardia* or *Entamoeba*. Instead, this energy-generating step is catalysed by acetate thiokinase (AT), a non-sedimentable enzymatic activity of likely cytosolic localisation (Lindmark 1980; Reeves et al. 1977). In *Cryptosporidium* neither ASCT nor AT activities have been detected and genes encoding these enzymes seem to be absent from the *C. parvum* genome (Abrahamsen et al. 2004). It is therefore unlikely that acetyl-CoA catabolism plays a role in the generation of biological energy in this organism. Together with the cytosolic nature of *Giardia* and *Entamoeba* AT these findings suggest that – even if PFO and PNT were found associated with mitosomes of *Cryptosporidium*, *Giardia* and *Entamoeba* – energy metabolism in these parasites is not compartmentalised.
3. A defining feature of hydrogenosomes is the evolution of molecular hydrogen resulting from the transfer of electrons from ferredoxin to $H^+$, a reaction catalysed by Fe-hydrogenase. The discovery of Fe-hydrogenase genes in the genomes of *Giardia* and *Entamoeba* prompted a reassessment of traditional studies which found no unequivocal evidence for the generation of molecular hydrogen in these organisms (Müller 1988; Reeves 1984). No hydrogenase activity has yet been demonstrated in these organisms but recently the evolution of a small amount of hydrogen gas was detected in axenically grown *Giardia* (Lloyd et al. 2002b); however, at only a fraction of the hydrogen produced by trichomonad hydrogenosomes, the amount is much too small to suggest the involvement of a fully functional hydrogenase enzyme. Although the cellular localisation of Fe hydrogenase has not been directly demonstrated in *Giardia* or in *Entamoeba*, epitope-tagged versions of the protein in both organisms display cytosolic localisation (Ghosh et al. 2000; Zourmpanou and Tovar, unpublished data). Phylogenetic analysis of Fe hydrogenases from protists and fungi and hydrogenase-like proteins found in other eukaryotes is problematic and at present does not satisfactorily resolve their evolutionary origins (Horner et al. 2000, 2002). However, their structure and phylogenetic affiliations suggest that eukaryotic Fe hydrogenases are not monophyletic in origin and that they have been subjected to frequent rearrangements and modifications during the course of evolution (Embley et al. 2003).

All of these observations suggest that mitosomes and hydrogenosomes have a common origin but that they have not followed identical evolutionary paths. Their biochemical diversity is likely the result of lineage-specific loss and/or relocation of enzymatic components and the recruitment of host proteins that in turn may have been acquired by lateral transfer from other organisms (Andersson et al. 2003a; Doolittle et al. 2003; Karlberg et al. 2000; Tielens et al. 2002). Mitosomes are without a doubt the most derived of mitochondrial organelles and it seems unlikely that the reductive evolutionary tendencies will stop there. Our current failure to identify a single bona fide amitochondrial eukaryote suggests either insufficient sampling, insufficient evolutionary time for complete loss to have occurred or that total loss of the mitochondrial organelle is a lethal event.

The bacterial ancestry of mitochondria, although suspected, was only demonstrated unequivocally by the sequencing and phylogenetic analyses of remnant mitochondrial genomes from a range of eukaryotes (Gray et al. 1998, 1999). Another remnant organellar genome, this time from a hydrogenosome-containing ciliate, has now been used to demonstrate beyond reasonable doubt the previously suspected evolutionary connection between mitochondria and hydrogenosomes (Boxma et al. 2005). Although no remnant mitosomal genomes have been identified in *Entamoeba*, *Giardia*, *Encephalitozoon* and *Cryptosporidium*, organellar DNA has recently been demonstrated in *B. hominis* (Abrahamsen et al. 2004; Katinka et al. 2001; León-Avila and Tovar 2004; Nasirudeen and Tan 2004). If these organelles prove to be true mitosomes, their remnant genome may be useful in mapping the evolutionary origins of these enigmatic organelles.

## 11.6 Conclusions

Much remains to be investigated about the physiology of mitosomes in diverse eukaryotic groups and about their distribution across the taxonomical spectrum. However, their apparent biochemical simplicity begs the question as to why mitosomes have been retained in parasitic protozoa. Current evidence suggests that FeS cluster metabolism may be the only common physiological function shared by these organelles (although, with the exception of *Giardia*, the mitosomal localisation of the proteins involved in this process remains to be directly demonstrated). Protecting these oxygen-labile enzymes from molecular oxygen present in the cytoplasm would be a good reason to keep them compartmentalised in mitosomes, but the same criteria should apply to other oxygen-sensitive enzymes such as PFO, which, at least in *Entamoeba*, appears to have been relocated to the cytoplasm. Whether or not protection of oxygen-sensitive enzymes/processes turns out to be the only reason for mitosome retention, the biological simplicity of mitosomes makes them ideal biological models for the study of the minimal molecular composition required for the replication and inheritance of an endosymbiosis-derived organelle.

Mitosome research will continue to be an important part of a wider effort aimed at understanding the evolutionary origins of the eukaryotic cell and indeed the evolutionary history of its endosymbiosis-derived organelles. Mitosomes, like mitochondria and hydrogenosomes, are heterogeneous in function and appearance but it is clear that they all share a common origin with mitochondria and hydrogenosomes. Mapping the precise routes by which the original mitochondrial endosymbiont evolved into aerobic and anaerobic mitochondria, hydrogenosomes and mitosomes is a matter of great interest and of considerable research activity. The remarkable similarities displayed by mitosomes, hydrogenosomes and mitochondria in terms of their morphology, biogenesis and biochemistry strongly support the view that mitosomes, like hydrogenosomes, represent vestigial anaerobic mitochondria.

*Acknowledgements:* I thank Graham Clark for critically reading the manuscript. Research on this topic in my laboratory is funded by the Wellcome Trust (059845) and by the BBSRC (111/C13820 and BB/C507145).

## References

Abrahamsen MS, Templeton TJ, Enomoto S, Abrahante JE, Zhu G, Lancto CA, Deng M, Liu C, Widmer G, Tzipori S, Buck GA, Xu P, Bankier AT, Dear PH, Konfortov BA, Spriggs HF, Iyer L, Anantharaman V, Aravind L, Kapur V (2004) Complete genome sequence of the apicomplexan, *Cryptosporidium parvum.* Science 304:441–445

Adam RD (2001) Biology of *Giardia lamblia.* Clin Microbiol Rev 14:447–475

Ali V, Shigeta Y, Tokumoto U, Takahashi Y, Nozaki T (2004) An intestinal parasitic protist, *Entamoeba histolytica,* possesses a non-redundant nitrogen fixation-like system for iron-sulfur cluster assembly under anaerobic conditions. J Biol Chem 279:16863–16874

Andersson JO, Sjogren AM, Davis LA, Embley TM, Roger AJ (2003a) Phylogenetic analyses of diplomonad genes reveal frequent lateral gene transfers affecting eukaryotes. Curr Biol 13:94–104

Andersson SG, Karlberg O, Canback B, Kurland CG (2003b) On the origin of mitochondria: a genomics perspective. Philos Trans R Soc Lond B Biol Sci 358:165–179

Arakawa H, Amaya Y, Mori M (1990) The NH2-terminal 14-16 amino acids of mitochondrial and bacterial thiolases can direct mature ornithine carbamoyltransferase into mitochondria. J Biochem (Tokyo) 107:160–164

Arisue N, Hashimoto T, Yoshikawa H, Nakamura Y, Nakamura G, Nakamura F, Yano TA, Hasegawa M (2002a) Phylogenetic position of *Blastocystis hominis* and of stramenopiles inferred from multiple molecular sequence data. J Eukaryot Microbiol 49:42–53

Arisue N, Sánchez LB, Weiss LM, Müller M, Hashimoto T (2002b) Mitochondrial-type hsp70 genes of the amitochondriate protists *Giardia intestinalis, Entamoeba histolytica* and two microsporidians. Parasitol Int 51:9–16

Bakatselou C, Kidgell C, Clark CG (2000) A mitochondrial-type hsp70 gene of *Entamoeba histolytica.* Mol Biochem Parasitol 110:177–182

Baldauf SL, Roger AJ, Wenk-Siefert I, Doolittle WF (2000) A kingdom-level phylogeny of eukaryotes based on combined protein data. Science 290:972–977

Bernas T, Dobrucki JW (2000) The role of plasma membrane in bioreduction of two tetrazolium salts, MTT, and CTC. Arch Biochem Biophys 380:108–116

Biderre C, Pages M, Metenier G, Canning EU, Vivares CP (1995) Evidence for the smallest nuclear genome (2.9 Mb) in the microsporidium *Encephalitozoon cuniculi*. Mol Biochem Parasitol 74:229–231

Boxma B, de Graaf RM, van der Staay GW, van Alen TA, Ricard G, Gabaldon T, van Hoek AH, Moon-van der Staay SY, Koopman WJ, van Hellemond JJ, Tielens AG, Friedrich T, Veenhuis M, Huynen MA, Hackstein JH (2005) An anaerobic mitochondrion that produces hydrogen. Nature 434:74–79

Bradley PJ, Lahti CJ, Plümper E, Johnson PJ (1997) Targeting and translocation of proteins into the hydrogenosome of the protist *Trichomonas*: similarities with mitochondrial protein import. EMBO J 16:3484–3493

Brown DM, Upcroft JA, Edwards MR, Upcroft P (1998) Anaerobic bacterial metabolism in the ancient eukaryote *Giardia duodenalis*. Int J Parasitol 28:149–164

Bui ETN, Bradley PJ, Johnson PJ (1996) A common evolutionary origin for mitochondria and hydrogenosomes. Proc Natl Acad Sci USA 93:9651–9656

Cavalier-Smith T (1983) A 6-kingdom classification and a unified phylogeny. In: Schwemmler W, Schenk HEA (eds) Endocytobiology. de Gruyter, Berlin, pp 1027–1034

Cavalier-Smith T (1987) Eukaryotes with no mitochondria. Nature 326:332–333

Cavalier-Smith T (1991) Archamoebae: the ancestral eukaryotes? Biosystems 25:25–38

Cavalier-Smith T (2002) The phagotrophic origin of eukaryotes and phylogenetic classification of Protozoa. Int J Syst Evol Microbiol 52:297–354

Chan KW, Slotboom DJ, Cox S, Embley TM, Fabre O, van der Giezen M, Harding M, Horner DS, Kunji ER, Leon-Avila G, Tovar J (2005) A novel ADP/ATP transporter in the mitosome of the microaerophilic human parasite *Entamoeba histolytica*. Curr Biol 15:737–742

Clark CG, Roger AJ (1995) Direct evidence for secondary loss of mitochondria in *Entamoeba histolytica*. Proc Natl Acad Sci USA 92:6518–6521

Coombs GH (1999) Biochemical peculiarities and drug targets in *Cryptosporidium parvum*: lessons from other coccidian parasites. Parasitol Today 15:333–338

Desagher S, Martinou JC (2000) Mitochondria as the central control point of apoptosis. Trends Cell Biol 10:369–377

Dolezal P, Smid O, Rada P, Zubacova Z, Bursac D, Sutak R, Nebesarova J, Lithgow T, Tachezy J (2005) *Giardia* mitosomes and trichomonad hydrogenosomes share a common mode of protein targeting. Proc Natl Acad Sci USA 102:10924–10929

Doolittle WF, Boucher Y, Nesbo CL, Douady CJ, Andersson JO, Roger AJ (2003) How big is the iceberg of which organellar genes in nuclear genomes are but the tip? Philos Trans R Soc Lond B Biol Sci 358:39–57

Dyall SD, Lester DC, Schneider RE, Delgadillo-Correa MG, Plumper E, Martinez A, Koehler CM, Johnson PJ (2003) *Trichomonas vaginalis* Hmp35, a putative pore-forming hydrogenosomal membrane protein, can form a complex in yeast mitochondria. J Biol Chem 278:30548–30561

Ellis JE, Setchell KD, Kaneshiro ES (1994) Detection of ubiquinone in parasitic and free-living protozoa, including species devoid of mitochondria. Mol Biochem Parasitol 65:213–224

Embley TM, Finlay BJ, Dyal PL, Hirt RP, Wilkinson M, Williams AG (1995) Multiple origins of anaerobic ciliates with hydrogenosomes within the radiation of aerobic ciliates. Proc R Soc Lond B Biol Sci 262:87–93

Embley TM, Horner DS, Hirt RP (1997) Anaerobic eukaryote evolution: hydrogenosomes as biochemically modified mitochondria? Trends Ecol Evol 12:437–441

Embley TM, van der Giezen M, Horner DS, Dyal PL, Foster P (2003) Mitochondria and hydrogenosomes are two forms of the same fundamental organelle. Philos Trans R Soc Lond B Biol Sci 358:191–203

Field AS, Marriott DJ, Milliken ST, Brew BJ, Canning EU, Kench JG, Darveniza P, Harkness JL (1996) Myositis associated with a newly described microsporidian, *Trachipleistophora hominis*, in a patient with AIDS. J Clin Microbiol 34:2803–2811

Finlay BJ, Fenchel T (1989) Hydrogenosomes in some anaerobic protozoa resemble mitochondria. FEMS Microbiol Lett 65:311–314

Germot A, Philippe H, Le Guyader H (1996) Presence of a mitochondrial-type 70-kDa heat shock protein in *Trichomonas vaginalis* suggests a very early mitochondrial endosymbiosis in eukaryotes. Proc Natl Acad Sci USA 93:14614–14617

Ghosh S, Field J, Rogers R, Hickman M, Samuelson J (2000) The *Entamoeba histolytica* mitochondrion-derived organelle (crypton) contains double-stranded DNA and appears to be bound by a double membrane. Infect Immun 68:4319–4322

Gray MW (2005) Evolutionary biology: the hydrogenosome's murky past. Nature 434:29–31

Gray MW, Burger G, Lang BF (1999) Mitochondrial evolution. Science 283:1476–1481

Gray MW, Lang BF, Cedergren R, Golding GB, Lemieux C, Sankoff D, Turmel M, Brossard N, Delage E, Littlejohn TG, Plante I, Rioux P, Saint-Louis D, Zhu Y, Burger G (1998) Genome structure and gene content in protist mitochondrial DNAs. Nucleic Acids Res 26:865–878

Hehl AB, Marti M (2004) Secretory protein trafficking in *Giardia intestinalis*. Mol Microbiol 53:19–28

Hirt RP, Logsdon JM Jr, Healy B, Dorey MW, Doolittle WF, Embley TM (1999) Microsporidia are related to Fungi: evidence from the largest subunit of RNA polymerase II and other proteins. Proc Natl Acad Sci USA 96:580–585

Horner DS, Hirt RP, Kilvington S, Lloyd D, Embley TM (1996) Molecular data suggest an early acquisition of the mitochondrion endosymbiont. Proc R Soc Lond B Biol Sci 263:1053–1059

Horner DS, Hirt RP, Embley TM (1999) A single eubacterial origin of eukaryotic pyruvate: ferredoxin oxidoreductase genes: implications for the evolution of anaerobic eukaryotes. Mol Biol Evol 16:1280–1291

Horner DS, Foster PG, Embley TM (2000) Iron hydrogenases and the evolution of anaerobic eukaryotes. Mol Biol Evol 17:1695–1709

Horner DS, Heil B, Happe T, Embley TM (2002) Iron hydrogenases – ancient enzymes in modern eukaryotes. Trends Biochem Sci 27:148–153

Jarvis JA, Ryan MT, Hoogenraad NJ, Craik DJ, Hoj PB (1995) Solution structure of the acetylated and noncleavable mitochondrial targeting signal of rat chaperonin 10. J Biol Chem 270:1323–1331

Karlberg O, Canback B, Kurland CG, Andersson SG (2000) The dual origin of the yeast mitochondrial proteome. Yeast 17:170–187

Katinka MD, Duprat S, Cornillot E, Metenier G, Thomarat F, Prensier G, Barbe V, Peyretaillade E, Brottier P, Wincker P, Delbac F, El Alaoui H, Peyret P, Saurin W, Gouy M, Weissenbach J, Vivares CP (2001) Genome sequence and gene compaction of the eukaryote parasite *Encephalitozoon cuniculi*. Nature 414:450–453

Keeling PJ (2001) Parasites go the full monty. Nature 414:401–402

Keeling PJ, Fast NM (2002) Microsporidia: biology and evolution of highly reduced intracellular parasites. Annu Rev Microbiol 56:93–116

Keithly JS, Langreth SG, Buttle KF, Mannella CA (2005) Electron tomographic and ultrastructural analysis of the *Cryptosporidium parvum* relict mitochondrion, its associated membranes and organelles. J Eukaryot Microbiol 52:132–140

Knight J (2004) *Giardia*: not so special, after all? Nature 429:236–237

LaGier MJ, Tachezy J, Stejskal F, Kutisova K, Keithly JS (2003) Mitochondrial-type iron-sulfur cluster biosynthesis genes (IscS and IscU) in the apicomplexan *Cryptosporidium parvum*. Microbiology 149:3519–3530

León-Avila G, Tovar J (2004) Mitosomes of *Entamoeba histolytica* are abundant mitochondrion-related remnant organelles that lack a detectable organellar genome. Microbiology 150:1245–1250

Lill R, Kispal G (2000) Maturation of cellular Fe-S proteins: an essential function of mitochondria. Trends Biochem Sci 25:352–356

Lindmark DG (1980) Energy metabolism of the anaerobic protozoon *Giardia lamblia*. Mol Biochem Parasitol 1:1–12

Lloyd D, Harris JC, Maroulis S, Wadley R, Ralphs JR, Hann AC, Turner MP, Edwards MR (2002a) The 'primitive' microaerophile *Giardia intestinalis* (syn. *lamblia, duodenalis*) has specialized membranes with electron transport and membrane-potential-generating functions. Microbiology 148:1349–1354

Lloyd D, Ralphs JR, Harris JC (2002b) *Giardia intestinalis*, a eukaryote without hydrogenosomes, produces hydrogen. Microbiology 148:727–733

Loftus B, Anderson I, Davies R, Alsmark UC, Samuelson J, Amedeo P, Roncaglia P, Berriman M, Hirt RP, Mann BJ, Nozaki T, Suh B, Pop M, Duchene M, Ackers J, Tannich E, Leippe M, Hofer M, Bruchhaus I, Willhoeft U, Bhattacharya A, Chillingworth T, Churcher C, Hanc, Z, Harris B, Harris D, Jagels K, Moule S, Mungall K, Ormond D, Squares R, Whitehead S, Quail MA, Rabbinowitsch E, Norbertczak H, Price C, Wang Z, Guillen N, Gilchrist C, Stroup SE, Bhattacharya S, Lohia A, Foster PG, Sicheritz-Ponten T, Weber C, Singh U, Mukherjee C, El Sayed NM, Petri WA Jr, Clark CG, Embley TM, Barrell B, Fraser CM, Hall N (2005) The genome of the protist parasite *Entamoeba histolytica*. Nature 433:865–868

Lucattini R, Likic VA, Lithgow T (2004) Bacterial proteins predisposed for targeting to mitochondria. Mol Biol Evol 21:652–658

Mai Z, Ghosh S, Frisardi M, Rosenthal B, Rogers R, Samuelson J (1999) Hsp60 is targeted to a cryptic mitochondrion-derived organelle ("crypton") in the microaerophilic protozoan parasite *Entamoeba histolytica*. Mol Cell Biol 19:2198–2205

Martin W, Müller M (1998) The hydrogen hypothesis for the first eukaryote. Nature 392:37–41

Martin W, Hoffmeister M, Rotte C, Henze K (2001) An overview of endosymbiotic models for the origins of eukaryotes, their ATP-producing organelles (mitochondria and hydrogenosomes), and their heterotrophic lifestyle. Biol Chem 382:1521–1539

Matouschek A, Pfanner N, Voos W (2000) Protein unfolding by mitochondria. The Hsp70 import motor. EMBO Rep 1:404–410

McArthur AG, Morrison HG, Nixon JE, Passamaneck NQ, Kim U, Hinkle G, Crocker MK, Holder ME, Farr R, Reich CI, Olsen GE, Aley SB, Adam RD, Gillin FD, Sogin ML (2000) The *Giardia* genome project database. FEMS Microbiol Lett 189:271–273

Meza I (1992) *Entamoeba histolytica*: phylogenetic considerations. A review. Arch Med Res 23:1–5

Miller DM, Delgado R, Chirgwin JM, Hardies SC, Horowitz PM (1991) Expression of cloned bovine adrenal rhodanese. J Biol Chem 266:4686–4691

Morrison HG, Roger AJ, Nystul TG, Gillin FD, Sogin ML (2001) *Giardia lamblia* expresses a proteobacterial-like DnaK homolog. Mol Biol Evol 18:530–541

Müller M (1988) Energy metabolism of protozoa without mitochondria. Annu Rev Microbiol 42:465–488

Müller M (2003) Energy metabolism. Part I: anaerobic protozoa. In: Marr J, Nilsen T, Komuniecki R (eds) Molecular medical parasitology. Academic, London, pp 125–139

Nakamura Y, Hashimoto T, Yoshikawa H, Kamaishi T, Nakamura F, Okamoto K, Hasegawa M (1996) Phylogenetic position of *Blastocystis hominis* that contains cytochrome-free mitochondria, inferred from the protein phylogeny of elongation factor 1 alpha. Mol Biochem Parasitol 77:241–245

Nasirudeen AM, Tan KS (2004) Isolation and characterization of the mitochondrion-like organelle from *Blastocystis hominis*. J Microbiol Methods 58:101–109

Nixon JE, Wang A, Field J, Morrison HG, McArthur AG, Sogin ML, Loftus BJ, Samuelson J (2002) Evidence for lateral transfer of genes encoding ferredoxins, nitroreductases, NADH oxidase, and alcohol dehydrogenase 3 from anaerobic prokaryotes to *Giardia lamblia* and *Entamoeba histolytica*. Eukaryot Cell 1:181–190

Ogbadoyi EO, Robinson DR, Gull K (2003) A high-order trans-membrane structural linkage is responsible for mitochondrial genome positioning and segregation by flagellar basal bodies in trypanosomes. Mol Biol Cell 14:1769–1779

Putignani L, Tait A, Smith HV, Horner D, Tovar J, Tetley L, Wastling JM (2004) Characterization of a mitochondrion-like organelle in *Cryptosporidium parvum*. Parasitology 129:1–18

Reeves RE (1984) Metabolism of *Entamoeba histolytica* Schaudinn, 1903. Adv Parasitol 23:105–142

Reeves RE, Warren LG, Susskind B, Lo HS (1977) An energy-conserving pyruvate-to-acetate pathway in *Entamoeba histolytica*. Pyruvate synthase and a new acetate thiokinase. J Biol Chem 252:726–731

Regoes A, Zourmpanou D, Leon-Avila G, van der Giezen M, Tovar J, Hehl AB (2005) Protein import, replication and inheritance of a vestigial mitochondrion. J Biol Chem 280:30557–30563

Riordan CE, Langreth SG, Sánchez LB, Kayser O, Keithly JS (1999) Preliminary evidence for a mitochondrion in *Cryptosporidium parvum*: phylogenetic and therapeutic implications. J Eukaryot Microbiol 46:52S–55S

Riordan CE, Ault JG, Langreth SG, Keithly JS (2003) *Cryptosporidium parvum* Cpn60 targets a relict organelle. Curr Genet 44:138–147

Rodríguez MA, García-Pérez RM, Mendoza L, Sánchez T, Guillén N, Orozco E (1998) The pyruvate:ferredoxin oxidoreductase enzyme is located in the plasma membrane and in a cytoplasmic structure in *Entamoeba*. Microb Pathog 25:1–10

Roger AJ, Clark CG, Doolittle WF (1996) A possible mitochondrial gene in the early-branching amitochondriate protist *Trichomonas vaginalis*. Proc Natl Acad Sci USA 93:14618–14622

Roger AJ, Svärd SG, Tovar J, Clark CG, Smith MW, Gillin FD, Sogin ML (1998) A mitochondrial-like chaperonin 60 gene in *Giardia lamblia*: evidence that diplomonads once harbored an endosymbiont related to the progenitor of mitochondria. Proc Natl Acad Sci USA 95:229–234

Rospert S, Looser R, Dubaquie Y, Matouschek A, Glick BS, Schatz G (1996) Hsp60-independent protein folding in the matrix of yeast mitochondria. EMBO J 15:764–774

Rotte C, Stejskal F, Zhu G, Keithly JS, Martin W (2001) Pyruvate:NADP+ oxidoreductase from the mitochondrion of *Euglena gracilis* and from the apicomplexan *Cryptosporidium parvum*: a biochemical relic linking pyruvate metabolism in mitochondriate and amitochondriate protists. Mol Biol Evol 18:710–720

Sigler PB, Xu Z, Rye HS, Burston SG, Fenton WA, Horwich AL (1998) Structure and function in GroEL-mediated protein folding. Annu Rev Biochem 67:581–608

Silberman JD, Sogin ML, Leipe DD, Clark CG (1996) Human parasite finds taxonomic home. Nature 380:398

Slapeta J, Keithly JS (2004) *Cryptosporidium parvum* mitochondrial-type HSP70 targets homologous and heterologous mitochondria. Eukaryot Cell 3:483–494

Sogin ML (1991) Early evolution and the origin of eukaryotes. Curr Opin Genet Dev 1:457–463

Sogin ML, Gunderson JH, Elwood HJ, Alonso RA, Peattie DA (1989) Phylogenetic meaning of the kingdom concept: an unusual ribosomal RNA from *Giardia lamblia*. Science 243:75–77

Soltys BJ, Gupta RS (1994) Presence and cellular distribution of a 60-kDa protein related to mitochondrial hsp60 in *Giardia lamblia*. J Parasitol 80:580–590

Sutak R, Dolezal P, Fiumera HL, Hrdy I, Dancis A, Delgadillo-Correa M, Johnson PJ, Müller M, Tachezy J (2004) Mitochondrial-type assembly of FeS centers in the hydrogenosomes of the amitochondriate eukaryote *Trichomonas vaginalis*. Proc Natl Acad Sci USA 101:10368–10373

Tachezy J, Sánchez LB, Müller M (2001) Mitochondrial type iron-sulfur cluster assembly in the amitochondriate eukaryotes *Trichomonas vaginalis* and *Giardia intestinalis*, as indicated by the phylogeny of IscS. Mol Biol Evol 18:1919–1928

Tetley L, Brown SM, McDonald V, Coombs GH (1998) Ultrastructural analysis of the sporozoite of *Cryptosporidium parvum*. Microbiology 144:3249–3255

Tielens AG, Rotte C, van Hellemond JJ, Martin W (2002) Mitochondria as we don't know them. Trends Biochem Sci 27:564–572

Tovar J, Fischer A, Clark CG (1999) The mitosome, a novel organelle related to mitochondria in the amitochondrial parasite *Entamoeba histolytica*. Mol Microbiol 32:1013–1021

Tovar J, Leon-Avila G, Sánchez LB, Sutak R, Tachezy J, van der Giezen M, Hernández M, Müller M, Lucocq JM (2003) Mitochondrial remnant organelles of *Giardia* function in iron-sulphur protein maturation. Nature 426:172–176

Townson SM, Upcroft JA, Upcroft P (1996) Characterisation and purification of pyruvate:ferredoxin oxidoreductase from *Giardia duodenalis*. Mol Biochem Parasitol 79:183–193

Van De Peer Y, Ben Ali A, Meyer A (2000) Microsporidia: accumulating molecular evidence that a group of amitochondriate and suspectedly primitive eukaryotes are just curious fungi. Gene 246:1–8

van der Giezen M, Tovar J (2004) Mitosomes, hydrogenosomes and mitochondria; variations on a theme? In: Hirt RP, Horner DS (eds) Organelles, genomes and eukaryotic phylogeny: an evolutionary synthesis in the age of genomics. CRC, Boca Raton, pp 289–308

van der Giezen M, Tovar J (2005) Degenerate mitochondria. EMBO Rep 6:525–530

van der Giezen M, Cox S, Tovar J (2004) The iron-sulphur cluster assembly genes *iscS* and *iscU* of *Entamoeba histolytica* were acquired by horizontal gene transfer. BMC Evol Biol 4:7

van der Giezen M, Tovar J, Clark CG (2005) Mitochondrion-derived organelles in protists and fungi. Int Rev Cytol 244:177–227

Vossbrinck CR, Woese CR (1986) Eukaryotic ribosomes that lack a 5.8S RNA. Nature 320:287–288

Vossbrinck CR, Maddox JV, Friedman S, Debrunner-Vossbrinck BA, Woese CR (1987) Ribosomal RNA sequence suggests microsporidia are extremely ancient eukaryotes. Nature 326:411–414

Weston CJ, Venning JD, Jackson JB (2002) The membrane-peripheral subunits of transhydrogenase from *Entamoeba histolytica* are functional only when dimerized. J Biol Chem 277:26163–26170

Wiedemann N, Frazier AE, Pfanner N (2004) The protein import machinery of mitochondria. J Biol Chem 279:14473–14476

Williams BAP, Hirt RP, Lucocq JM, Embley TM (2002) A mitochondrial remnant in the microsporidian *Trachipleistophora hominis*. Nature 418:865–869

Zierdt CH (1991) *Blastocystis hominis* – past and future. Clin Microbiol Rev 4:61–79

Zierdt CH (1993) Taxonomic status of *Blastocystis hominis*: reply. Parasitol Today 9:18

Zierdt CH, Donnolley CT, Müller J, Constantopoulos G (1988) Biochemical and ultrastructural study of *Blastocystis hominis*. J Clin Microbiol 26:965–970

# Index

MIX
Papier aus verantwortungsvollen Quellen
Paper from responsible sources
FSC® C105338

If you have any concerns about our products,
you can contact us on
**ProductSafety@springernature.com**

In case Publisher is established outside the EU,
the EU authorized representative is:
**Springer Nature Customer Service Center GmbH**
**Europaplatz 3, 69115 Heidelberg, Germany**

Printed by Libri Plureos GmbH
in Hamburg, Germany